PEST MANAGE

PEST MANAGEMENT

Editor

Tahir Alfarsi

scitus
academics

Pest Management

Edited by **Tahir Alfarsi**

This Book is Printed on Acid-Free Paper. ♾

ISBN: 9781681179162
Library of Congress Control Number: 2017948945

SCITUS Academics LLC,
616, Corporate Way, Suite 2, 4766,
Valley Cottage, NY 10989

www.scitusacademics.com

Notice

Printed in United States of America.

Preface

Integrated Pest Management (IPM) is an effective and environmentally sensitive approach to pest management that relies on a combination of common-sense practices. IPM programs use current, comprehensive information on the life cycles of pests and their interaction with the environment. This information, in combination with available pest control methods, is used to manage pest damage by the most economical means, and with the least possible hazard to people, property, and the environment. In crop protection, chemical control rapidly revealed its limitations, as well as its possibilities, and alternative solutions to pest management problems have been recommended since at least the 1960s. A new strategy was developed under the rubric 'integrated control', envisaging the employment of a range of different control measures, constrained by their compatibility and the requirement for minimizing noxious effects on the wider environment. Despite these difficulties, a biological, then ecological, orientation has underlain the development of crop protection over the last 50 years. This process has been marked by multiple and diverse interpretations of the concept of IPM

This book 'Pest management' covers current and comprehensive information on the life cycles of pests and their relations with the environment. This information is used to manage pest damage by the most cost-effective means and with the least possible hazard to people, property, and the environment. The book chapters detail about historical review of IPM concepts; strategies and some experiences in applications of IPM; pest control in organic agricultural system; and the use of entomopathogenic and molluscoparasitic nematodes, insect pheromones, semiochemicals, detergents, and soaps as a part of IPM scheme. It also delivers how agroecological concepts may contribute to sustainable pest management.

Table of Contents

Chapter 1 Integrated Pest Management 1

Chapter 2 Pest Control in Organic Systems 13

Chapter 3 Companion Planting and Insect Pest Control 59

Chapter 4 Biochemical Insecticide Resistance in Tea Pests 87

Chapter 5 Semiochemicals and Their Potential Use in Pest Management 129

Chapter 6 Use of Banker Plant System for Sustainable Management of the most Important Insect Pest in Rice Fields in China 153

Chapter 7 Insect Pathogenic Bacteria in Integrated Pest Management 169

Chapter 8 Insect Pests of Sweetpotato in Uganda: Farmers' Perceptions of Their Importance and Control Practices 189

Chapter 9 A Modelling Methodology to Assess the Effect of Insect Pest Control on Agro-Ecosystems 207

Chapter 10 **The Rise and Demise of Integrated Pest Management in Rice in Indonesia** **223**

Chapter 11 **Development of a Microbial-Based Integrated Pest Management Program for *Helicoverpa* spp. (Lepidoptera: Noctuidae) and Beneficial Insects on Conventional Cotton Crops in Australia** **261**

Chapter 12 **Improving Strawberry Irrigation with Micro-Sprinklers and Their Impact on Pest Management** **285**

Chapter 13 **Arthropod Pest Control for UK Oilseed Rape – Comparing Insecticide Efficacies, Side Effects and Alternatives** **297**

Chapter 14 **A Fundamental Step in IPM on Grapevine: Evaluating the Side Effects of Pesticides on Predatory Mites** **327**

Chapter 15 **Pest Control and Resistance Management Through Release of Insects Carrying a Male-Selecting Transgene** **341**

Chapter 16 **Efficient Management of Fruit Pests by Pheromone Nanogels** **371**

Chapter 17 **Developing the Arsenal Against Pest and Vector Dipterans: Inputs of Transgenic and Paratransgenic Biotechnologies** **393**

Index **419**

CHAPTER 1

Integrated Pest Management

C.O. Ehi-Eromosele[1], O.C. Nwinyi[2] and O.O. Ajani[1]

[1] Chemistry Department, School of Natural and Applied Sciences, College of Science and Technology, Covenant University, Canaan Land, Ota, Ogun State, Nigeria

[2] Department of Biological Sciences, School of Natural and Applied Sciences, College of Science and Technology, Covenant University,, Canaan Land, Ota, Ogun State, Nigeria

1. INTRODUCTION

Integrated Pest Management (IPM) has had a varied history, with different definitions. It has been implemented under an array of different connotations (Lewis *et al.*, 1997). The term was earlier used as "integrated control" by Bartlett (Bartlett, 1956) and was further elaborated on by Stern and co-workers (Stern *et al.*, 1959). In reference to the concept of integrating the use of biological and other controls in complementary ways, the term was later broadened to embrace coordinated use of all biological, cultural, and artificial practices (van den Bosch and Stern, 1962). The term "IPM," under various authors have advocated for the principle of incorporating, the full array of pest management practices with production objectives in a total systemic approach. Nonetheless, there is no universally agreed definition of IPM.

Alastair (2003) conceived IPM as a method of rationalizing pesticide use to prevent or delay the resurgence of pest populations, that had become resistant to pesticides, and to protect beneficial insects. Today, concerns about pesticide residues in the food chain and in the environment have led to alternative definitions that exclude the use of conventional pesticides. Nevertheless, there are broad agreements on the core principles of IPM. These include IPM being:

- An integrated scheme due to the methods of control that are seen as component technologies rather than alternatives.

- emphasizes pest management, within a balanced system whereas control strategy suggests direct intervention with little concern for sustainability (Alastair, 2003).

In principle, IPM may be defined as a flexible and holistic system. This views the agro-ecosystem as an interrelated whole, that utilizes a variety of biological, cultural, genetic, physical, and chemical techniques that hold pests below economically damaging levels with a minimum disruption to the cropping ecosystem and the surrounding environment (Malena, 1994). In the definition of FAO (2012), integrated pest management (IPM) is an ecosystem approach to crop production and protection that combines different management strategies and practices to grow healthy crops and minimize the use of pesticides. IPM therefore, utilizes the best mix of control tactics for a given pest problem when compared with the crop yield, profit and safety of other alternatives (Kenmore *et al.,* 1985). Other definitions of IPM according to the United States Environment Protection Agency (2012), involves an effective and environmentally sensitive approach to pest management that relies on a combination of common-sense practices. IPM could be a broad based ecological approach to structural and agricultural pest control that integrates pesticides/herbicides into a management system incorporating a range of practices for economic control of a pest. In addition, IPM, attempts to prevent infestation, to observation of patterns of infestation when they occur, and to intervene (without poisons) when one deems necessary. Finally, IPM is the intelligent selection and use of pest control actions that ensures favourable economic, ecological and sociological consequences (Sandler, 2010).

1.1. Succinct History of Ipm

IPM is not a new philosophy. The concept has been around since the 1920's when cotton pest management program was developed. Under this scheme, insect control was "supervised" by qualified entomologists, and insecticide applications were based on conclusions reached from periodic monitoring of pest and natural-enemy populations. This was viewed as an alternative to calendar-based insecticide programs. Supervised control is based on a sound knowledge of the ecology and analysis of the projected trends in pest and natural-enemy populations. In supervised control, (integrated control) the best mix of chemical and biological controls is sought and identified for a given insect pest. The chemical insecticides are used in a manner that is least disruptive to biological control. The chemical controls are applied only after regular monitoring indicates that a pest population had reached an economic threshold level. Thus, such treatment is required to prevent the population from reaching an economic injury level where economic losses would exceed the cost of the artificial control measures.

Typically, the main aim of IPM programmes is on agricultural insect pests (IPM Guidelines, 2009). Although, originally developed for agricultural pest management, IPM programmes are now developed to encompass diseases, weeds, and other pests that may interfere with the management objectives of sites such as residential and commercial structures, lawn and turf areas, and home and community gardens. IPM programs use current, comprehensive information on the life cycles of pests and their interaction with the environment. This information, in combination with available pest control methods, is used to manage pest damage by the most economical means, and with the least possible hazard to people, property, and the environment.

The IPM approach can be applied to both agricultural and non-agricultural settings, such as the home, garden, and workplace. IPM takes advantage of all appropriate pest management options including, the judicious use of pesticides.

2. IPM – A PHILOSOPHY

2.1. A Pest Management Strategy

Integrated Pest Management (IPM) is a philosophy that involves the management of a pest instead of controlling or eradicating a pest. It requires a greater knowledge of the pest, crop and the environment. Therefore, its strategy focuses on harnessing inherent strengths within ecosystems and directing the pest populations into acceptable bounds rather than toward eliminating them. This strategy avoids undesirable short term and long term ripple effects and will ensure a sustainable future (Lewis *et al.*, 1997).

IPM programs should be operated with "pest management objectives" rather than "pesticide management objectives". Integrated pest management is a comprehensive long term pest management program based on knowledge of an ecosystem that weighs economic, environmental, and social consequences of interventions (Flint and van den Bosch, 1981). The foundation for pest management in agricultural systems should be an understanding and shoring up of the full composite of inherent plant defenses, plant mixtures, soil, natural enemies, and other components of the system. These natural "built in" regulators are linked in a web of feedback loops that are renewable and sustainable. The use of pesticides and other "treat-the-symptoms" approaches are unsustainable and should be the last option rather than the first line of defense. A pest management strategy should always start with the question "Why is the pest a pest?". It should also seek to address underlying weaknesses in ecosystems and/or agronomic practice(s) that have allowed organisms to reach pest status (Lewis *et al.*, 1997).

2.2. An Integrated Process

Integration or compatibility among pest management tactics is central to Integrated Pest Management. Simply mixing different management tactics does not constitute IPM. Mixing the tactics arbitrarily may actually aggravate pest problems or produce other unintended effects. IPM recognizes there is no "cure-all" in pest control (dependence on any one pest management method will have undesirable effects). Reliance on a single tactic will favor pests that are resistant to that practice. In IPM, integrated control seeks to identify the best mix of chemical and biological controls for a given insect pest. The term "integrated" is thus synonymous with "compatibility."

2.3. Understanding Pest Biology and Ecology

The determination of the correct cause of pest problem (understanding pest biology) and ecology is essential in manipulating the environment to the crop's advantage and to the detriment of the pest.

2.4. Acceptable Pest Levels

IPM recognizes that eradication of a pest is seldom necessary or even desirable, and generally not possible. The primary objective in pest management is not to eliminate a pest organism but to bring it into acceptable bounds (Lawal *et al.*, 1997). The emphasis is on control, not eradication. IPM holds that wiping out an entire pest population is often impossible, and the attempt can be expensive and environmentally unsafe. IPM programmes initial task is to establish acceptable pest levels, called action thresholds, and apply controls where the thresholds are crossed. These thresholds are pest and site specific, meaning that it may be acceptable at one site to have for instance a weed such as white clover, but at another site it may not be acceptable. By allowing a pest population to survive at a reasonable threshold, selection pressure is reduced. This stops the pest gaining resistance to chemicals produced by the plant or applied to the crops. If many of the pests are killed, then any that have resistance to the chemical will form the genetic basis of the future, more resistant, population. By not killing all the pests there are some un-resistant pests left that will dilute any resistant genes that appear (Wikipedia, 2011).

2.5. Ipm A Continuum, Not an End

Agriculture is a dynamic system that continually changes to changing crop production practices. IPM must continually change to meet pest management challenges. IPM is a continuum that will change with time. Every farmer practices some type of IPM, as long as they make progress to better its management. As new pest control techniques are discovered, the producer and crop advisor must adapt their pest control program to reflect these changes. What is considered a good IPM program today may be considered a chemical intensive program in a few years. Additionally, some good advice to the producer and crop advisor is to try the new changes on a limited scale, while becoming comfortable with the suggested practices before wide-scale changes are made.

3. IPM PROCESS

IPM is applicable to all types of agriculture and sites such as residential and commercial structures, lawn and turf areas, and home and community gardens. The process includes:

3.1. Proper Identification Pest Damage and Responsible Pests

Identification must be the first objective. When the identity of a pest is not known, then, a strategy built to control the pest cannot be transferred from one site to another, primarily, because the pest species or strain (biotype) might behave differently. Thus, a solid foundation must be built on systematic, taxonomy, etiology, and spatial distribution (Irwin, 1999). Cases of mistaken identity may result in ineffective actions. If plant damage is due to over-watering, it could be mistaken for fungal infection, since many fungal and viral infections arise under moist conditions. This could lead to spray costs, but the plant would be no better off.

3.2. Pest and Host Life Cycles Biology

Understanding crop growth and development is an underlying principle of IPM. We cannot just focus on the pest. The interactions between crop and pest (as well as the environment) are very important. To deplore an efficient IPM programme, literature and other data sources about the pest, the pest's life cycle, host range, distribution, movement, and basic biology will have to be researched. At the time you see a pest, it may be too late to do much about it except maybe spray with a pesticide (Metcalf and Luckmann, 1994). Often, there is another stage of the life cycle that is susceptible to preventative actions. For example, weeds reproducing from last year's seed can be prevented with mulches and pre-emergent herbicide. Also, learning what a pest needs to survive allows you to remove these.

3.3. Monitor or Sample The Environment for Pest Populations

After the pest has been correctly identified, monitoring must begin before it becomes a problem. Sampling and monitoring methodologies must be designed and tested to provide the ability for assessing instantaneous and dynamic aspects of the pest's density, activity, or incidence (Irwin, 1999). Understanding how the environment affects pest and crop development is very important. Understanding interactions with the environment allows crop advisors to react to changing conditions. Environmental influences like drought stress influences pest management recommendations. When a crop is under stress it can be less capable of dealing with the stress caused by insects that extract plant sap (e.g. aphids, leafhoppers) and this stress may slightly lower the economic threshold. Weed populations which would not normally cause an economic loss may do so under drought conditions when they compete with the crop for limited water.

The weather is notorious for affecting pest development and survival. Certain weather patterns may affect weed seed germination and explain why certain weeds are more abundant during wet fall or springs.

3.4. Establish Action Threshold (Economic, Health, and Aesthetic)

The question here is: how many are too many or how much can be tolerated? In some cases, there are standardized numbers of pests that can be tolerated. Soybeans are quite tolerant of defoliation, so if there are a few caterpillars in the field and their population may not be increasing dramatically; thus, no urgent action may be necessary. Conversely, there is a point at which an action must be taken to control cost. For instance the farmer can control cost at the point when the cost of damage by the pest is more than the cost of control. This is an economic threshold. Tolerance of pests varies according to the health hazard (low tolerance) or merely a cosmetic damage (high tolerance in a non-commercial situation). Different sites may also have varying requirements based on specific areas. For instance, white clover may be perfectly acceptable on the sides of a tee box on a golf course, but unacceptable in the fairway where it could cause confusion in the field of play (Purdue University, 2006).

3.5. Choose an Appropriate Combination of Management Tactics

The word 'integrated' in IPM initially referred to the simultaneous use or integration of any number of tactics in combination, with focus on maintaining a single pest species below its economic injury level. Although, in theory a single strategy results from the simultaneous integration of several tactics, in practice, the integration actually occurs in a step-wise, time-delayed fashion. Several of the tactics are compatible, but some are not. Certainly the tactics of biological control, habitat manipulation, and legal control go alongside. The tactic of host resistance can stand alone or be combined with the other tactics just mentioned. Chemical control is generally compatible with host resistance. Thus, a management strategy integrates one or several compatible tactics into a single package (Irwin, 1999).

3.6. Evaluate and Record Results

Evaluation is often one of the most important steps in Integrated Pest Management (Bennett *et al.*, 2005). It is the process of reviewing an IPM program and the results it has generated. Asking the following questions is useful: Did the steps one took effectively control the population? Was this method safe enough? Where there any expected side effects? What is the next step? Understanding the effectiveness of the IPM program allows the site manager to make modifications to the IPM plan prior to pests reaching the action threshold and requiring action again.

4. PEST MANAGEMENT TACTICS

There are different pest management tactics to suppress pests. They include host resistance, chemical, biological, cultural, mechanical, sanitary and mechanical controls. The primary pest management tactic involves maximization of built-in pest reduction features of an ecosystem. Molecular or genetic mechanisms are potentially manifested in a number of these more specific tactics. Each category, discussed below, employs a different set of mechanisms for suppressing populations.

4.1. Chemical Control

The therapeutic approach of killing pest organisms with toxic chemicals has been the prevailing pest control strategy for over 50 years. Safety problems and ecological disruptions continue to ensue (Wright, 1996), and there are renewed appeals for effective, safe, and economically acceptable alternatives (Benbrook, 1996). Synthetic chemical pesticides are the most widely used method of pest control. The four major problems encountered with conventional pesticides are toxic residues, pest resistance, secondary pests, and pest resurgence (Lewis, 1997). The use of natural pesticides and organophosphates that are more environmentally friendly are encouraged and synthetic pesticides should only be used as a last resort or only used as required and often only at specific times in a pests life cycle.

4.2. Biological Control

This involves the use of other living things that are enemies of a pest in order to control it. Sometimes, the term "biological control" has been used in a broad context to encompass a full spectrum of biological organisms and biologically based products including pheromones, resistant plant varieties, and autocidal techniques such as sterile insects. IPM is mainly aimed at developing systems based on biological and non-chemical methods as much as possible.

4.3. Host Plant Resistance

This involves breeding varieties with desirable economic traits, but less attractive for pests, for egg laying and subsequent development of insect, disease or nematode. It also involves withstanding the infestation/infection or the reduction of pests to level that they are not large numbers during the plant growth period (Sharma, 2007).

4.4. Cultural Measures

This involves practices that suppress pest problems by minimizing the conditions that favour their existence (water, shelter, food). Some of these factors are intrinsic to crop production while making the environment less favourable for survival, growth and reproduction of pest species. If followed in an appropriate manner, the cultural practices can provide significant relief from pests. The selection of appropriate site for the cultivation of field crops and fruit trees can reduce future infestation from insect pests. The culture should be selected in such a manner that it should be suitable for growing in the area and tolerant to important pests diseases of the area.

4.5. Mechanical Control

This is the use of machinery and other tools to control pests. It involves agricultural practices like tillage, slash and burn, and hand weeding. The pruning of infested parts of fruits and forest trees and defoliation in certain crops help reduce the pest population. Chaffing of sorghum/maize stalks and burning of stubbles kills maize borer.

4.6. Sanitary Control

Preventive practices are important part of an IPM programme. These include cleaning field equipment (i.e., tillage equipment, haying equipment, etc.), planting certified seeds and quarantine of infested crops or farmlands. These are methods used to prevent the introduction of a pest into the field.

4.7. Natural Control

Natural control involves the enhancement of naturally occurring pest management methods to combat pests like using beneficial insects and diseases. Here, insecticides will only be used when they are economically feasible and it is apparent that natural enemies will not control the pests.

5. IPM: A MULTI-DISCIPLINARY APPROACH

IPM is a management intensive philosophy which stresses a multidisciplinary approach. Pests interact with each other, the crop, and the environment. Similarly, pest and crop management disciplines must work together to develop control recommendations that reflect these interactions.

For example, management of the Soybean aphid includes entomologists who study the insect and their damage to soybean, agronomists that identify crop stage which are most vulnerable to soybean aphid damage, plant pathologists who study the viruses transmitted by aphid feeding, and soil scientists who study the aphids interaction with nutrient deficiencies.

6. BENEFITS OF AN IPM PROGRAMME

The benefits of Integrated Pest Management are immense directly to farming and indirectly to society.

a. Integrated Pest Management (IPM) protects environment through elimination of unnecessary pesticide applications. In IPM, pesticides are used at the smallest effective dose when other methods of pest control have failed. Also, they are used in bringing a pest organism to acceptable bounds with as little ecological disruption as possible.

b. IPM improves profitability. Since IPM programme applies the most economical management pest tactics, profitability is ensured for the grower or farmer.

c. It reduces risk of crop loss by a pest. Applying pest management and monitoring tactics will also ensure the reduction of crop loss or damage by pests.

d. Long term sociological benefits of IPM would also emerge in areas of employment, public health, and well being of persons associated with agriculture.

7. DISADVANTAGES OF AN IPM PROGRAMME

In spite of the numerous benefits of IPM stated so far, there are also some drawbacks to it:

7.1. An Ipm Program Requires a Higher Degree of Management

Making the decision not to use pesticides on a routine or regular basis requires advanced planning and therefore, a higher degree of management. This planning includes attention to field histories to anticipate what the pest problems might be, selecting crop varieties which are resistant or tolerant to pest damage, choosing tillage systems that will suppress anticipated pest damage while giving the crop the greatest yield potential.

7.2. Ipm can be more Labor Intensive

Consistent, timely and accurate field scouting takes time. However, it is this information that is necessary and is the corner stone of IPM programs.

Without this information you cannot make intelligent management decision.

7.3. Success can be weather Dependant

Weather can complicate IPM planning. For example you might want to lower herbicide rates and use row cultivation to manage weed pressure. However an extended wet period may reduce (or eliminate) the effectiveness of row cultivation. Therefore, good IPM planners will have a alternate plan for when these problems arise.

REFERENCES

1. Alastair, O. (2003) Integrated Pest Management for Resource-Poor African Farmers: Is the Emperor Naked? World Development Vol. 31, No. 5, pp. 831–845
2. Barlett, B. R. (1956) Natural predators. Can selective insecticides help to preserve biotic control Agric. Chem. 11, 42–44.
3. Benbrook, C. M. (1996) Pest Management at the Crossroads (Consumer's Union, Yonkers, NY).
4. Bennett, G.W., Owens, J.M., Corrigan, R.M. (2005) Truman's Scientific Guide to Pest Management Operations, 6th Edition, Purdue University, Questex, USA. pp. 10-12,
5. Ehler, L. E.; Bottrell D. G. (2000) The illusion of integrated pest management. Issues in Science and Technology 16, 61-64.
6. Flint, M. L. and van den Bosch, R. (1981) Introduction to Integrated Pest Management (Plenum, New York).
7. Food and Agricultural Organisation of the United Nations http://www.fao.org/agriculture/crops/core-themes/theme/pests/ipm/en/
8. IPM Guidelines . UMassAmherst: Integrated Pest Management, Agriculture and Landscape Program 2009 http://www.umass.edu/umext/ipm/ publications /guidelines/index.html.
9. Irwin, M.E. (1999) Agricultural and Forest Meteorology 97: 235–248
10. Kenmore, P.E., Heong, K.L., Putter, C.A (1985) Political, social and perceptual aspects of integrated pest management programmes. In:Lee, B.S., Loke, W.H., Heong, K.L. (Eds.), Integrated Pest Management in Asia. Malaysian Plant Protection Society, Kuala Lumpur, pp. 47- 66
11. Lewis, W. J., van Lenteren, J. C., Sharad C. Phatak, and Tumlinson, J. H. (1997) Proc. Natl. Acad. Sci. USA 94: 12243–12248
12. Malena, C. (1994) Gender issues in IPM in African agriculture. Socio-economic series no. 5. Chatham: Natural Resources Institute.
13. Purdue University Turf Pest Management Correspondence Course, Introduction, 2006
14. Robert L. M., William H. L. (1994). Introduction to Insect Pest Management. New York: John Wiley and Sons, Inc. pp. 266.
15. Sandler, Hilary A. (2010). "Integrated Pest Management". Cranberry Station Best Management Practices 1 (1): 12–15.
16. Stern, V. M., Smith, R. F., van den Bosch, R. and Hagen, K. S. (1959) The integration of chemical and biological control of the spotted alfalfa aphid (the integrated control concept). Hilgardia 29:81-101.
17. Sharma, B.K. (1997) Environmental Chemistry, Krishna Ltd., Delhi pp. soil-114

18. United States Protection Agency "Pesticides and Food: What Integrated Pest Management Means". 2011. http://www.epa.gov/pesticides/food/ipm.htm.
19. United States Environmental Protection Agency "Integrated Pest Management (IMP) Principles". 2012. http://www.epa.gov/pesticides/factsheets/ipm.htm.
20. van den Bosch, R and Stern, V. M. (1962) The Integration of Chemical and Biological Control of Arthropod Pests Annu. Rev. Entomol. 7: 367–387.
21. Vereijken, P., Edwards, C., El Titi, A., Fougeroux, A. and Way., M. (1986) Proceedings of Workshop on Integrated Farming. Bull. I. O. B. C./W. P. R. S./IX 2, 1–34.
22. Wikipedia, The Free Encyclopaedia http://en.wikipedia.org/wiki/ Integrated_ pest_management
23. Wijnands, F. G. and Kroonen-Backbier, B. M. A. (1993) In Modern Crop Protection: Developments and Perspectives, ed. Zadoks, J. C. (Wageningen Pers, Wageningen, The Netherlands), pp. 227–234.
24. Wright, L. 1996. Silent sperm. (A reporter at large.) The New Yorker (Jan. 15):42,48,50-53,55.
25. www.cias.wisc.edu/curriculum/.../Notes_for_IPM_powerpoint.doc

CHAPTER 2

Pest Control in Organic Systems

Vasile Stoleru[1]* and Vicenzo Michele Sellitto[2]

ABSTRACT

Conventional agriculture techniques applied in the latest decades have had undesired consequences on the environmental sustainability, carried out to the soil erosion, the degradation of the ecological system, changing the balance between beneficial and harmful pests, and contamination of soil, water, and agricultural products by heavy metals and pesticides. Thus, in organic agriculture, using synthetic chemicals for pest control is prohibited, assigning to the diversity a major role. The study provides to the reader many important practical data, judiciously documented, which are useful for the researchers and farmers from the world. Pest control in organic agriculture can be obtained through prevention and curative measure, but modern agriculture must be focused on the prevention.

Keywords: organic agriculture, pest control, preventive and curative methods

1. INTRODUCTION

Organic agriculture (OA) farming aims to achieve sustainable, diversified, and balanced systems, with the purpose of protecting the environment for present and future generations. In the same way, OA provides on the food market, products of a certain nutritional quality, suitable in terms of lower contaminants.

The organic product is governed by some well defined principles, aimed at ensuring environmental and crop sustainability.

1.1. Circumstances of Pest Control in Organic Systems

Being a type of sustainable agriculture the purpose of OA can be expressed by a mini–max function, maximizing production and minimizing the negative agricultural activities on the environment [1].

OA stimulates the activity of useful microorganisms, flora and fauna. Soils under crops are increasingly lifeless and infested with weeds, diseases, as well as pests. This situation is determined by current agricultural practices that excel in monoculture and short crop rotations, of 2–3 years, much delayed and bad quality soil tillage and plant care, burning plant debris, etc.

Biodiversity management. The soil's biological resources are vital to the economic and social development of all humanity. That is why, it is more and more frequently recognized that biological diversity is universal asset, of inestimable value for future generations. Biological (ecologic, organic) agriculture generally uses a greater number of cultivated species, to explore their suitability and ecological plasticity. Non-using synthetic herbicides, and instead using milder solutions for weed destruction, ensures the coexistence of weeds together with the crop.

Protecting the natural landscape. Elevation diversity, as well as flora and fauna variability, is inseparable to the applied vegetable growing systems, the most aggressive ones being of the intensive type, often causing deterioration.

Many cultivation techniques applied in the past decades have had undesired consequences on the environment, contributing to soil erosion, the degradation of the ecological system, contamination of ground water and crops with pesticides and nitrates.

Organic agriculture aims to preserve the environment unaltered, using organic fertilizers and also less soluble mineral fertilizers, organic fertilizers, such as composts and green fertilizers, avoiding to use products that can have harmful effects [2].

The use of synthetic herbicides and pesticides are prohibited, and only products that are harmless for the plant are allowed, products based on simple minerals (Cu, S, Na, silicate, etc.) or plant extracts (pyrethrum), including the application of physical (thermal) methods.

In organic agriculture, the emphasis is laid on the quality of human intervention over nature, which is non-aggressive, compared to conventional agriculture.

1.2. Standards and Regulations Regarding Organic Farming

After 2010, OA can be considered a period of consolidation for standards and the regulations, which aimed and still aims to facilitate international trade with organic products in order to reduce legislative gaps which exist among the various certification types, such as the EC Regulations [3, 4], the USA (NOP), Australia (AS 6000-2009), Japan (JAS), and Switzerland (Bio Swiss). Thus, the EC Regulation of organic agriculture [3] has been improved in the least years with new regulations, targeting aquaculture and organic wine production.

The number of the certification bodies, in 2013, was at 569, increasing from 2010 when there were 532. Most certification bodies are found in the European Union, Japan, the United States of America, South Korea, China, Canada, India, and Brazil [5].

Organic farming (biological, ecological) is currently one of the most dynamic forms of agriculture. This affirmation is mainly supported by the expansion of agricultural areas, currently occupying 40.2 of the surface in Oceania, 26.6 in Europe, and 15.3% in Latin America. There are also cases of countries, such as Argentina, Spain, and USA, in which the area increased in 2013 compared to 2011 with over 185,000 ha.

Around the world, at the end of 2013, the organically certified area covered more than 78 million ha. Organically certified agricultural areas covered over 43 million ha (1% from total arable land), including the same land under its conversion period, but excluded wild collection and aquaculture. From these data, it appears that the organically administered surface has had a growth rate of over 14.94% compared with 2012 (approx. 37.4 million ha). Europe and Oceania recorded the fastest land expansion rhythm in 2013, compared to 2011, which shows that the expansion of the areas is supported by an intensive marketing of organic products [5].

Compared to 2012, the organically certified area in the world increased by over 5.6 million ha, which means a growth rate of the arable production from the total agricultural area of 0.1%.

At the end of 2013, the situation of the organic agricultural area distributed on categories of land use highlighted that 63% was permanent grassland, 18% was arable land (cereals, green fodder, oilseed, vegetable, and protein crops), and 7% was permanent crop (coffee, olives, nuts, grapes, and cocoa) and the rest with other crops [5].

Of course, in some countries, the conversion areas or the cultivated ones are decreasing, especially due to legislation and government support, which differ from country to country (UK).

Global sales of organic food and drinks reached more 72 billion dollars at the end of 2013. Compared to 2009, this sector revenue increased almost five times. Europe and North America made a big contribution to cover these specific sectors. Asia, Latin America, and Africa have become really important producers of organic crops for this market. About 43% from this market is covered by the United States followed by Europe at percent 40% [5].

In 2013, the countries with the largest organic markets were the USA (24.3 billion €), Germany (7.6 billion €), and France (4.4 billion €) [5].

2. PEST CONTROL MEASURES

Organic farming (OF) is a system-based agricultural production system working with rather than against natural systems [2].

The major differences that have been made in terms of technology between organic and conventional cultivation of plants are as follows: soil fertility, weeds, pathogens, and pest control.

Pest control in organic agriculture can be obtained through prevention and curative measure but must be focused on the preventive infestation of pests [2]. Measures to prevent infestation by pests refers to: phytosanitary quarantine (special for seed and planting materials used for establishing crops); monitoring pest infestation (used in general agro-expert stations or traps); choice of cultivars according to the criterion of resistance and ecological plasticity; seed conditioning; destruction of problematic weeds; solarization; and hygienic conditions.

2.1. Prevention Pests in an Organic System

The fundamental principle of controlling pests in organic systems (OS) should consider the mechanism of adjusting its biocenosis (total community of organisms from o biotope), through the correlation and interdependence between the cultivated species, pathogens, weeds, pests, technology, and the environment. Protecting plants from pests and diseases probably has the greatest impact on achieving an organic vegetable crop, due to the very large spectrum of pathogens and pests from these crops. The first major attempt to reduce chemical treatments took place even before 1970, when the concept of integrated control was promoted [6, 7]. According to this concept, all technical methods are allowed to maintain the populations of pests and pathogens under a certain degree of impairment, which does not affect the yields from an economic point of view.

This concept is approved by the International Organization for Biological Control (IOBC), but first of all natural factors must be used, together with other methods appropriate for the economic, ecological, and toxicological requirements [8].

In organic farming, the principles of the integrated pest control are perfectly applicable in substantializing the mechanisms for fighting pests, diseases, but most chemical means are forbidden; instead, new unconventional methods have been used, like some biodynamic preparations.

The strong attack of some pests may be favored by some technical mistakes, in general, or mistakes in the environmental context such as the following: improper choice of the place of culture; using seeds or plants that are weakly developed; mistakes in crop association; practicing monocultures without using proper crop rotation; incorrectly executed soil tillage; unilateral or excessive fertilization, without organic fertilizers; insufficient fertilization; extreme weather conditions; and improper choice of the sowing period [1, 9].

2.1.1. Phytosanitary quarantine

The quarantine is a complex of preventive measures taken to stop the penetration of diseases, pests, or weeds from other countries and to limit their spread. Overall, export products between countries shall be binding accompanied by a phytosanitary document certifying that the seeds or agricultural materials for setting up the crop (seeds, cuttings, tubers, bulbs, seedlings, shrubs, or trees) are free from pest quarantine.

There are numerous species (mites, insects), generally in polyphagus that are considered extremely dangerous and huge efforts have been made to limit their expansion, for example: *Leptinotarsa decemlineata* (Colorado beetle), *Tetranychus urticae (*red spider mite*), Myzus persicae (*green peach aphid*), Bemisia tabaci (*silverleaf whitefly*), Trialeurodes vaporariorum* (greenhouse whitefly), *Liriomyza trifolii (*leaf miner flies*), Tuta absoluta* (tomato leaf miner), *Spodoptera litura (*Oriental leaf worm moth*), Frankliniella intonsa red (*red thrips*), Diabrotica virgifera virgifera* (western corn rootworm), or others [10–12].

2.1.2. Maintenance of biodiversity

Synthetic pesticides are not permitted in organic farming which serves to preserve and enhance biodiversity within the system. Natural enemies of pest species are therefore able to thrive, exerting control on pest populations. Conservation and improvement of natural features of the landscape, such as hedgerows and ponds and the construction of beetle banks and sown flower strips, have also enabled communities of predators to flourish.

In agriculture, in general, farmers work with biological organisms, which behave differently under the action of nature's biotic or abiotic factors [13].

The pests are very adaptive to the changes of production systems, especially from the transfer from conventional to organic farms (in conversion).

In OA, pest problems are influenced by three major components of farming systems, such as: crop species and cultivar, agro-ecosystem structure, and technology production (**Figures 1** and **2**).

Figure 1. Management of land for organic agriculture (photograph by Stoleru Vasile).

Researchers developed flowering strips that are tailored to requirements of the specific complex of natural enemies within a cropping system. So, any experiments identified selective plant species that would improve the longevity and parasitization rate of the parasitoid wasps (*Microplitis mediator*, *Diadegma fenestrale*, and *D. semiclausum*) on the *Mamestra brassicae*.

Figure 2. Sea buckthorn hedge on an organic farm (photograph by Stoleru Vasile).

Comparing the effects of floral and extrafloral nectar of different plants, beneficial effects of *Fagopyrum esculentum* (floral nectar), *Centaurea cyanus* (floral and extrafloral nectar), and non-flowering *Vicia sativa* (extrafloral nectar) on parasitoids were found. Extensive plant screening is essential to achieve plant selectivity and to maximize biological control. *F. esculentum*, *C. cyanus* and *V. sativa* are recommended as selective plant species to enhance parasitoids of *M. brassicae* [14].

2.1.3. Selection of cultivars according to the resistance and ecological plasticity criteria

The cultivar is perhaps the most important factor that productivity and quality depend on. Because of its biological and technological potential, it will be expressed in terms of appropriate measures [15].

In order to choose the most suitable cultivar for OA, the farmer should take into account main criteria: consumer preferences regarding appearance, taste [2], etc.; climate and soil conditions, adaptation to extreme environmental conditions; extreme temperatures, the length of the photoperiod, tolerance to high concentrations of salts, and economic use of fertilizers; resistance or tolerance to diseases and pests; cultivation technology (field, greenhouse, tunnels, time of sowing, planting and the harvesting period, irrigated regime or less, mechanization) [16]; and product destination: fresh consumption and industrialization (canning, freezing, dehydration, etc.);

A cultivar cannot meet all these requirements, but, depending on the destination of the products and both the consumers' requirement and farmers' preferences, the most suitable biological material will be chosen under the given conditions [17].

There are very different requirements from the growers regarding variety characteristics, depending on the size of the surfaces and the destination of the products. Thus, for small gardens, created by amateurs for their own consumption, large fruit species can be cultivated, as they are more sensitive to transport and storage. OA can be used as varieties, hybrids, local populations, and clones [3, 4], but not accepted genetically modified organisms.

Choosing varieties and hybrids with resistance to pathogens and pests is necessary both for protected crops and for early field crops, because the investment is often large, so risks and loss must be eliminated [18–20].

For many crops (tomatoes, cucumbers, eggplants, bushes, or trees) grafted method may be used that causes plant vigor and thus resistance to nematode (**Figure 3**).

In **Table 1** are presented any cultivars with resistance or tolerance to the attack of different pests, especially for nematode control, in temperate climate conditions.

Figure 3. Fado hybrid grafted on the Rezistar rootstock for an attack on nematodes (photograph by Stoleru Vasile).

Table 1. Varieties created with resistance or tolerance to various pests.

Species	Cultivar	Pest resistant or tolerance
Tomato	Getina F1 Gloria F1, Splendid F1, Solara F1, Nemarom F1	*Meloidogyne incognita Chitw.*, *Meloidogyne hapla Chitw.*
Dianthus	Sooty	*Meloidogyne arenaria Neal.*,
Pineapple	Turiacu	*Meloidogyne arenaria Neal.*,
Zucchini	Amalthee	*Meloidogyne* spp.
Cucumber	Dasher II	*Meloidogyne* spp.
Soybean	Huasteca 300	*Tamaulipas state*

Recent research on the outside cabbage crop in the temperate climate highlighted, Timpurie de Vidra cultivar (cv) of early cabbage is most resistant to the cabbage fly (8.5% degree of attack) in comparison with the Golden acre cv., where the degree of attack was 14.2%, during two study years [21].

The reaction of cultivars resistant to pests and the nematode default may be determined by its presence in the plant silica [22], iron [23] genes that provide resistance [18, 24, 25], or protein presence in bean or cowpea [23, 26, 27].

2.1.4. Seed conditioning

Numerous pests, especially from the coleopteran order, can be found between the seeds or inside them during sowing, as they feed within their endosperm, endangering seed germination or weakening the newly sprouted plant [8]. The larvae and adults of nematodes (*Ditylenchus dipsaci, Tylenchorhynchus cylindricus*) attack both the garlic and onion bulbs but also the roots of the vegetables, making the plant die dry [28, 29].

2.1.5. Crop rotation

Effective crop rotations are fundamental to pest control in OS. Correct rotations provide an obstacle to the pest life cycles by removing host crops for prolonged periods of time. They also help in supporting a more diverse and stable agro-ecosystem to assist with natural pest suppression.

In areas where the climate permits, two or three crops can be grown during on the year on the same area, both in greenhouses or tunnels. From this point of view, it must be considered as species that succeed have no common pests (**Table 2**).

Table 2. Plot design for successive crops in greenhouse/tunnel.

No.	Crop	Sowing period	Planting	End of the crop
1	Lettuce, anticipated	20 VIII–10 IX	20 IX–10 X	15–30 III
2	Sweet pepper	10–15 I	1–15 IV	20–30 IX
3	Green onion	25–30 IV	1–15 X	25 III–5 IV

For the outdoor crops, in **Table 3**, some design rotation successive crops are presented. For these designs bear in mind that the crops that are grown on the same land area must belong to the same botanical family, have no common diseases and pests, and have different growing seasons.

Table 3. Plot design for successive crops in the field.

Type of crop	Crop	Sowing period	Planting	End of the crop
Vegetable design (1)	Radish	10–20 III	–	10–15 IV
	Tomato	10–15 II	15–20 IV	25–30 IX
Vegetable design (2)	Peas pods	1–15 III	–	10–20 VI
	Late cabbage	10–15 V	20–30 VI	15–20 X
Mixed crop design (1)	Barley	20–30 IX	–	10–15 VI
	Cauliflower	10–15 V	25–30 VI	1–10 X
Mixed crop design (2)	Wheat	10–20 X	–	10–20 VI
	Cucumber	1–10 VII	–	1–5 X

2.1.6. Crop Monitoring

Monitoring insects is fundamental in organic farming systems (OFS). Correct identification of insects and insect biology knowledge when they colonize crops is one of the main activities of management decisions that lead to optimal moment. This can be done by simply checking the crop (aphids, spider mites) or by using pheromone traps (thrips, cydia, white fly, rose fly, carrot fly and cabbage moth).

Pests of agricultural crops can cause damage directly (lower leaf surface, destroying fruit), (**Figure 4a**) or indirectly (gale or gates run for various soil diseases, such as *Rhizoctonia* sp. or *Fusarium* sp.), because many pests performing the biological cycle in soil (**Figure 4b**).

Prognosis and warning are performed by the centers dealing with plant protection, and they establish, at the right moment, the imminent danger of setting off massive pest attacks.

(a) (b)

Figure 4. (a) Eggplant leaf affected by *Thrips tabaci* (photograph by Stoleru Vasile). (b) Tomato root affected by *Thrips tabaci* (photograph by Stoleru Vasile).

2.1.7. Management Practices When It Comes To Pest Control

Cultural activities in organic farming may be considered as specific as crop production practices that implemented in the initial stages of the organic farm plan to reduce the likelihood of insect pest infestation. These measures are based on disrupting the biological cycle of the pest as follows: an unavailable crop to pests in space and time; unacceptable crop to pests by interfering with location; reducing the pest on the crop by natural enemies, etc.

Cultural practices are among the oldest techniques used for pest suppression, and many of the practices used in conventional and organic farming today have their roots in traditional agriculture. Effective deployment of cultural tactics is information intensive; it requires knowledge of pest–crop interactions and about the natural enemies of the pests.

2.1.8. Intercropping System

Intercropping is the practice of growing two or more crops (usually different families) in the same area. Strip cropping is a derivation of intercropping and is the practice of growing two or more crops in alternating strips across a field. Both practices serve to increase biodiversity and make the habitat less suitable for pest development (**Figure 5a, b**).

(a) (b)

Figure 5. (a) Intercropping management of the runner bean with maize (photograph by Hamburda Silvia). (b) Intercropping management of the runner bean with sunflower (photograph by Hamburda Silvia).

2.1.9. Tillage Management

Much of the pest population from both soil and foliar can be influenced through tillage practices. Tillage systems reduce insect pressure in succeeding crops. Fields are usually tilled in the fall or early spring when many kinds of insects are in the overwintering stage within the soil or in crop residues. Direct destruction of the insect or its overwintering chamber, removal of the protective cover, elimination of food plants, and disruption of the insect life cycle generally kills many of the insects through direct contact, starvation or exposure to predators, and weather.

Crop irrigation by sprinkler reduces the number of pests in crops [30]. Irrigation by culverts reduces the number of galas in the soil and thus causes interruption of the biological cycle of soil insects.

2.1.10. Mulches

Mulch is a layer film of material applied to the soil surface for the following reasons: to conserve moisture, to improve the fertility and health of the soil, to reduce weed growth, and to pressure soil land crop infestation with different pests [31].

Mulch is usually but not exclusively organic in nature (**Figure 6a**). It may be non-biodegradable (e.g., plastic sheeting) or biodegradable (e.g., bark chips). It may be applied to bare soil or around existing plants. Mulch consisting of manure or compost is incorporated naturally into the soil through the activity of worms and other organisms [32].

(a) (b)

Figure 6. (a) Organic cabbage mulched with phacelia (photograph by Stoleru Vasile). (b) Organic cabbage mulched with biodegradable plastic (photograph by Stoleru Vasile).

All mulch types suppress insects in comparison with bare soil. Different colors of plastic have been tested; clear, white, yellow, or aluminum (reflective) colors may provide some additional suppression of aphids and whiteflies [33]. Blue and yellow may bring in more pests. Plastic can be painted the desired color (**Figure 6b**). Before choosing a mulch type, farmers should check with their certifier bodies to see whether the practice is allowable by organic regulations [34].

2.1.11. Optimum Crop Health

The driving force behind the sustainability and environmental preservation derived through organic farming comes through healthy living soil. Microbes in the soil process organic matter to provide a balance of minerals and nutrients which are utilized by plants to achieve healthy, vigor crop growth. When this balance is achieved, the associated health of the crop gives it a heightened ability to withstand pest and disease attack. Good crop husbandry and hygiene also make a significant contribution to the health of the crop and the prevention of pest problems.

2.1.12. "Host Weed" Removal

Numerous dangerous species find favorable conditions for the summer or winter diapause on the spontaneous vegetation from the forest skirt, the borderline of strip ground, roads, or railways or the less cared for agricultural crops. So, the cabbage aphid has as host plant the cole, and the Colorado beetle has as host plant the black nightshade—*Solanum nigrum* [8].

Storing crops in hygienic conditions generally represents an additional source of pest infestation. (e.g., the bean weevil (*Acanthoscelides obsoletus*), pea weevil—*Bruchus pisorum*). They can be fought against either by storing the products in refrigerated storerooms for a certain period of time or by vacuuming the products in a special room [35].

2.2. Curative Measures

Curative care or curative measure is the health care given for environmental conditions where a measure is considered achievable, or even possibly so, and directed to this end. Curative care differs from the preventive method, which aims at preventing the appearance of pests, which concentrates on reducing the degree of the attack.

2.2.1. Physical–Mechanical Methods

According to specific regulation (EU 834/2007), in OA, it can be used following measures: thermotherapy, heliotherapy, radiotherapy, ultrasounds, nets, fences, or traps.

Thermotherapy is recommended only if the vegetal remains are highly infested with pests and, as much as possible, after collecting and removing the remains from the cultivated area. In OA according to EU Regulation 834/2007, this method is restrictive and can be applied only in problematic crops. If this is not possible, in situ burning may be used, but only after a thorough investigation of the opportunity of such a measure and registering it in the farm register and announcing the local organization of environmental protection (EU 889/2008).

Heliotherapy. The method is very simple and has been the subject of thorough research studies carried out at the Central Food Technological Research Institute in India [1]. This method consists of exposing the infested seeds to a temperature of 60°C for 10 min [17]. In order to do so, seeds are put in a dark color polyethylene bag with high molecular and density weight, which, at its turn, is tightly covered by another transparent low density polyethylene bag. The entire operation is carried out on a plane surface exposed to sun. The two foils act as a condenser making the temperature inside the seed bag quickly increase leading to the pests' death.

Radiotherapy is used for sterilizing males with the aid of X-rays and gamma radiations. Achieving the dominant lethal mutations has led to obtaining a biological method called autocide.

Figure 7. Ultrasound for pest control (photograph by Stoleru Vasile).

Scientifically, literature mentioned the effects of X-ray irradiation applied on six floriculture insect pests (*Tetranychus urticae, Myzus persicae, Bemisia tabaci, Liriomyza trifolii, Spodoptera litura*, and *Frankliniella intonsa*) placed in the bottom sections of rose and chrysanthemum pots. After irradiation with an X-ray dose of 150 Gy, the development of nymphs and adults of *M. persicae* and eggs, nymphs, and adults of *B. tabaci* was prevented at every position in the pots. *T. urticae* nymphs irradiated at 200 Gy newly emerged adults laid eggs in the bottom section of rose boxes only. *L. trifolii* adults irradiated at 200 Gy were completely inhibited. Radiotherapy method depends on dose of X-ray irradiation, insects, and crops [10].

Other physical or mechanical methods refer to installing various barriers, such as: nets for carrot fly, ultrasounds for soil insects (**Figure 7**), metallic fences for snails (**Figure 8**), layers for aphids and Lepidoptera's insects (**Figure 9**), traps or rollers (carrot fly, thrips) (**Figure 10**), flooding, and crushing the eggs of caterpillars or even the adults.

Figure 8. Metallic fence for protection against the snail (photograph by Stoleru Vasile).

Flooding provides better results in fighting against underground pests (mice, moles, crickets, etc.) by flooding their galleries. The impossibility of knowing the exact side of their galleries reduces the method's practical value and limits its use [36].

Figure 9. Early crops protected with Agryl P_{17} (photograph by Stoleru Vasile).

2.2.2. Biotechnical Methods

Installing food bait traps. They can consist of parts of plants, fruits, tubercles, or feed and are placed on the ground or in storehouses. After collecting the pests, traps are removed, soaked in boiling water or burnt [31, 37].

Installing pheromone traps. Pheromones are chemical substances secreted and spread outside the body and determine a response only from the individuals of the same species (**Figure 11**). There are multiple types of pheromones, according to the role they fulfill: sexual, alarm, aggregation, path marking, recognition, and social regulation (e.g., ATRAGAM and ATRAPOM are a sexual pheromone used for *Autographa gamma* and *Cydia pomonella*) [8, 38].

Figure 10. Thrips and whitefly plaque applied in tomato crops (photograph by Stoleru Vasile).

Table 4 presents other products that can be applied in organic farming, based on the pheromones.

Natural enemies (predators and parasites). This category includes methods in order to attract animals that eat insects and other harmful living animals.

The effect of control pest in OA is to increase functional biodiversity, that is, to use wild flowers to attract parasitoids into the cabbage field—or to retain them if we release them—to increase natural pest control, directly through the added plants and the organisms that use them as resources and indirectly through the reduction of pesticides.

Figure 11. Attractive traps for pests control in tomato crops (photograph by Stoleru Vasile).

Table 4. .Pheromones and attractants for organic farming.

Name of product	Pest	Crop	Pheromone/attractant	Application
Codling Moth Pagoda Trap with lure®	Codling moth	Apples	Pheromone attracts male moths, for monitoring only	8 traps/ha
Rollertrap®	Range of insects	Various	Two sided sticky trap	Yellow and blue
Pheromone trap®	Butterflies and moths	Protected and field crops	Monitors butterfly and moth population	5–8 traps/ha
Agralan Envirofleece®	Various pests	Protected and field crops	Polypropylene fleece, physical barrier to pests	17–30 g/m^2

Creating proper shelters and feed for the useful fauna (frogs, green lizards, snakes, insectivore insects, and mammals), including their artificial breeding, have positive effects for farmers. Snakes can be used against rodents; hedgehogs counteract the attack of shell-less snails, mice, mole crickets, and also the Colorado beetles [39].

Predators catch and eat their prey. Some common predatory arthropods include ladybird beetles, carabid (ground) beetles, staphylinid (rove) beetles, syrphid (hover) flies, lacewings, minute pirate bugs, nabid bugs, big-eyed bugs, and spiders.

Entomophagy predators are species of animals which consume other animals, pests in particular.

The main species of insects and nematodes used for fighting against harmful insects are presented in **Table 5**. This method of biological control is widely used in horticulture, especially in protected areas, such as flower, orchard, and vegetables crops (**Figures 12–17**).

Table 5. Parasites and predators permitted for organic pest control.

Name of products	Pests controlled	Crops	Parasites/predators	Application/dose
Aphipar®	Aphids (cotton aphid, peach and potato aphid, tobacco aphid	Protected crops	*Aphidius colemani* parasitic wasp	Preventive = 0.25 ex. m^2, curative light = 1 ex. m^2, curative heavy = 2 ex. m^2 (7 days interval, 3–6 application/year)
Ervipar®	Aphids (potato aphid, glasshouse potato aphid)	Protected crops	*Aphidius colemani* parasitic wasp	Preventive = 0.25 ex. m^2, curative light = 0.5 ex. m^2, curative heavy = 2 ex. m^2 (7 days interval, continuously application)
Aphidend®	Aphids	Protected crops	*Aphidoletes aphidimyza* (gall midge)	Curative light = 1 ex. m^2, curative heavy = 10 ex. m^2 (continuously application)

Chrysoperla®	Aphids, whitefly, various thrips, caterpillars	Various	*Chrysoperla carnea* lacewing	ex./m², curative light = 1 ex./m², curative heavy = 5 ex./m² (continuously application)
Aphilin®	Aphids	Protected crops	*Aphelinus abdominalis* parasitic wasp	0.1–0.5 adult/m² for preventive use
Ervibank®	Aphids	Protected crops	*Aphidius ervi* parasitic wasp	Preventive = 0.5 ex./m² on interval of 14 days
Tricho-strip®	Caterpillars (lepidopteran eggs)	Protected crops	*Trichogramma brassicae* parasitic wasp	Preventive: min. 8 × 5 ex./m² each 7 days, curative light: min. 8 × 10 ex./m² each 7 days, curative heavy: min. 8 × 20/m² each 7 days
Fightacat®	Caterpillars (lepidopteran eggs)	Various	*Trichogramma evanescens* parasitic wasp	Preventive: min. 8 × 5 ex./m² each 7 days, curative light: min. 8 × 10 ex./m² each 7 days, curative heavy: min. 8 × 20/m² each 7 days
Anagrus®	Leafhopper	Protected	*Anagrus atomus* parasitic wasp	Preventive = 0.1 ex./m², curative heavy = 0.5 ex./m² (7 days interval)
Minex® Fightamine® Minusa®	Leaf miners	Protected crops	*Dacnusa sibirica* and *Diglyphus isaea* parasitic wasps	Preventive = 0.25 ex./m², curative light = 0.5 ex./m², curative heavy = 2 ex./m² (continuously application)
Miglyphus®	Leaf miners	Protected crops	*Diglyphus isaea* parasitic wasps	Preventive = 0.1 ex./m², curative heavy = 1 ex./m² (continuously application)
Cryptolaemus®	Mealy bug	Protected crops	*Cryptolaemus* sp. Australian ladybird	*Greenhouses*: 5 beetles per infested plant, *outdoors*: 1250–12,500 beetles per hectare *orchards*: 2500–5000 beetles per hectare

Spidex® Spidex-T® Fightamite A®	Mites (two spotted spider mite and carmine spider mite	Protected crops	*Phytoseiulus persimilis* predatory mite	Curative light = 0.5 ex./m^2, curative heavy = 2 ex./m^2 (continuously application every week)
Fightamite B®	Mites (two spotted spider mite)	Protected crops	*Feltiella acarisuga* (*Therodiplosis persicae*) predatory midge	250 adults/1000 m^2
Typhlodromus®	Mites (red spider mite, two spotted spider mite, strawberry mite, broad mite, fruit tree spider mite)	Protected and outdoor crops	*Typhlodromus pyri* predatory mite	250–500 adults/1000 m^2
Fightascale®	Soft scale insect	Protected crops	*Metaphycus helvolus* parasitic wasp	1 adult/m^2
Entonem® Nemasys®	Sciarid flies	Various crops,	*Steinernema feltiae* nematode	200–400 ex./m^2
Entomite®	Soil-living insects, thrips, collembola, nematodes, sciarid files	Various	*Hypoaspis aculeifer* or *Hypoaspis miles* predatory mites	Preventive = 100 ex./m^2, curative light = 200 ex./m^2, curative heavy = 500 ex./m^2 (one application)
Nemaslug® Slugsure®	Slugs	Various	*Phasmarhabditis hermaphrodita* nematode	500–1000 ex./m^2
Thripex® Fightathrip®	Thrips (various), spider mites	Protected crops	*Amblyseius cucumeris* predatory mite	Preventive = 50 ex./m^2, curative light = 100 ex./m^2(application at 14 days), curative heavy = 100 ex./m^2 (one application/week)
Thripor® Fightabug®	Thrips (various)	Protected crops	*Orius laevigatus*, *Orius insidiosus* or *Orius majusculus* predatory bug	Preventive = 0.5 ex./m^2, curative light = 1 ex./m^2, curative heavy = 10 ex./m^2 (one application/14 days)
Larvanem® Nemasys H®	Vine weevil	Various	*Heterorhabditis megidis* nematode	Curative light = 500.000/m^2, curative heavy = 1.000.000/m^2 (one application)
En-strip®	Whitefly	Protected crops	*Encarsia formosa* parasitic wasp	Preventive = 1.5–3 ex./m^2, curative light = 3–6 ex./m^2, curative heavy = 9 ex./m^2 (one applic./week)

Fightafly B®	Whitefly, leafhopper, leaf miner, spider mite	Protected crops	*Macrolophus caliginosus* predatory bug	Curative light = 10 ex./m^2, curative heavy = 50 ex./m^2 (one applic./14 days)

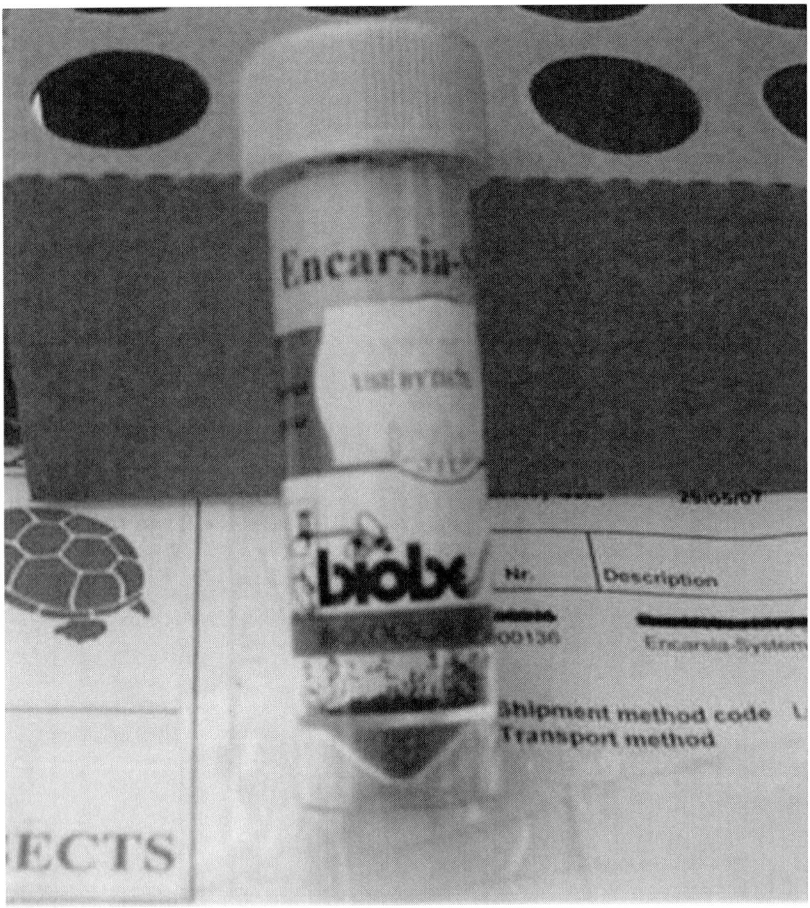

Figure 12. *Encarsia formosa* for greenhouse crops (photograph by Stoleru Vasile).

Figure 13. Applying the parasite wasp to a cucumber crop (photograph by Stoleru Vasile).

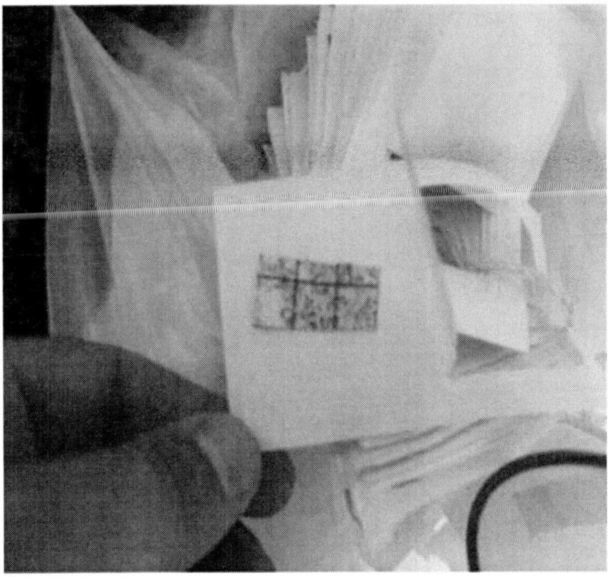

Figurc 14. *Trichogramma* eggs plaques made in a laboratory (photograph by Stoleru Carmen).

Figure 15. Application of *Trichogramma* plaques for white butterfly eggs (photograph by Stoleru Carmen).

Figure 16. Tomato fruit damage by *Helicoverpa armigera* (photo by Deleanu Florina).

Figure 17. Whitefly in on the flower crop (photograph by Stoleru Vasile).

Biological methods. Biological control consists of using organisms and products against other living beings. The methods correspond to the future approaches; they are characterized by high selectivity and improbability levels regarding the fact of inducing the pest resistance phenomena, as well as a good capacity of self-perpetuation.

Economically speaking, these methods are more expensive, at least initially, when they have to be projected and produced, or when special installations are necessary and they require a lot of manual work for operation or for the uphill works. But in the end, does not the environment's health and ours implicitly deserve a bonus from the beneficiary?

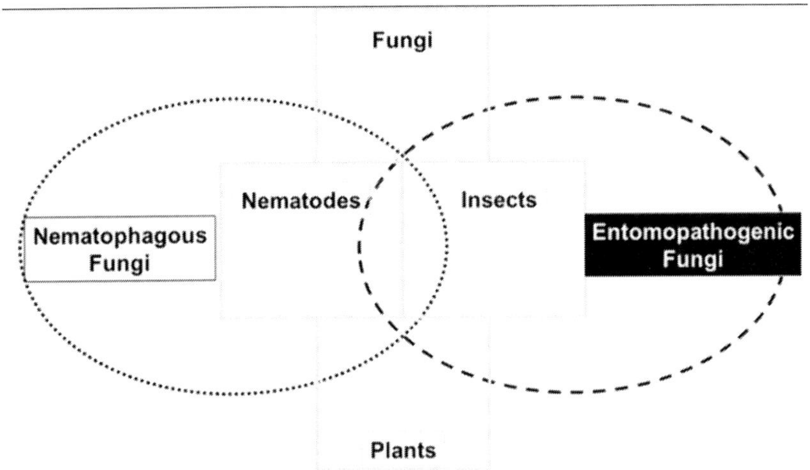

Figure 18. Multitrophic lifestyle of fungal parasites [38].

Microbiological control is a modern, efficient method but still quite expensive; it consists of using certain preparations based on living organisms (viruses, bacteria, fungi) that parasites and kill some of the pests.

Nowadays, more than 500 species of insect parasite fungi are known. Their advantage is that they spread out easily through spores and they are resistant to unfriendly conditions for long periods of time (**Figure 18**). In general, the relation between pests and their parasites are affected by global change, abiotic and biotic stresses to crops [40].

Beauveria sp. and *Metarhizium* sp. are two pathogenic fungi for insects which can penetrate the host insect through its exoskeleton due to its production of chitinolytic enzymes (**Figures 19** and **20**). Once inside the host, the fungus develops and feeds, causing its host's death.

Figure 19. *B. bassiana* parasitism for *Bemisia tabaci* control (photograph by Sellitto Michele).

The infested insects, still living, experience limited motion ability and the incapacity to feed themselves; moreover, they represent a source of infection for other insects [37].

Figure 20. *B. bassiana* on palm carbide (*Rhynchophorus ferrugineus*) (photograph by Sellitto Michele).

Different studies have shown that *Beauveria* sp. and *Metarhizium* sp. actively control species from the following genera Coleoptera (*Melolontha* sp., *Diabrotica* sp.), Lepidoptera (*Tuta absoluta*), or Orthoptera [aphids, greenhouse whitefly, thrips, [41, 42] etc].

Pochonia sp. is a hyphomycete that acts as a parasite of nematode eggs. Its antagonistic activity is related to the production of proteolytic and chitinolytic enzymes that degrade the cellular structure of nematodes, especially that of eggs and females in the early stage (**Figures 21** and **22**).

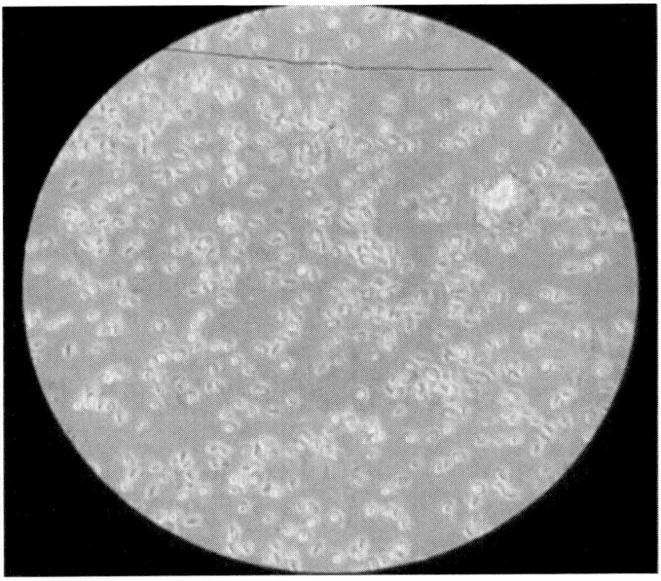

Figure 21. Conidi of *Pochonia chlamydospores* (photograph by Sellitto Michele).

Figure 22. Tubers of a potato attacked by nematodes (photograph by Aurelio Ciancio).

Arthrobotrys sp. is a fungus that parasitizes nematodes. The nematodes' biocontrol activity is related to the production of ring-like structures which swallow when a nematode pass by and catches it. Afterwards, the nematode is degraded by enzymes and used by the fungus as feed.

The combination between *Pochonia* sp. and *Arthrobotrys* sp. represents the most effective biological control method for the nematodes from a genera *Meloidogyne* sp., *Globodera* sp., and *Heterodera* sp. (**Figures 23** and **24**).

The literature dealing with this subject mentions tests that proved that the use of these fungi, on soils sterilized using chemical products and solarization and steam, has maintained the soil and the level of nematodes below the damaging threshold for many years, compared to the soils where these fungi were not present [43].

Figure 23. Adult of a nematode parasited by *Pochonia chlamydospores* (photograph by V.M. Sellitto, 2014).

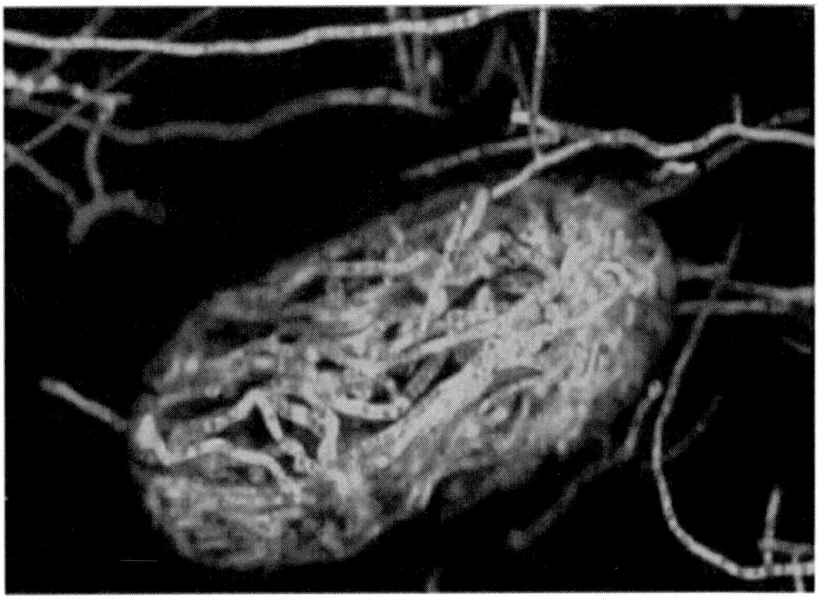

Figure 24. Egg of a nematode parasited by *Pochonia chlamydospores* (photograph by L. Lopez-Llorca, 2015).

Lecanicillium lecanii is a pathogenic fungus for numerous species of insects. This fungus acts as follows: the fungus spores lie and remain on the insects' exoskeleton, and then, they germinate and mechanically penetrate the insects' exoskeleton, due to their production of chitinolytic enzymes. From the industrial products containing entomopathogenic fungi, we mention the following: Muscardin M 45® and *Beauveria* spores (from *B. bassiana*), Boverin® (from *B. densa*), and Mitecidin® (from *Streptomyces aureus*), which act against the Colorado beetle and other coleopters (**Table 6**). Applying myco-insecticides, Naturalis-L® (*Beauveria bassiana*) and PreFeRal®WG (*Paecilomyces fumosoroseus*), were applied against adult *Rhagoletis cerasi* (Diptera: Tephritidae). In the first case, *B. bassiana* significantly reduced the number of damaged fruit (efficacy: 69–74%), whereas damage was not significantly reduced with PreFeRal®WG (efficacy: 27%) [44].

Table 6. Biological control agents used in organic farming.

Name of product	Pest	Crop	Microorganisms	Dose/application
Vertalec®	Aphids	Protected crops	*Verticillium lecanii* fungal spores	2 g/L
Mycotal®	Whitefly, thrip larvae	Protected crops	*Verticillium lecanii* fungal spores	curative light = 0.1% (2–3 applic.), curative heavy = 0.1% (3–4 applic.)
Novosol FC® Dipel WP® Bactura WP®	Caterpillars	Vegetables, fruit, ornamentals	*Bacillus thuringiensis* var. kurstaki	1–1.6 kg/ha, depending of crop
Thuricide®	Caterpillars	Various	*Bacillus thuringiensis* var. *kurstaki*	2–4 tsp/gal water
Bactospeine®	Caterpillars	Various	*Bacillus thuringiensis* wettable powder	1–1.6 kg/ha, depending of crop

Once the fungus is in, it develops and digests the insect from the inside until it kills it. Infested insects die in 4–6 days and are then covered with a whitish efflorescence, depending on the fungus sporulation. Thus, these insects become a source of infection for other insects. In addition, *Lecanicillium lecanii* can colonize certain tissues of the host plant, achieving an induced systemic resistance.

Many studies have shown that *L. lecanii* controls aphids, whitefly, and Thripidae genus. Other studies have proven that *Lecanicillium* sp. also controls certain nematode species as well as certain plant diseases, such as the gray mold (**Table 7**).

Table 7. Microbiological products to control pests from vegetable crops.

No.	Commercial name	Biological composition	Dose
1.	Pochar Linia Greenpower®	*Glomus* sp., *Pochonia* sp., *Arthrobotrys* sp.	2–3 l/ha in time and after transplantation
2.	Lecan Linia Greenpower®	*Glomus* sp., *Lecanicillium* sp.	2 l/ha or 0.2% foliar
3.	Metab Linia Greenpower®	*Glomus* sp., *Metarhizium* sp., *Beauveria* sp.	2 l/ha at the root or 0.2% foliar

From the bacteria used to fight against insects, *Bacillus thuringiensis* (**Figure 25**) and *B. subtillis* are the most popular (**Figure 26**).

Figure 25. Bacillus thuringiensis var. kurstaki (photograph by V.M. Sellitto).

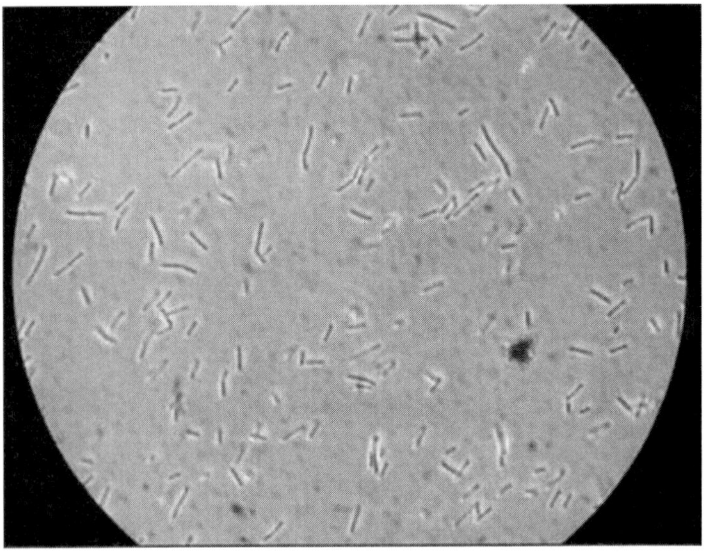

FIGURE 26. *Bacillus subtillis* (photograph by V.M. Sellitto).

During the last years, strains of *B. thuringiensis* were studied for their effect on the insect, through different toxins (**Table 8**).

Table 8. Toxins produced by strains of *Bacillus thuringiensis.*

No.	Toxins	Activities
1	Cry toxins	Pore formation on cell membrane; cytolysis activity
2	Vip toxins	Wide spectrum of insect activity
3	Thuricin	Bacteriocin
4	Hemolysin	Lysis of vertebrate red blood cells
5	Beta-exotoxins	Inhibition of RNA polymerase
6	Phospholipase-C	Cell membrane alteration

It laid at the basis of the process of obtaining numerous commercial products: Agritol®, Dipel®, Thuricide®, Novodor 3FC®, Vectobac®, Bactospeine®, Thuringine®, Entobakterin®, Thurintox®, or Foray®. These products are highly efficient in counteracting the larvae of certain butterflies from vegetables crops [37].

Out of more than 300 viruses that cause diseases for more than 175 species of insects, polyhedric viruses are the most known; they are used at obtaining certain preparations industrially, such as Biotrol VHZ® and VSE®, Vitex® (against caterpillars), and Virin-ENS® (recommended in fighting against the cabbage moth). Nuclear polyhedrosis viruses (NPV) and granulosis viruses (GV) are available to get rid of some caterpillar pests (*Mamestra brassicae, Helicoverpa armigera, Autographa gamma, Pieris brassicae, and Euproctis chrysorrhoea*) [45] (**Figure 27**).

Figure 27. Uninfected (bottom) beet armyworm (*Spodoptera exigua*) and beet armyworm killed by the nuclear polyhedrosis virus. Photograph credit: David Nanace, USDA ARS.

Genetic methods. The works of ameliorating plants have as their main objective the production of cultivars endowed with greater resistance. This is why the forms providing higher mechanical resistance are promoted (with thicker cuticle or suber, with a waxy protective layer or with abundant porosity), physiological or chemical (by growing the content of substances with repellent or insecticide effect).

Several aphid species can proliferate in winter lettuce crops, such as *Nasonovia ribisnigri* (Mosley), *Myzus persicae* (Sul.), *Aulacorthum solani* (Kalt.), *Macrosiphum euphorbiae* (Th.), and *Hyperomyzus lactucae* (L.). *N. ribisnigri* is the most damaging one because it preferentially develops in the lettuce heart [46, 47]. In addition to feeding damage and the loss of product quality due to their presence when the lettuce is marketed, aphids are also vectors of viruses, such as the lettuce mosaic virus. Finally, slugs (*Deroceras* sp. and *Arion* sp.) and snails can also cause feeding damage to lettuce in winter.

Complete resistance to the aphid *N. ribisnigri* and partial resistance to *M. persicae* are conferred by a dominant gene called Nr, which has been introduced in many European cultivars [48]. However, this resistance was recently bypassed by a new *N. ribisnigri* biotype named Nr:1 [49].

2.2.3. Using plants to fight against pests

This method relies on certain plants' feature of secreting in the earth or in the air certain substances with repulsive or destructive effects on pests. By and large, these plants can be cultivated in the field, as border or associated with the crops. The important species with insecticide effect are presented in **Table 9**.

Biochemical methods. The products used for protecting plants against harmful insects can be classified according to the raw material used, into two categories: vegetal insecticides and mineral insecticides.

Table 9. Plants used in organic farming with a repellent effect.

Species	Controlled pests
Yarrow (*Achillea millefolium*)	Aphids, mites, psyllids, thrips
Queen of poisons (*Aconitum* sp.)	Coleopteran larvae
Sweet flag (*Acorum calamus*)	White cabbage butterfly
Onion (*Allium cepa*)	Mites, ants, storehouse pests
Garlic (*Allium sativum*)	Thrips, storehouse pests
Birthwort (*Aristolochia clematitis*)	Bed bug
Absinthium (*Artemisia absinthium*)	Nematodes, caterpillars, fleas
Mugwort (*Artemisia vulgaris*)	Fleas, Colorado beetle
Lamb's quarters (*Chenopodium album*)	Colorado beetle, white butterfly
Hemlock (*Conium maculatum*)	Coleopteran larvae
Coriander (*Coriandrum sativum*)	Aphids, spiders, Colorado beetle (repellent effect)
Spurge (*Euphorbia* sp.)	Caterpillars, aphids
White sweet clover (*Melilotus albus*)	Colorado beetle
Mint (*Mentha* sp.)	Colorado beetle
Tobacco (*Nicotiana tabacum*)	Aphids, mites, Colorado beetle

Black nightshade (*Solanum nigrum*)	Aphids, mites, Colorado beetle, cabbage butterfly
Yew (*Taxus baccata*)	Various insects
Field penny-cress (*Thlaspi arvense*)	Bed bug (repellent)
Common nettle (*Urtica dioica*)	Aphids, mites
Mullein (*Verbascum phlomoides*)	Colorado beetle

Vegetal insecticides. Insecticides of natural origin are substances which can cause the death of insects interfere in the development or reproduction being responsible to attract or repel them. Today, worldwide, there are more than 1450 species of plants with insecticide effects, from which only approximately 50 are useful [1]. As far as our country is concerned, too little from the 200 species credited with this action have been or are being effectively used in this purpose, and even fewer have been studied from this point of view.

Stinging nettle (*Urtica dioica*). Action: it stimulates plant growth, it slows down the attack of certain insects, counteracts aphids, and spiders before the formation of leaves and flowers [37].

Fern (*Dryopteris filix-mas*). Leaf purine and decoction, undiluted, are used against shell-less snails (every time needed). At the same time, this product, diluted 10 times with water, is used for the late spring treatments against aphids.

Wormwood (*Artemisia absinthium*). This plant can be used as an undiluted purine (caterpillars, lice), cold extract diluted twice for Solanaceae against the larvae of the Colorado beetle [37], or decoction is used undiluted against the cabbage fly [2].

Tansy (*Tanacetum vulgare*) is used as an undiluted infusion every time it is needed against ants, aphids, fleas and other insects.

Wild garlic (*Allium ursinum*). Wild garlic infusion is used undiluted, by repeatedly aspersing the plants every 3 days against aphids and mites. Purine is also used undiluted against the carrot fly (*Psila rosae*), but only during its flight period.

Garlic (*Allium sativum*). It can be used in the treatment of mites and also in seed treatments. Garlic in its natural state is eventually cultivated in rows, has a nematode effect (*Meloidogyne* sp.), and drives away the striped field mouse.

Table 10. The action spectrum of the *Chrysanthemum cinerariaefolium* extract.

Scientific name	Common name	Scientific name	Common name
Leptinotarsa decemlineata	Colorado beetle	*Acyrthosiphon pisum*	Pea aphid
Mamestra brassicae	Cabbage moth	*Pieris brassicae*	Large white
Pieris rapae	Small white	*Trialeurodes vaporariorum*	Greenhouse whitefly
Gnorimoschema lycopersicella	Tomato pinworm		

Pyrethrum (Chrysanthemum cinerariaefolium, Pyrethrum cinerariaefolium). Pyrethrum is a contact insecticide having paralyzing effect and a wide range of actions. The great advantage, ecologically speaking, is that it completely decomposes into harmless compounds in only 48 h after application [50]. Pyrethrum is noticed on a large number of insects and mites with a soft body or when they are still in a larval stage, as a solution with concentration of 0.1% (**Table 10**). The extract of pyrethrum cannot recommend mixture with alkaline products, Bordeaux mixture [1, 39].

Derris powder (*Derris* sp.). Derris powder is applied to a large number of aphids, nematodes, and insects, more vulnerable as their ingestion capacity is larger (larvae). Its toxicity for warm blooded animals is null, while for the other ones, it is lethal, used as decoct of ground fresh or dried roots, in a solution of 0.01%.

Gliricidia (*Gliricidia sepium*). Action: repellent, parasitic, rodenticide, mixed with grain, left from place to place on a field or put in warehouses; in a few days, it kills the rodents [9, 51].

Neem (*Azadirachta indica*). It is a repellent, hormonal disruptive (it blocks the larval metamorphosis process), nematocide and antimicotic. Azadirachtin is extracted from this plant's seeds, the active substance of NeemAzal T/S®.

The preparations destroy the eggs, larvae, and adults of more than 200 species of field or storehouse pests in the case of beans, cereals, tomatoes, and field plants from the most various classes: nematodes, ants, bed bugs, grasshoppers, etc. Neem oil is used in fighting against certain pests on plants, and ground marc has a nematode effect [33].

Bitter wood (*Quassia amara*). The active substances of this preparation act as contact and ingestion insecticide but are slower than pyrethrum. It is used in fighting against many pests: aphids, flies, cabbage aphids, etc.

Decoct is made from 100 to 150 g chips of bitter wood at 10 l water. The bitter wood decoction can be improved by adding an equal amount of solution of potassium soap in a concentration of 1–2.5% [51].

Traditionally, in organic fruit growing, the apple sawfly *Hoplocampa testudinea* Klug is controlled by the use of extracts of bitter wood of 6 g/ha/in 500 l. For a good efficiency, the bitter extract can be mixed with Nemmazal T/S® [52, 53].

Table 11. Commercial products permitted to use in organic farming.

Name of product	Pests	Crops	Agent for control	Dose/concentration
Savona®	Whiteflies, thrips, aphids, mealy bugs, leafhoppers	Various	Fatty acid	1–2%, one applic./week
Liquid Derris®	Various biting and sucking insects	Various	Rotenone derived from *Lonchocarpus utilis and L. urucu*	0.8–1 l ha
Bug-Me-Not Bloom and Leaf Astringent Spray (CP) ®	Various insects	Various	Insect repellent based on neem extract	Insect repellent. 4–6 tsp./10 l
Bug-Me-Not Root and Soil Granules (CP) ®	Various insects	Various	Insect repellent based on neem extract	Insect repellent. 4–6 tsp./10 l
AquaPy®	For insects in grain stores	Store	Natural pyrethrum, organic products (e.g., grain) must be removed before use	1 l 3000 m^3
Jet 5®	For cleaning glasshouses polytunnels	Store	Peroxyacetic acid	0.2%

2.2.4. Repellent Mineral Products

Potassium alum. This preparation is used as solution with a concentration of 0.4% with good efficacy against lice and caterpillars. At the same time, aspersing the soil with this solution is quite efficacious against shell-less snails. Basalt flour. It is used as a powder. Its action against pests is explained because of a change of the pH at the surface of aerial organs from weak acid (preferred by most pests) to weak alkaline or mechanical action on the insect's body, their eyes, and trachea [2].

2.2.5. Insecticide Mineral Preparations

Potassium soap is successfully used against mites (red spider) and the cabbage aphid. The treatment is applied alone or in combination with other products (horsetail extract) by repeatedly aspersing the plants with various solution types: 200–300 g soap at 10 l water (lice); 200–300 g soap + 0.5 l alimentary alcohol + 1 table-spoonful of lime and 1 table-spoonful of cooking salt at 10 l of water, against the red spider and the larvae of the Colorado beetle [39].

The preparation is used as solution with concentration of 1–2% with good efficacy against lice and leaf fleas, found under the name Neudosan® or Savona® [9], like as other products presented in **Table 11**.

3. ACKNOWLEDGEMENTS

The authors wish to thank Mr. Dean Hufstetler for reviewing article.

REFERENCES

1. Toncea I. Practical Guide for organic agriculture. Organic technologies for land cultivation. Cluj-Napoca: Publisher Academic Press 2002.
2. Stoleru V, Munteanu N, Sellitto VM. New approach of organic vegetable systems. Aracne Editrice; Rome 2014.
3. EC Regulation 834. Organic production and labelling of organic products and repealing Regulation (EEC) No 2092/91, (2007).
4. EC Regulation 889. Laying down detailed rules for the implementation of Council Regulation (EC) No 834/2007 on organic production and labelling of organic products with regard to organic production, labelling and control, (2008).
5. Willer H. The World of Organic Agriculture. Statistics and Emerging. Trends 2015. Frick: FIBL; 2015. 306 p.
6. Knipling EF. Entomology and the Management of Man's Environment. Australian Journal of Entomology. 1972;11(3):153–67.
7. Hatman M, Bobes I, Lazar A, Gheorghies C, Glodeanu C, Severin V, Tusa C, Popescu I, Vonica I. Phytopatology. Bucharest: Editura Didactica si Pedagogica; 1989. 468 p.

8. Georgescu T. Horticultural entomology: Dosoftei Publisher; 2006. 426 p.
9. Stoleru V, Albert IO. Vegetable growing using organic measures. Risoprint; Cluj-Napoca 2007.
10. Yun S-H, Koo H-N, Kim HK, Yang J-O, Kim G-H. X-ray irradiation as a quarantine treatment for the control of six insect pests in cut flower boxes. Journal of Asia-Pacific Entomology. 2016;19(1):31–8.
11. Osouli S, Ziaie F, Haddad Irani Nejad K, Moghaddam M. Application of gamma irradiation on eggs, active and quiescence stages of *Tetranychus urticae* Koch as a quarantine treatment of cut flowers. Radiation Physics and Chemistry. 2013;90:111–9.
12. Boiteau G, Heikkilä J. Chapter 12 - Successional and Invasive Colonization of the Potato Crop by the Colorado Potato Beetle: Managing Spread. In: Alyokhin A, Giordanengo CV, editors. Insect Pests of Potato. San Diego: Academic Press; 2013. p. 339–71.
13. Zeleke KT, Nendel C. Analysis of options for increasing wheat (*Triticum aestivum* L.) yield in south-eastern Australia: The role of irrigation, cultivar choice and time of sowing. Agricultural Water Management. 2016;166:139–48.
14. Pfiffner L, Wyss E. Use of sown wildflower strips to enhance natural enemies of agricultural pests. In: Gurr GM, Wratten SD, Altieri MA, editor. Ecological Engineering for Pest Management: Advances in Habitat Manipulation for Arthropods. Collingwood, VIC, Australia: CSIRO Publishing; 2004. p. 167–88.
15. Munteanu N. Tomatoes, peppers and eggplants. Iasi: Ion Ionescu de la Brad; 2003. 214 p.
16. Indrea D, Apahidean S, Apahidean M, Maniutiu D, Sima R. Vegetable growing Editor. Bucuresti: Ceres; 2012. 624 p.
17. Ciofu R, Stan N, Popescu V, Chilom P, Apahidean S, Horgos A, Berar V, Lauer KF, Atanasiu N. Tratat de legumicultura. Bucuresti: Editura Ceres; 2004. 1165 p.
18. López-Pérez J-A, Le Strange M, Kaloshian I, Ploeg AT. Differential response of Mi gene-resistant tomato rootstocks to root-knot nematodes (*Meloidogyne incognita*). Crop Protection. 2006;25(4):382–8.
19. Schneider JHM, s'Jacob JJ, van de Pol PA. Rosa multiflora 'Ludiek', a rootstock with resistant features to the root lesion nematode *Pratylenchus vulnus*. Scientia Horticulturae. 1995;63(1–2):37–45.
20. King SR, Davis AR, Zhang X, Crosby K. Genetics, breeding and selection of rootstocks for Solanaceae and Cucurbitaceae. Scientia Horticulturae. 2010;127(2):106–11.
21. Stoleru V, Munteanu N, Stoleru CM, Rotaru L. Cultivar selection and pest control techniques on organic white cabbage yield. Notulae Botanicae Horti Agrobotanici. 2012;40(2):190–6.
22. Vilela M, Moraes JC, Coelho M, Françoso J, Santos-Cividanes TMd, Sakomura R. Response of moderate pest resistant and susceptible cultivar of sugarcane to silicon application. American Journal of Plant Sciences. 2014;5(26):3823–8.
23. Abdel-Sabour AG, Obiadalla-Ali HA, AbdelRehim KA. Genetic and chemical analyses of six cowpea and two Phaseolus bean species differing in resistance to weevil pest. Journal of Crop Science and Biotechnology. 2010;13(1):53–60.
24. López-Gómez M, Flor-Peregrín E, Talavera M, Verdejo-Lucas S. Suitability of zucchini and cucumber genotypes to populations of *Meloidogyne arenaria*, *M. incognita*, and *M. javanica*. Journal of Nematology. 2015;47(1):79–85.
25. Andersson SC, Johansson E, Baum M, Rihawi F, El-Bouhssini M. New resistance sources to Russian wheat aphid (*Diuraphis noxia*) in Swedish wheat substitution and translocation lines with rye (Secale cereale) and Leymus mollis. Czech Journal of Genetics and Plant Breeding. 2015;51(4):162–5.
26. Silva SZd, Alves LFA, Coelho SRM, Tessaro D. Technological and nutritional quality of cowpea (*Vigna unguiculata* (L.) Walp.) cultivars infested by *Callosobruchus maculatus* (Fabr.) (Coleoptera: Bruchidae). Journal of Food, Agriculture and Environment. 2014;12(2):731–4.

27. Maldonado Moreno N, Ascencio Luciano G, Gill Langarica HR. Huasteca 300, a new soybean cultivar for the south of Tamaulipas state. Agricultura Técnica en México. 2009;35(4):481–5.
28. Lopez-Llorca LV, Olivares-Bernabeu C, Salinas J, Jansson H-B, Kolattukudy PE. Pre-penetration events in fungal parasitism of nematode eggs. Mycological Research. 2002;106(4):499–506.
29. Ciancio A, Bonsignore R, Vovlas N, Lamberti F. Host records and spore morphometrics of *Pasteuria penetrans* group parasites of nematodes. Journal of Invertebrate Pathology. 1994;63(3):260–7.
30. Sabareanu-Stoleru C-M. Research on the technology for white cabbage growing (*Brassica oleracea* L., var. *capitata* L., f. *alba* DC) in ecological vegetable system, in conditions of Iasi county [Doctoral thesis]. Iasi: University of Agricultural Sciences and Veterinary Medicine; 2010.
31. Gill HK, McSorley R. Impact of different organic mulches on the soil surface arthropod community and weeds. International Journal of Pest Management. 2012;58(1):33–40.
32. Gill HK, McSorley R, Treadwell DD. Comparative performance of different plastic films for soil solarization and weed suppression. HortTechnology. 2009;19(4):769–774.
33. Brian Caldwell ES, Abby S, Anthony S, Christine S. Resource Guide for Organic Insect and Disease Management. Ithaca, New York: Arnold Printing Corp; 2013. 201 p.
34. Linker HM, Orr DB, Barbercheck ME. Insect management on organic farms. Center for Environmental Farming Systems. 2009:1–37.
35. Stan N, Munteanu N, Stan T. Vegetable Growing. Iasi: Ion Ionescu de la Brad 2003. 316 p.
36. Lampkin N. The Principles of Organic Farming. U.K.: Farming Press, Miller Freeman; 2001.
37. Calin M. The guide of the recognition and pest control of vegetable in biological agriculture. Bacau: Tipoactiv; 2005. 376 p.
38. Lind K, Lafer G, Schloffer K, Innerhofer G, Meister H. Organic Fruit Growing. Wallingfors: CABI Publishing; 2003. 282 p.
39. Stoleru V. The management of organic vegetable systems. Iasi: Ion Ionescu de la Brad; 2013. 250 p.
40. Lopez-Llorca L, editor Fungal parasites of invertebrates: useful tools for adapting crops to global change. International Congress Soil and Food, Resources for a Healthy Life; 2015; Iasi: UASVM Iasi.
41. Pires L, Marques E, Wanderley-Teixeira V, Teixeira A, Alves L, Alves S. Ultrastrucure of *Tuta absoluta* parasitized eggs and the reproductive potential of females after parasitism by *Metharhizium anisopliae*. Micron. 2009;40:255–261.
42. Lecuona R, Riba G, Cassier P, Clement JL. Alterations of insect epicuticular hydrocarbons during infection with *Beauveria bassiana* or *B. brongniartii*. Journal of Invertebrate Pathology. 1991;58(1):10–18.
43. McSorley R, Wang KH, Rosskopf EN, Kokalis-Burelle N, Hans Petersen HN, Gill HK, Krueger R. Nonfumigant alternatives to methyl bromide for management of nematodes, soil-borne diseases, and weeds in production of snapdragon (*Antirrhinum majus*). International Journal of Pest Management. 2009;55(4):265–273.
44. Daniel C, Wyss E. Field applications of Beauveria bassiana to control the European cherry fruit fly *Rhagoletis cerasi*. Journal of Applied Entomology. 2010;134(9-10):675–81.
45. Kalha CS, Singh PP, Kang SS, Hunjan MS, Gupta V, Sharma R. Entomopathogenic viruses and bacteria for insect-pest control. Integrated Pest Management. Academic Press, Elsevier 2014; pp. 225–244.
46. Scorsetti AC, Maciá A, Steinkraus DC, López Lastra CC. Prevalence of Pandora neoaphidis (Zygomycetes: Entomophthorales) infecting *Nasonovia ribisnigri*

(Hemiptera: Aphididae) on lettuce crops in Argentina. Biological Control. 2010;52(1):46–50.

47. Liu YB. Distribution and population development of *Nasonovia ribisnigri* (homoptera: Aphididae) in iceberg lettuce. J Econ Entomol. 2004;97:883–90.

48. Shrestha G, Enkegaard A, Steenberg T. Laboratory and semi-field evaluation of *Beauveria bassiana* (Ascomycota: Hypocreales) against the lettuce aphid, *Nasonovia ribisnigri* (Hemiptera: Aphididae). Biological Control. 2015;85:37–45.

49. Ten Broeke CJM, Dicke M, Van Loon JJA. Feeding behaviour and performance of different populations of the black currant-lettuce aphid, *Nasonovia ribisnigri*, on resistant and susceptible lettuce. Entomologia Experimentalis et Applicata. 2013;148:130–41.

50. Stan N, Munteanu N. Legumicultura. Iasi: Ion Ionescu de la Brad; 2001. 242 p.

51. Munteanu N, Stoleru V, Filipov F, Teliban G, Lorena-Diana P. To assess the climatic potential of biological vegetable in Iasi county. Grant 31/2006, 2006 financed by UEFISCDI.

52. Sjöberg P, Swiergiel W, Neupane D, Lennartsson E, Thierfelder T, Tasin M, Rämert B. Evaluation of temperature sum models and timing of *Quassia amara* (Simaroubaceae) wood-chip extract to control apple sawfly (*Hoplocampa testudinea* Klug) in Sweden. Journal of Pest Science. 2015;88(2):301–10.

53. Kienzle J, Kopp B, Schulz C editor. Control of the apple sawfly (*Hoplocampa testudinea* Klug) with extracts from *Quassia amara* L.: Quality and combination. 10th International Conference on Cultivation Technique and Phytopathological Problems in Organic Fruit-Growing; 2002. Weinsberg, Germany: Proceedings to the Conference from 31st January.

CHAPTER 3

Companion Planting and Insect Pest Control

Joyce E. Parker[1], William E. Snyder[2], George C. Hamilton[1] and Cesar Rodriguez-Saona[1]

[1] *Department of Entomology, Rutgers University, New Brunswick, NJ, USA*

[2] *Department of Entomology, Washington State University, Pullman, WA, USA*

1. INTRODUCTION

There is growing public concern about pesticides' non-target effects on humans and other organisms, and many pests have evolved resistance to some of the most commonly-used pesticides. Together, these factors have led to increasing interest in non-chemical, ecologically-sound ways to manage pests [1]. One pest-management alternative is the diversification of agricultural fields by establishing "polycultures" that include one or more different crop varieties or species within the same field, to more-closely match the higher species richness typical of natural systems [2, 3]. After all, destructive, explosive herbivore outbreaks typical of agricultural monocultures are rarely seen in highly-diverse unmanaged communities.

There are several reasons that diverse plantings might experience fewer pest problems. First, it can be more difficult for specialized herbivores to "find" their host plant against a background of one or more non-host species [4]. Second, diverse plantings may provide a broader base of resources for natural enemies to exploit, both in terms of non-pest prey species and resources such as pollen and nectar provided by the plant themselves, building natural enemy communities and strengthening their impacts on pests [4]. Both host-hiding and encouragement of natural enemies have the potential to depress pest populations, reducing the need for pesticide applications and increasing crop yields [5, 6]. On the other hand, crop diversification can present management and economic challenges for farmers, making these schemes difficult to implement. For example, each of two or more crops in a field could require quite different management practices (e.g., planting, tillage and harvest all might need to occur at different times for the different crops), and growers must have access to profitable markets for all of the different crops grown together.

"Companion planting" is one specific type of polyculture, under which two plant species are grown together that are known, or believed, to synergistically improve one another's growth [7]. That is, plants are brought together because they directly mask the specific chemical cues that one another's pests use to find their hosts, or because they hold and retain particularly effective natural enemies of one another's pests. In this chapter we define companion plants as interplantings of one crop (the companion) within another (the protection target), where the companion directly benefits the target through a specific known (or suspected) mechanism [8, 9]. Companion plants can control insect pests either directly, by discouraging pest establishment, and indirectly, by attracting natural enemies that then kill the pest. The ideal companion plant can be harvested, providing a direct economic return to the farmer [2] in addition to the indirect value in protecting the target crop. However, "sacrificial" companion plants which themselves provide no economic return can be useful when their economic benefit in increased yield of the target exceeds the cost of growing the companion [10, 11].

Companion planting has received less attention from researchers than other diversification schemes (such as insectary plants and cover crops), but this strategy is widely utilized by organic growers [8, 9]. Generally, recommendations on effective companion-target pairings come from popular press articles and gardening books, which make claims of the benefits of bringing together as companions aromatic herbs, certain flowers [12], or onions (*Allium* L. spp.) [13]; nearly always, vegetables are the protection target. However, these recommendations most-commonly reflect the gut-feeling experiences of particular farmers that these pairings are effective, rather than empirical data from replicated trials demonstrating that this hunch is correct. Indeed, more-rigorous examinations of companion-planting's effectiveness have yielded decidedly mixed evidence [e.g. 9, 14 and 15]. Here, we first review companion plants that disrupt host-location by the target's key pests, and then those that operate by attracting natural enemies of the protection target's pests. For companions operating through either mechanism, we discuss case-studies where underlying mechanisms have been examined within replicated field trials, highlighting evidence for why each companion-planting scheme succeeded or failed.

2. COMPANIONS THAT DISRUPT HOST LOCATION BY PESTS

Herbivorous insects use a wide variety of means to differentiate between host and non-host plants. Consequently, host-finding behavior of the target's pests plays a key role in selecting an effective companion plant. Typically, host plant selection by insects is a catenary process involving sequences of behavioral acts influenced by many factors [16]. These can include the use of chemical cues, assessment of host plant size, and varying abilities to navigate and identify hosts among the surrounding vegetation. Therefore, both visual and chemical stimuli play key roles in host plant location and eventual acceptance. At longer distances, host-location often is primarily through the detection and tracking of a chemical plume [17]. At this scale, abiotic factors may play a strong role. For example, an odorous plume can be influenced not just by plant patch size, but also by temperature and wind speed, which can change the plume's spatial distribution and concentration [17]. As the insect draws near to the host plant, visual cues can increase in importance [17]. Visual indications that a suitable host has been located can include the size, shape and color of the plant [18]. Therefore, based on the dual roles of chemical and visual cues in host-location by herbivores, to be effective disruptors of host-location by the target's pests, companion plants would need to: (1) disrupt the ability of the pest to detect or recognize the target's chemical plume; (2) disrupt or obscure the visual profile of the target; or (3) act simultaneously through both chemical and visual disruption of host location.

Furthermore, ecological differences among pest species are likely to impact the effectiveness of companion planting. For example, specialist herbivores appear to be relatively strongly dissuaded from staying in diverse plantings where their host is just one component of the plant community, whereas generalist herbivores sometimes prefer diverse to simple plantings [19, 20]. Presumably this is because diverse plantings provide relatively few acceptable hosts per unit area for a specialist, but (potentially) several different hosts acceptable to a generalist. Likewise, the size/mobility of the pest is likely to be important. Potting et al. (2005) in reference [21] suggested that smaller sized arthropods such as mites, thrips, aphids and whiteflies that can be passively transported by wind currents, have limited host detection ability. Of course, when a pest moves haphazardly through the environment there is no active host-location behavior for a companion plant to disrupt! Apparently because insects that travel passively with wind currents may cause them to bypass trap crops leading to companion plant failure. Conversely, larger sized insects capable of direct flight have good sensory abilities that allow them to perform oriented movement and thus represent good candidates for control by companion planting [21].

2.1. Companions That Draw Pests Away From The Protection Target

Trap crops are stands of plants grown that attract pest insects away from the target crop [11, 22] (Figure 1).

Figure 1. A trap crop of Pacific gold mustard (companion plant) is flanked on both sides by broccoli (target crop). The symbols (+) represent the principal mechanism at work. Here, the trap crop, designated with two (+) signs, are more attractive than the protection target-broccoli. The mustard trap crop is used to attract pest insects away from broccoli.

Once pests are concentrated in the trap crop the pests can be removed by different means, such as burning or tilling-under the trap crop [11] or by making insecticide applications to the trap. A highly-effective trap crop can bring a relatively large number of pests into a relatively small area, such that pest management within the trap crop requires coverage of less ground than if the entire planting of the protection target had to be treated. Even if left unmanaged through other means, pests feeding within the trap are not damaging the protection target. Because trap crops are more attractive to the pest, they are usually rendered unmarketable due to pest damage. This means that, to be economically-viable, the cost of establishing and maintaining the trap crop must equal or exceed the value of crop-protection within the protection target.

There are many successful examples of trap cropping. For example, in California the need to spray for *Lygus* Hahn in cotton was almost completely eliminated due to the success of alfalfa trap crops [23-25]. In soybeans, Mexican bean beetles can be controlled using a trap crop of snap beans [26]. Similarly, for over 50 years in Belorussia early-planted potato trap crops have been used to protect later plantings of potatoes from Colorado potato beetle attack [27]. Even though many successful examples of trap crops have been reported, several studies have also demonstrated contradictory results with many declaring unsuccessful [28-31] to unreliable control of pests [32]. For example, Luther et al. (1996) in reference [29] explored trap crops of Indian mustard and Tastie cabbage to control diamondback moth and *Pieris rapae* L. in Scorpio cabbage and discovered that these trap crops were effective at attracting these pests; however, the distance between the trap crop and the protection target allowed for pests to spillover back into the protection target. In another experiment using Indian mustard as a trap crop, Bender et al. (1999) in reference [30] intercropped Indian mustard with cabbage to control lepidopterous insects and found that Indian mustard did not appear to preferentially attract these insect pests. Overall, the relative effectiveness of the trap crop depends on the spatial dimensions of the trap crop and protection target, the trap crop and protection target species and pest behavior.

The need to control pests in the trap crop can be avoided when "dead-end" traps are deployed. Dead-end trap cropping utilizes specific plants that are highly preferred as ovipositional sites, but incapable of supporting development of pest offspring [33, 34]. For example, the diamond back moth (*Plutella xylostella* L.), a pest of *Brassica* crops, is highly attracted for oviposition to the G-type of yellow rocket (*Barbarea vulagaris* R. Br.), but the larvae are not able to survive on this host plant [35]. This inability to survive has been attributed to a feeding deterrent, monodesmosidic triterpenoid saponin [36] and so larvae cannot complete development. Similarly, potato plants genetically engineered to express *Bacillus thuringiensis* (Bt) proteins that are deadly to the Colorado potato beetle, and planted early in the season, can act as dead-end traps that kill early-arriving potato beetles [37].

Trap crop effectiveness can be enhanced by incorporating multiple plant species simultaneously. Diverse trap crops include plants with different chemical profiles, physical structures and plant phenologies, therefore, diverse trap crops may provide for a more attractive trap crop. For example, in Finland, mixtures of Chinese cabbage, marigolds, rape and sunflower were used successfully as a diverse trap crop to manage the pollen beetle (*Melighetes aeneus* F.) in cauliflower [38]. Furthermore, Parker (2012) in reference [39] conducted experiments exploring the use of simple and diverse trap crops to control the crucifer flea beetle (*Phyllotreta cruciferae* Goeze) in broccoli (*Brassica oleracea* L. var. *italica*). The trap crops included monocultures and polycultures of two or three species of Pacific gold mustard (*Brassica juncea* L.), pac choi (*Brassica rapa* L. subsp. *pekinensis*) and rape (*Brassica napus* L.). Results indicated that broccoli planted adjacent to diverse trap crops containing all three trap crop species attained the greatest dry weight suggesting that the trap crops species were not particularly effective when planted alone, however, provided substantial plant protection when planted in multi-species polycultures. Thus, diverse trap crops consisting of all three trap crop species (Pacific gold mustard, pac choi and rape) provided the most effective trap crop mixture.

The success of trap crops depends on a number of variables, such as the physical layout of the trap crop (e.g., size, shape, location) and the pests' patterns of movement behavior [40]. For example perimeter trap crops, trap crops sown around the border of the main crop [41], have been used to disrupt Colorado potato beetle (*Leptinotarsa decemlineata* Say) colonization of potato fields from overwintering sites that ring the field [42-44]. However, depending on the pest targeted for control and the cropping system, perimeter trap crops may not be the most effective physical design. For example, a perimeter trap crop may not impede pest movement if the pest descends on a crop from high elevations. In reference [11] Hokkanen (1991) has recommended an area of about 10% of the main crop area be devoted to the trap. A smaller trap crop planting leaves more farm ground available for planting marketable crops.

In general, throughout trap cropping literature, trap crops are most effective when they are attractive over a longer period of time than the target crop, and when trap crops target mobile pests that can easily move among the trap and protection-target plantings [11]. References [11] and [41] reported trap crop success particularly with larger beetles [11] and tephritid flies [41], insects generally capable of direct flight.

2.2. Plants That Repel

Plants with aromatic qualities contain volatile oils that may interfere with host plant location, feeding, distribution and mating, resulting in decreased pest abundance [45-47] (Figure 2).

Figure 2. Intercroppings of spring onions (companion plant) are implemented to protect broccoli (target crop) from pest attack. Here, spring onions are used as a repellent to push pest insects away from broccoli. The symbols represent their potential attractive (+) and repellent (-) properties.

Moreover, certain plants contain chemical properties which can repel or deter pest insects and many of these products are used to produce botanical insecticides. For example, pyrethrum obtained from dried flower of the pyrethrum daisy (*Tanacetum cinerariaefolium* L.), neem extracted from seeds of the Indian neem (*Azadirachta indica* A. Juss.) and essential oils extracted from herbs such as rosemary, eucalyptus, clove, thyme and mint have been used for pest control [48]. Generally, aromatic herbs and certain plants are recommended for their supposed repellent qualities. For example, herbs such as basil (*Ocimum basilicum* L.) planted with tomatoes have been recorded to repel thrips [49] and tomato hornworms [50]. Plants in the genus *Allium* (onion) have been observed to exhibit repellent properties against a variety of insects and other arthropods including moths [51], cockroaches [52], mites [53] and aphids [54]. These examples represent a wide array of arthropods that respond to repellent odors and demonstrate the potential repellent plant properties can have on pest control.

Furthermore, many studies have reported a wide variety of companion plants to contain repellent properties against pests of *Brassica* crops. *Brassica* species are an economically important crop throughout the world [55], sometimes comprising up to 25% of the land devoted to vegetable crops [56]. These companion plants included sage (*Salvia officinalis* L.), rosemary (*Rosemarinus officinalis* L.), hyssop (*Hyssop officinalis* L.), thyme (*Thymus vulgaris* L.), dill (*Anethum graveolens* L.), southernwood (*Artemisia abrotanum* L.), mint (*Menta* L. spp.), tansy (*Tanacetum vulgare* L.), chamomile (several genera), orange nasturtium (*Tropaeolum Majus* L.) [57], celery and tomatoes [57, 58]. Similarly, intercropping tomatoes with cabbages has been suggested to repel the diamond back moth [59] and ragweed (*Ambrosia artemisifolia* L.) has been used to repel the crucifer flea beetle (*Phyllotreta cruciferae*) from collards (*Brassica oleracea* L. var. *acephala*) [60], both widespread pests of *Brassica* crops.

Not all studies using repellent companion plants have reported positive results. Early data have suggested no scientific evidence that odors from aromatic plants can repel or deter pest insects [61]. In reference [62] Latheef and Irwin (1979) found no significant differences in the number of eggs, larvae, pupae, or damage by cabbage pests between companion plants; French marigold (*Tagetes patula* L.), garden nasturtium pennyroyal (*Mentha pulegium* L.), peppermint (*Mentha piperita* L.), garden sage, thyme and control treatments. Furthermore, French marigolds (*Tagetes patula* L.) intercropped in carrots did not repel the carrot fly (*Psila rosae* F.) [47]. Even reports of frequently recommended companion herbs did not always improve pest control. For example, there were no differences in diamond back moth oviposition between Brussels sprouts (*B. oleracea*) intercropped with sage (*S. officinalis*) and thyme (*T. vulgaris*) [61]. Sage and thyme represent two common companion plants noted for their pungent odors [9]. Billiald et al. (2005) in reference [63] and Couty et al. (2006) in reference [64] concluded that if these highly aromatic plants were truly repellent, insects would not land on non-host companion plants.

Indeed, other mechanisms other than repellent odors might have a prominent role in plant protection. In reference [61] Dover (1986) noted reduced oviposition by the diamond back moth caused by contact stimuli and not repellent volatiles of sage and thyme. Therefore, sage and thyme were still protecting the target crop; however, this protection was caused by alternative mechanisms other than repellent odors. Similarly, research has demonstrated that aromatic plants such as marigolds (*Tagetes erecta* L.) and mint (*Mentha piperita* L.) did not repel the onion fly (*Delia antiqua* Meigen) or the cabbage root fly (*D. radicum* L.), but instead disrupted their normal chain of host plant selecting behaviors [16, 65, 66].

The response to a repellent plant will vary depending on the behavior of the insect and the plant involved. As a result, a repellent plant that can be effective for one pest might not provide effective control for another [67]. Finally, many experiments to determine plant's repellent capabilities were carried out in laboratory settings and do not necessarily represent field conditions [9].

2.3. Plants That Mask

Companion plants may release volatiles that mask host plant odors [59, 60, 68] interfering with host plant location (Figure 3).

Figure 3. Marigolds (companion plant) are intercropped with broccoli (target crop) to interfere with host plant location. Here, several mechanisms may be involved in protecting broccoli including masking host plant odors or visually camouflaging broccoli making it less apparent. Here, the symbol (+) is shaded to represent a less apparent target crop-broccoli.

For example, host location by the cabbage root fly (*D. radicum*) was disrupted when host plants were surrounded by a wide variety of plants including weeds and marketable crops [69, 70] such as spurrey (*Spergula arvensis* L.) [71], peas (*Pisum sativum* L.) [72], rye-grass (*Lolium perenne* L.) [72] or clover [73, 74]. However, Finch and Collier (2000) in reference [9] suggested that even though these diverse companion plants contain different chemical profiles, it is unlikely that all would be able to mask host plant odors. Further research has demonstrated that in a wind tunnel, cabbage root fly move toward *Brassica* plants surrounded by clover just as much as *Brassica* plants grown in bare soil indicating that odors from clover did not mask those of the *Brassica* plants [75].

In addition to hiding odors emitted by the protection target, companion plants have also been reported to alter the chemical profile of the protection target. For example, certain companion plants can directly affect adjacent plants by chemicals taken up through its roots [76]. African marigolds (*Tagetes* spp.) produce root exudates which can be absorbed by neighboring plants [77] and may help to explain the reports of African marigold reducing pest numbers [9]. African marigolds also release thiopene, which acts as a repellent to nematodes [78]. Similarly, studies exploring various barley cultivars discovered that airborne exposure of certain combinations of undamaged cultivars caused the receiving plant to become less acceptable to aphids [79-81] and this was also confirmed in field settings [80]. Thus, volatile interactions between odors of host and non-host plants and even single species with different cultivars can affect the behavior of pest insects.

2.4. Plants That Camouflage or Physically Block

In addition to protecting crops with olfactory cues, companion plants may also physically and visually camouflage or block host plants [9, 14, 15, 20, 47, 60, 82]. The 'appropriate/inappropriate landing' theory proposed that green surfaces surrounding host plants may disrupt host plant finding [9]. The 'appropriate/inappropriate landing' theory was originally inspired from studies exploring the oviposition behavior of cabbage herbivores and found that reduced damage in intercropping systems were attributed to a disruption of oviposition behavior [9]. This can occur when insects land on a companion plant instead of the target crop before or during oviposition [83]. For example, Atsatt and O'Dowd (1976) in reference [84] demonstrated that *Delia radicum* (L.) (cabbage root fly) spent twice as much time on a non-host plant after landing on it compared to a host plant. This demonstrated that companion plants can disrupt and arrest *D. radicum* on inappropriate hosts (companion plants). Consequently, *D. radicum* will start its oviposition process from the beginning which may reduce the total number of eggs layed on the target crop. Studies have found similar post-alighting behavior of *Delia floralis* Fallén (turnip root fly) and the decision to oviposit after landing on host and non-host plants [85, 86].

Companion plants may visually (Figure 3) or physically (Figure 4) obstruct host plant location rendering host plants less apparent [87].

Figure 4. Dill (companion plant) is used as a physical barrier to protect broccoli (target crop) from pest attack. Here, the height of the dill can impede pest movement.

For example, host plant location in the crucifer flea beetle (*Phyllotreta cruciferae* Goeze) is disrupted when non-host plant foliage, either visual or hidden, is present [60]. Similarly, Kostal and Finch (1994) in reference [72] and Ryan et al. (1980) in reference [88] both showed that artificial plant replicas made from green card or green paper could disrupt host plant location. Companion plant height is also an important factor in pest suppression. Tall plants can impede pest movement within a cropping system [89]. For example, maize has been used to protect bean plants from pest attack [90] and dill has been used as a vegetative barrier to inhibit pest movement in organic farms (personal observation). Frequently recommended companion plants used as physical barriers include sunflowers, sorghum, sesame and peal-millet [91]. In addition, companion plant barriers may also be used to reduce the spread and transmission of insect vectored viruses [92].

Nevertheless, these mechanisms may not rely solely on physical obstruction [93]. For example, the presence of desiccated clover plants (brown in color), which retained the same architecture as living plants (green in color), but only differed in their appearance from living plants, did not reduce the number of cabbage root fly (*D. radicum*), diamond back moth (*P. xylostella*) and the large white butterfly (*Pieris brassicae* L.) eggs when compared to the target crop on bare ground [93]. However, when live clover surrounded the target crop, the numbers of eggs laid were reduced suggesting that the physical presence of clover alone was not enough to prevent a reduction in oviposition [93]. Therefore, the size, shape, color and chemical profiles of companion plants may interact together reducing pest numbers making it is difficult to tease apart specific mechanisms which may be contributing to pest control.

2.5. Combinations of Companion Planting Techniques

In some systems, different companion planting methods have been combined to work synergistically and improve pest control. For example, in Kenya trap crops have been combined with repellent plants and implemented successfully in a 'push-pull' system [94] to control spotted stem borer (*Chilo partellus* Swinhoe) in maize (*Zea mays* L.) [95, 96]. The repellent plants included a variety of non-host plants such as molasses grass (*Melinis minutiflora* P. Beauv.), silverleaf desmodium (*Desmodium uncinatum* Jacq.) or green leaf desmodium (*Desmodium intortum* Mill.) and the trap crop plantings included Napier grass (*Pennisetum purpurerum* Schumach) or Sudan grass (*Sorghum vulgare sudanense* Hitchc.) [94]. Here, the 'push' (repellent companion plants) drives the pest insect away from the target crop while the 'pull' (trap crop) simultaneously lures the pests toward the trap crop. Kahn and Pickett

(2003) in reference [96] have reported thousands of farmers in east Africa to utilize the push-pull strategies to protect maize and sorghum. In addition, Komi et al. (2006) in reference [97] suggested that maize-legumes or maize-cassava intercrops can provide a 'push' for push-pull systems incorporating Jack-bean (*Canavalia ensiformis* L.) as a highly attractive trap crop 'pull'. The goal of the push-pull strategy aims to minimize negative environmental consequences and maximize pest control, sustainability and crop yield [94].

3. PLANTS THAT ENHANCE CONSERVATION BIOLOGICAL CONTROL

While the previous theories explored bottom-up forces in which companion plants improved pest control, Root (1973) 'enemies hypothesis' in reference [4] explored top-down mechanisms. He proposed that natural enemy populations are greater in polycultures because diverse habitats provide a greater variety of prey and host species that become available at different times. Furthermore, a greater diversity of prey and host species allows natural enemy populations to stabilize and persists without driving their host populations to extinction [4]. Altogether these theories present processes which may contribute to the lower abundance of pest insects in mixed cropping systems. Not surprisingly, companion plants may provide pest control by one or several of these mechanisms.

Pest populations can be managed by enhancing the performance of locally existing communities of natural enemies [98]. This can be accomplished by incorporating non-crop vegetation, such as flowering plants also known as insectary plants, into a cropping system (Figure 5).

Companion plants can provide essential components in conservation biological control by serving as an alternative food source and supplying shelter to natural enemies [99]. Many natural enemies including predators and parasitoids require non-prey food items in order to develop and reproduce [100-102]. For example, adult syrphids whose larvae are voracious predators of aphids, feed on both pollen and nectar [103]. Pollen and nectar are essential resources for natural enemies which satisfy different health requirements. Nectar is a source for carbohydrates and provides energy, while pollen supplies nutrients for egg production [103-106]. In wheat fields in England and in horticultural and pastoral habitats in New Zealand over 95% of gravid female syrphids were found with pollen in

Figure 5. Flowering companion plants are incorporated into this mixed vegetable farm to enhance the efficacy of natural enemies and improve pest suppression.

their gut [103]. As a result, flowering plants can increase the fecundity and longevity of parasitic hymenoptera [107-109] and predators [110, 111]. In addition to increasing natural enemy fitness, improved nutrition may also enhance foraging behavior [e.g. 112, 113] and increase the female-based sex ratio of parasitoid offsprings [114]. A wide variety of natural enemies utilize non-prey food sources. For example, pollen and nectar have been demonstrated to be highly attractive to variety of predators including syrphids [103, 115, 116], coccinellids [117-119], and lacewings [117].

One method to increase natural enemy density using companion plants includes incorporating certain flowering plants into a cropping system. This is often accomplished by planting flowering strips or border plantings in crop fields. Plants in the family Apiaceae are highly attractive to certain beneficial insect populations and are generally recommended as insectary plants [120]. This can be attributed to their exposed nectaries and the structure of their compound inflorescence which creates a "landing platform" [121, 122]. In addition, natural enemies are attracted to the field by the color and odor of companion plants [123]. Another commonly used insectary plant is *Phacelia tanacetifolia* Benth, which has been employed in borders of crop fields because it produces large amounts of pollen and nectar [124, 125]. For example, White et al. (1995) in reference [116] incorporated plantings of *P. tanacetifolia* near cabbage (*B. oleracea*) to increase syrphid densities to control aphids. Similarly, MacLeod (1992) in reference [126] and Lövei et al. (1993) in reference [127] demonstrated that syrphids are highly attracted to the floral resources provided by coriander and buckwheat. Companion plants may work simultaneously influencing both top-down and bottom-up mechanisms. For example, while some studies have demonstrated dill to improve pest control by containing repellent properties, other studies have indicated that dill may also increase predator populations. Patt et al. (1997) in reference [128] found reduced survivorship and populations of Colorado potato beetle (*L. decemlineata*) when dill was intercropped with eggplant and attributed the lower pest numbers to improved biological control.

Flowering companion plants have been used in different cropping systems to enhance the impact of natural enemies. For example, in organic vineyards, [110, 111] increased natural enemies by supplying access to nectar-producing plants such as alyssum (*Lobularia maritima* L.). Other various herbs have also been used this way in Europe [126, 129, 130] and in New Zealand [115, 127]. Overall, flowering companion plants have been implemented in a variety of crops including cereals, vegetable crops and fruit orchards [99, 131-137] to improve conservation biocontrol. In addition to food resources, companion plants can provide shelter from predators and pesticides as well as favorable microclimates [138, 139] including overwintering sites [140]. Furthermore, companion plants can also influence the spatial distribution of natural enemies in and around crops [141, 142] improving pest control.

Indeed, the advantages of plant based resources for natural enemies have only recently been recognized by major reviews [99, 143- 146], and the growing empirical evidence has demonstrated their importance in pest suppression. However, the interactions between the companion plant, target pests and their natural enemies are complex. For example, incorporating companion plants may not necessarily improve biological control if the flowering does not coincide with the activity of natural enemies [147], or if natural enemies do not move from the companion plants to the target crop [117, 148]. Moreover, plant structures, such as the corolla, may obstruct feeding by natural enemies [128] and diverse habitats may complicate prey location by predators and parasitoids [143,149, 150]. Just as pest insects may react differently to the same companion plant, predators within the same family can also respond to similar companion plants in different ways. For example, certain syrphids are highly specialized feeders, while others are generalist [151] influencing companion plant selection. However, the possible obstructions to conservation biocontrol can be diminished. One way to improve the effectiveness of companion plants in conservation biocontrol is to select plants that benefit key natural enemies [152]. Again, this highlights the importance of implementing "careful diversification" as a pest management method [144, 153-155]. Overall, incorporating companion plants to enhance biological control holds promise for managing pests in crops.

Companion plants have also been used as banker plants. Banker plants are usually non-crop species that are deliberately infested with a non-pest insect and improves biological control by providing natural enemies with alternative prey [e.g. 156-158, but see 159, 160] even in the absence of pests [e.g. 156, 159, 161]. This allows natural enemy populations to reproduce and persists throughout the season. Banker plants have been used in both conservation and augmentative biological control programs. Many studies have used banker plants consisting of wheat or barley to sustain populations of the bird cherry-oat aphid (*Rhopalosiphum padi* L.) because this aphid feeds only on members of the Poaeceae family and does not pose a threat for vegetable and ornamental production [162]. However, success can be variable. Jacobson and Croft (1998) in reference [163] compared wheat, rye and corn as banker plants in its ability to sustain the bird cherry-oat aphid parasitoid (*Aphidius colemani* Viereck) and found that control was dependent on banker plant density, release rate and season. One successful example was implemented in apple orchards. To control the rosy apple aphid (*Dysaphis plantaginea* Passerini) in apple orchards, Bribosia et al. (2005) in reference [164] used Rowan trees (*Sorbus aucuparia* L.) as banker plants to maintain densities of Rowan aphids (*Dysaphis sorbi* L.) which served as an alternate host for the braconid parasitoid *Ephedrus persicae* Froggatt.

4. CONSTRAINTS AND CHALLENGES

Incorporating companion plants into pest management strategies is not without challenges. Farmers often face logistical constraints when incorporating companion plants into their field designs. For example, modern agriculture techniques and equipment are not conducive to growing multiple crops in one field [165]. Furthermore, companion plants may hinder crop yield and reduce economical benefits [166, 167]. Beizhou et al. (2011) in reference [168] reported an outbreak of secondary pests and reduced yield in an orchard setting. Decreased yields can often be attributed to competition for resources by incorporating inappropriate companion plants [169]. In certain cases, vegetational diversification can diminish the impacts of biological control. Generally, greater habitat diversity leads to a greater abundance of prey and host species. For instance, improved diversity can lead to reduced biological control by generalist predators which can be influenced by the greater diversity and abundance of alternative prey [123]. Straub et al. (2008) in reference [152] reviewed findings from natural enemy diversity experiments and found that results can range from negative (reduced control) to positive (improved control) due to effects from intraguild predation and species complementarity.

Therefore, choosing which type of companion plant to incorporate in a diversification scheme is challenging. For example, plant phenology, attractiveness and accessibility of the flowers to natural enemies [128] and pest species will play a key role in plant selection. However, it is possible to minimize the reductions in economic returns within companion planting schemes. It is important to use plants that can provide a satisfactory economic return, if possible, as compared to the target crop planted in monoculture [170]. In conservation biocontrol, to reduce negative impacts from biocontrol antagonists or the targeted pest, Straub et al. (2008) in reference [152] suggested using specific resources that can selectively benefit key natural enemies. Overall, whether companion plants control pests through bottom-up or top-down mechanism, their impact will depend on companion plant selection. This emphasizes the significance of finding the "right type" of diversity that combines species that complement one another in ecologically-relevant ways [67].

Designing companion planting schemes pose several impending issues. For instance, optimal distances between the companion plant and the target crop needs to be determined before specific recommendations can be made. The distance to which an insect is attracted to a source has proven to be variable and is a key area in companion plant success. Evans and Allen-Williams in reference [171] demonstrated that attraction can occur at distances of up to 20 m. Judd and Borden (1989) in reference [172] showed attraction of up to 100 m, however, other researchers have shown distances of only a few centimeters [173-176]. Therefore, adjusting the design depending on the insect's behavior and movement [83], the insect's search mode [177, 178] and diet breadth [20] may be necessary for companion plant success. Furthermore, an insect's feeding behavior will affect the success of companion plants in pest management strategies. For example plant structure can affect herbivory. Rape (*B. napus*) can be composed of trichomes that are nonglandular and simple or unbranched [179] and in some cases act as physical barriers that complicate feeding [180].

5. CONCLUSIONS

Many examples of companion plant to reduce pest numbers have been demonstrated; fewer diamondback moths were found on Brussels sprouts when intercropped with malting barley, sage or thyme [61]. Similarly, lower numbers of striped flea beetles were observed when Chinese cabbage (*Brassica chinensis* L.) was intercropped with green onions (*Allium fistulosum* L.) [181], while Mutiga et al. (2010) in reference [182] recorded significantly lower numbers of the cabbage aphid (*Brevicoryne brassicae* L.) when spring onion (*Allium cepa* L.) was intercropped with collard (*B. oleracea* var. *acephala*). However, the mechanisms through which companions protect the target are not well understood [183]. Many studies have suggested that chemical properties in the plant can repel insects [94], while others have suggested that companion plants are considered chemically neutral [66]. For example, Finch et al. (2003) in reference [66] demonstrated that commonly grown companion plants used for their repellent properties, marigolds and mint, did not repel the onion fly or the cabbage root fly (*D. radicum*), but rather interrupted their host finding and selecting behaviors [16, 65]. Thus, even though the companion plants did not repel pests, they were still able to disrupt host plant finding through alternative mechanisms. Overall, the effectiveness of companion plants to reduce pest numbers is not being disputed, but rather the mechanisms in which they work.

Caution should be taken before using companion plants in pest management as results can be mixed. For example, experiments conducted by Held et al. (2003) in reference [12] explored several putative companion plants in their ability to deter Japanese beetles (*Popillia japonica* Newman) from damaging roses and concluded that companion plants were unlikely to help. Diversifying cropping schemes is an essential step in the future of pest management. Companion planting represents just one of many areas in which a single farmer can incorporate diversifying schemes to reduce pest densities in an in-field approach. However, relatively subtle factors may determine whether crop-diversification schemes succeed or fail in improving pest suppression and crop response. Therefore, further research is needed on understanding the interactions between plant selection, mechanisms of benefit and patterns in time and crop phenology. Ultimately, cultural control strategies like companion planting can conserve species diversity, reduce pesticide use and enhance pest control.

REFERENCES

1. Denholm I, Devine G J and Williamson M S. Evolutionary genetics. Insecticide resistance on the move. Science 2002;297: 2222-2223.
2. Vandermeer J H. The ecology of intercropping. Cambridge University Press, New York, New York, USA; 1989.
3. Altieri M A and Nicholls C. Biodiversity and pest management in agroecosystems. New York, USA, Haworth Press; 1994.
4. Root R B. Organization of a plant-arthropod association in simple and diverse habitats: fauna of collards (Brassica oleracea). Ecological Monographs 1973;43: 95-120.
5. Pimentel D. Species diversity and insect population outbreaks. Annals of the Entomological Society of America 1961;54: 76-86.
6. van Emden H F and Williams G F. Insect stability and diversity in agroecosystems. Annual Reviews of Entomology 1974;19: 455-475.
7. Franck G. Companion planting: Successful gardening the organic way. Thorsons Pub. Group , Wellingborough, Northamptonshire, England; 1983.
8. Cunningham S J. Great garden companions: a companion planting system for a beautiful, chemical-free vegetable garden. Rodale, Emmaus, PA; 1998.
9. Finch S and Collier R H. Host-plant selection by insects – a theory based on 'appropriate/inappropriate landings' by pest insects of cruciferous plants. Entomologia Experimentalis 2000;96: 91-102.
10. Altieri M A. The ecological role of biodiversity in agroecosystems. Agriculture, Ecosystem and Environment 1999;74: 19-31.
11. Hokkanen H M T. Trap cropping in pest management. Annual Review of Entomology 1991;36: 119-138.
12. Held D W, Gonsiska P and Potter D A. Evaluating companion planting and non-host masking odors for protecting roses from the Japanese Beetle (Coleoptera: Scarabaeidae). Journal of Economic Entomology 2003; 96: 81-87.
13. Yepsen R B. The encyclopedia of natural insect and disease control. Rodale, Emmaus, PA; 1984.
14. Thiery D and Visser J H. Masking of host plant odour in the olfactory orientation of the Colorado potato beetle. Entomologia Experimentalis et Applicata 1986;41: 165-172.
15. Thiery D and Visser J H. Misleading the Colorado potato beetle with an odor blend. Journal of Chemical Ecology 1987;13: 1139-1146.
16. Atkins M D. Introduction to Insect Behaviour. Macmillan, New York, NY, USA; 1980.
17. Schoonhoven L, van Loon J J A and Dicke M. Insect-Plant Biology. Oxford University Press, New York; 2005.
18. Prokopy R J and Owens E D. Visual detection of plants by herbivorous insects. Annual Review of Entomology 1983;28: 337-364.
19. Bach C E. Effects of plant density and diversity on the population dynamics of a specialist herbivore, the striped cucumber beetle, Acalymma vitata (Fab.) Ecology 1980;61: 1515-1530.
20. Andow D A. Vegetational diversity and arthropod population response. Annual Review of Entomology 1991;36: 561-586.
21. Potting R, Perry J, and Powell JNW. Insect behavioural ecology and other factors affecting the control efficacy of agroecosystem diversification strategies. Ecological Modeling 2005;182: 199-216.
22. Shelton A M and Badenes-Perez F R. Concepts and applications of trap cropping in pest management. Annual Review of Entomology 2006;51: 285-308.
23. Stern V M, Mueller A, Sevacherian V and Way M. Lygus bug control in cotton through alfalfa interplanting. California Agriculture 1969;23: 8-10.

24. Sevacherian V and Stern V M. Host plant preferences of Lygus bugs in alfalfa-interplanted cotton fields. Environmental Entomology 1974;3: 761-766.
25. Stern V M. Environmental control of insects using trap crops, saniation, [prevention and harvesting. in Handbook of Pest Management in Agriculture. D. Pimentel (ed). Vols. 1,2. Boca Raton, FL; 1981.
26. Rust R W. Evaluation of trap crop procedures for control of Mexican bean beetle in soybeans and lima beans. Journal of Economic Entomology 1977;70: 630-632.
27. Dorozhkin N A, Bel'skaya S I and Meleshkevich A A. Combined protection of potatoes. Zasch. Rast 1975;7: 6-8.
28. Silva-Krott I U, Singh P, Lali T S and Maniappan R. Development of a trap cropping system for cabbage in Guam. Pest Management Horticultural Ecosystems 1995;1: 27-35.
29. Luther G C, Valenzuela H R and DeFrank J. Impact of cruciferous trap crops on Lepidopteran pests of cabbage in Hawaii. Environmental Entomology 1996;25: 39-47.
30. Bender D A, Morrison W P and Frisbie R E. Intercropping cabbage and Indian mustard for potential control of lepidopterous and other insects. Hortscience 1999;34: 275-279.
31. Shelton A M and Nault B A. Dead-end trap cropping: a technique to improve management of the diamondback moth, Plutella xylostella (Lepdioptera: Plutellidae). Crop Protection 2004;23: 497-503.
32. Musser F R, Nault B A, Nyrop J P and Shelton A M. Impact of a glossy collard trap crop on diamondback moth adult movement, oviposition, and larval survival. Entomologia Experimentalis et Applicata 2005;117: 71-81.
33. Thompson J N. Evolutionary ecology of the relationship between oviposition preference and performance of offspring in phytophagous insects. Entomologia Experimentalis et Applicata 1988;47: 3-14.
34. Thompson J N and Pellmyr O. Evolution of oviposition behavior and host preference in Lepidoptera: Annual Review of Entomology 1991;36: 65-89.
35. Badenes-Perez F R, Shelton A M and Nault B A. Evaluating trap crops for diamondback moth, Plutella xylostella (Lepidoptera: Plutellidae). Journal of Economic Entomology 2004;97: 1365-1372.
36. Shinoda T, Nagao T, Nakayama M, Serizawa H, Koshioka M, Okabe H and Kawai A. Identification of a triterpenoid saponin from a crucifer, Barbarea vulgaris, as a feeding deterrent to the diamondback moth, Plutella xylostella. Journal of Chemical Ecology 2002;28: 587-599.
37. Hoy C W. Colorado potato beetle resistance management strategies for transgenic potatoes. American Journal of Potato Research 1999;76: 215-219.
38. Hokkanen H M T. Biological and agrotechnical control of the rape blossom beetle Meligethes aeneus (Coleoptera, Nitidulidae). Acta Entomologica Fennica 1989;53:25-29.
39. Arker JE. Diversity by design. Exploring trap crops and companion plants to control Phyllotreta cruciferae, the crucifer flea beetle, in broccoli. Washington State University (Doctoral Dissertation); 2012.
40. Hannunen S. Modeling the interplay between pest movement and the physical design of trap crop systems. Agricultural and Forest Entomology 2005;7: 11-20.
41. Boucher T J, Ashley R, Durgy R, Sciabarrasi M and Calderwood W. Managing the pepper maggot (Diptera: Tephritidae) using perimeter trap cropping. Journal of Economic Entomology 2003;96: 420-432.
42. Hoy C W, Vaughn T T and East D A. Increasing the effectiveness of spring trap crops for Leptinotarsa decemlineata. Entomologia Experimentalis et Applicata 2000;96: 193-204.
43. Hunt DW and Whitfield AG. Potato trap crops for control of Colorado potato beetle (Coleoptera: Chrysomelidae) in tomatoes. Canadian Entomologist 1996;128: 407-412.

44. Weber D C, Ferro D N, Buonaccorsi J, Hazzard R V. Disrupting spring colonization of Colorado potato beetle to nonrotated potato fields. Entomologia Experimentalis et Applicata 1994;73: 39-50.
45. Lu W, Hou M L, Wen J H and Li J W. Effects of plant volatiles on herbivorous insects. Plant Protection 2007;33: 7-11.
46. Stan F, Helen B and Rosemary H C. Companion planting – do aromatic plants disrupt host-plant finding by the cabbage root fly and the onion fly more effectively than non-aromatic plants? Entomologia Experimentalis et Applicata 2003;109: 183-195.
47. Uvah I I I and Coaker T H. Effect of mixed cropping on some insect pests of carrots and onions. Entomologia Experimentalis et Applicata 1984;36: 159-167.
48. Isman B. Botanical insecticides, deterrents, and repellents in modern agriculture and increasingly regulated world. Annual Review of Entomology 2006;51: 45-66.
49. Anon. Companion planting. 2004a [Online]. Available: http://www.ghorganics.com/page2.html.
50. Anon. Organic vegetable IPM guide. 2004b [Online]. Available: http://msucares.com/pubs/publications/p2036.pdf.
51. Landolt P J, Hofstetter R W and Biddick L L. Plant essential oils as arrestants and repellents for neonate larvae of the codling moth (Lepidoptera: Tortricidae). Environmental Entomology 1999;28: 954-960.
52. Scheffler I and Dombrowski M. Behavioural responses of Blattella germanica L. (Orthopt., Blattelidae) induced by repellent plant extracts. Journal of Applied Entomology 1993;115: 499-505.
53. Dabrowski Z T and Seredynska U. Characterisation of the two-spotted spider mite (Tetranychus urticae Koch, Acari: Tetranychidae) response to aqueous extracts from selected plant species. Journal of Plant Protection Research 2007;47: 113-124.
54. Amarawardana L, Bandara P, Kumar V, Pettersson J, Ninkovic V and Glinwood R. Olfactory response of Myzus persicae (Homoptera: Aphididae) to volatiles from leek and chives- potential for intercropping with sweet pepper. Acta Agriculturæ Scandinavica B 2007;57: 87-91.
55. Harvey C T and Eubanks M D. Effect of habitat complexity on biological control by the red imported fire ant (Hymenoptera: Formicidae) in collards. Biological Control 2004;29: 348-358.
56. Talekar N S, Yang J C and Lee S T. Introduction of Diadegma semiclausum to control diamondback moth in Tawian, p. 263-270. in Diamondback moth and other crucifer pests N.S. Talekar (ed.). Proc. 2nd International Workshop, Shunhua, Taiwan. Asian Vegetable Research and Development Center; 1992.
57. Little B. Companion planting in Australia. Reed Books Pty Ltd: Frenchs Forest, New South Wales; 1989.
58. Philbrick H and Gregg R B. Companion plants and how to use them. The Devin-Adair Company. Old Greenwich, Connecticut; 1982.
59. Buranday R P and Raros R S. Effects of cabbage-tomato intercropping on the incidence and oviposition of the diamond back moth, Plutella xylostella (L.) Phillippine Entomology 1975;2: 369-375.
60. Tahvanainen J O and Root R B. The influence of vegetational diversity on population ecology of a specialize herbivore, Phyllotreta cruciferae (Coleoptera: Chrysomelidae) Oecologia 1972;10: 321-346.
61. Dover J W. The effects of labiate herbs and white clover on Plutella xylostella oviposition. Entomologia Experimentalis Applicata 1986;42: 243-247.
62. Latheef M A and Irwin R D. The effect of companionate planting on lepidopteran pests of cabbage. The Canadian Entomologist 1979;111: 863-864.
63. Billiald H, Finch S, Collier R H and Elliott M. Disrupting host-plant finding by pest insects of cruciferous plants using aromatic and non-aromatic plants. IOBC/WPRS Bulletin 2005;28 (4): 53-59.
64. Couty A, van Emden H, Perry J N, Hardie J, Pickett J A and Wadhams L J. The role of olfaction and vision in host-plant finding by the diamondback moth, Plutella xylostella. Physiological Entomology 2006;31: 134-145.

65. Zohren E. Laboruntersuchungen zu Massenanzucht, Lebensweise, Eiablage and Eiablageverhalten der Kohlfliege, Chortophila brassicae (Bouché) (Diptera, Anthomyiidae). Zeitschrift für angewandte Entomologie 1968;62: 139-188.
66. Finch S, Billiald H and Collier R H. Companion planting-do aromatic plants disrupt host-plant finding by the cabbage root fly and the onion fly more effectively than non-aromatic plants? Entomologia Experimentalis et Applicata 2003;109: 183-195.
67. Poveda K, Gomez M I and Martinez E. Diversification practices: their effect on pest regulation and production. Revista Colombiana de Entomologia 2008;34: 131-144.
68. Perrin R M and Phillips M L. Some effects of mixed cropping on the population dynamics of insect pests. Entomologia Experimentalis et Applicata 1978; 24: 385-393.
69. Dempster J P. Some effects of weed control on the numbers of the small cabbage white (Pieris rape L.) on Brussels sprouts. Journal of Applied Ecology 1969;6: 339-345.
70. Smith J G. Influence of crop backgrounds on aphids and other phytophagous insects on Brussels sprouts. Annals of Applied Biology 1976;83: 1-13.
71. Theunissen J, and den Ouden H. Effects of intercropping with Spergula arvensis on pests of Brussels sprouts. Entomologia Experimentalis et Applicata 1980;27: 260.268.
72. Kostal V and Finch S. Influence of background on host plant selection and subsequent oviposition by the cabbage root fly (Delia radicum). Entomologia Experimentalis et Applicata 1994;70: 153-163.
73. Theunissen J, Booij C J H, Schelling G and Noorlander J. Intercropping white cabbage with clover. IOBC/WPRS Bulletin XV/ 1992.;4: 104-114
74. Finch S and Edmonds G H. Undersowing cabbage crops with clover-the effects on pest insects, ground beetles and crop yield. IOBC/WPRS Bulletin 1994;17(8): 159-167.
75. Tukahirwa E M and Coaker T H. Effect of mixed cropping on some insect pests of brassicas; reduced Brevicoryne brassicae infestations and influences on epigeal predators and the disturbance of oviposition behavior in Delia brassicae. Entomologia Experimentalis et Applicata 1982;32: 129-140.
76. Theunissen J. Effects of intercropping on pest populations in vegetable crops. IOBC/WPRS Bulletin 1994;17 (8): 153-158.
77. Rovira A A. Plant root exudates. Botanical Review 1969;35: 35-39.
78. Matsumoto K and Kotulai J R. Field tests on the effectiveness of Azadirachta companion planting as a shoot borer repellent to protect mahogany. Japan International Research Center for Agricultural Sciences 2002;10: 1-8.
79. Pettersson J, Ninkovic V and Ahmed E. Volatiles from different barley cultivars affect aphid acceptance of neighbouring plants. Acta Agriculturae Scandinavica B 1999;49: 152-157.
80. Ninkovic V, Olsson U and Pettersson J. Mixing barley cultivars affects aphid host plant acceptance in field experiments. Entomologia Experimentalis et Applicata 2002;102: 177-182.
81. Glinwood R T, Gradin T, Karpinska B, Ahmed E, Jonsson L and Ninkovic V. Aphid acceptance of barley exposed to volatile phytochemicals differs between plants exposed in daylight and darkness. Plant Signaling Behavior 2007;2: 205-210.
82. Atsatt P R and O'Dowd D J. Plant defense guilds. Science 1976;193: 24-29.
83. Björkman M, Hambäck P A and Rämert B. Neighbouring monocultures enhance the effect of intercropping on the turnip root fly (Delia floralis). Entomologia Experimentalis et Applicata 2007;124: 319-326.
84. Morley K, Finch S and Collier R H. Companion planting-behavior of the cabbage root fly on host plants and non-host plants. Entomologia Experimentalis et Applicata 2005;117: 15-25.
85. Hopkins R J, Wright F, McKinley R G and Birch A N E. Analysis of sequential behavior patterns: the oviposition decision of the turnip root fly, Delia floralis. Entomologia Experimentalis et Applicata 1996;80: 93-96.

86. Hopkins R J, Wright F, Birch A N E and McKinlay E R G. The decision to reject an oviposition site: sequential analysis of the post-alighting behavior of Delia floralis. Physiological Entomology 1999;24: 41-50.

87. Feeny P. Plant apparency and chemical defense, pp. 1-10. in Biochemical interaction between plants and insects, J. Wallace and R. Mansell (eds.). Recent Advances in Phytochemistry 1976;10: 1-40.

88. Ryan J, Ryan M F and McNaeidhe F. The effect of interrow plant cover on populations of cabbage rootfly, Delia brassicae (Wied.). Journal of Applied Ecology 1980;17: 31-40.

89. Perrin R M. Pest management in multiple cropping systems. Agro-ecosystems 1977;3: 93-118.

90. Altieri M A and Doll J D. Some limitations of weed biocontrol in tropical ecosystems in Colombia. In: T.E. Freeman (ed.), Proceedings IV International symposium on Biological Control of Weeds. University of Florida, Gainesville pp. 74-82; 1978.

91. Thresh M. Cropping practices and virus spread. Annual Review of Phytopathology 1982;20: 193-218.

92. Toba H H, Kishaba A N, Bohn G W and Hield H. Protecting muskmelon against aphid-borne viruses. Phytopathology 1977;67: 1418-1423.

93. Finch S and Kienegger M A. behavioural study to help clarify how undersowing with clover affects host plant selection by pest insects of brassica crops. Entomologia Experimentalis et Applicata 1997;84: 165-172.

94. Cook S M, Khan Z R and Pickett J A. The use of "push-pull" strategies in integrated pest management. Annals Review Entomology 2007;52: 375-400.

95. Khan Z R, Ampong-Nyarko K, Chilishwa P, Hassanali A, Kimani S, Lwande W, Overholt W A, Pickett J A, Smart L E, Wadhams L J and Woodcock C M. Intercropping increases parasitism of pests. Nature 1997;388: 631-632.

96. Khan Z R and Pickett J A. The 'push-pull' strategy for stemborer management: a case study in exploiting biodiversity and chemical ecology. Ecological Engineering for Pest Management-Advances in Habitat Manipulation for Arthropods (ed. by GM Gurr, SD Wratten and MA Altieri), CABI Publishing, Wallingford, UK; 2004; pp. 155-164.

97. Komi A, Gounou S and Tamò M. The role of maize-legumes-cassava intercropping in the management of maize ear borers with special reference to Mussidia nigrivenella Ragonot (Lepidoptera: Pyralidae).Annales de la Société entomologique de France 2006;42: 495-502.

98. Van den Bosch, R, and Telford, A D. Environmental modification and biological control. In Biological Control of Pests and Weeds, ed. P DeBac, pp. 459-488. New York: Reinhold; 1964.

99. Landis D A, Wratten S D and Gurr G M. Habitat management to conserve natural enemies of arthropod pests in agriculture. Annual Review of Entomology 2000;45: 175-201.

100. Coll M and Guershon, M. Omnivory in terrestrial arthropods: mixing plant and prey diets. Annual Review of Entomology 2002;47: 267-297.

101. Patt J M, Wainright S C, Hamilton G C, Whittinghill D, Bosley K, Dietrick J and Lashomb J H. Assimilation of carbon and nitrogen from pollen and nectar by a predaceous larva and its effects on growth and development. Ecological Entomology 2003;28: 717-728.

102. Wäckers F L, van Rijn P C J, Bruin J. Plant-provided food for carnivorous insects: a protective mutualism and its applications. Cambridge: Cambridge University Press; 2005.

103. Hickman J M and Wratten S D. Use of Phacelia tanacetifolia strips to enhance biological control of aphids by hoverfly in cereal fields. Journal of Economic Entomology 1996;89: 832-840.

104. Schneider F. Beitrag zur Kenntnis der Generationsverhaltinisse und diapause raberischer Schwebfligen. Mitt. Schweiz. Entomol. Ges 1948;21: 249-285.

105. Jervis, MA., Lee, JC. and Heimpel, GE. 2004. Conservation biological control using arthropod predators and parasitoids – the role of behavioural and life-history studies. In: Gurr G, Wratten SD, Altieri M, editors. Ecological engineering for pest management: advances in habitat manipulation of arthropods. Melbourne: CSIRO Press. p. 69-100.
106. Lee J C, Heimpel G E and Leibee G L. Comparing floral nectar and aphid honeydew diets on the longevity and nutrient levels of a parasitoid wasp. Entomologia Experimentalis et Applicata 2004;111: 189-199.
107. Baggen L R, Gurr G M and Meats A. Flowers in tri-trophic systems: mechanism allowing selective exploitation by insect natural enemies for conservation biological control. Entomologia Experimentalis et Applicata 1999;91: 155-161.
108. Gurr G M and Nicol H I. Effect of food on longevity of adults of Trichogramma carverae Oatman and Pinto and Trichogramma carverae Oatman and Pinto and Trichogramma nr brassicae Bezdenko (Hymenoptera: Trichogrammatidae). Australian Journal of Entomology 2000;39: 185-187.
109. Bickerton M W and Hamilton G C. Effects of Intercropping with flowering plants on predation of Ostrinia nubilalis (Lepidoptera: Crambidae) eggs by generalist predators in bell peppers. Environmental Entomology 2012;41: 612-620
110. Begum M, Gurr G M, Wratten S D, Hedberg P and Nicol H I. The effect of floral nectar on the grapevine leafroller parasitoid Trichogramma carverae. International Journal of Ecology and Environmental Sciences 2004;30: 3-12.
111. Begum M, Gurr G M, Wratten S D, Hedberg P R and Nicol H I. Using selective food plants to maximize biological control of vineyard pests. Journal of Applied Ecology 2006;43: 547-554.
112. Lavandero B I, Wratten S D, Didham R K and Gurr G. Increasing floral diversity for selective enhancement of biological control agents: A double-edged sward? Basic and Applied Ecology 2006;7: 236-243.
113. Wade M R and Wratten S D. Excised or attached inflorescences? Methodological effects on parasitoid wasp longevity. Biological Control 2007;40: 347-354.
114. Berndt L A, Wratten S D and Hassan P G. Effects of buckwheat flowers on leafroller (Lepidoptera: Tortricidae) parasitoids in a New Zealand vineyard. Agriculture and Forest Entomology 2002;4: 39-45.
115. Carreck N L and Williams I H. Observations on two commercial flower mixtures as food sources for beneficial insects in the UK. Journal of Agricultural Science 1997;128: 397-403.
116. White A J, Wratten S D, Berry N A and Weighmann U. Habitat manipulation to enhance biological control of Brassica pests by hover flies (Diptera: Syrphidae). Journal of Economic Entomology 1995;88: 1171-1176.
117. Freeman Long R, Corbett A, Lamb C, Reberg-Horton C, Chandler J and Stimmann M. Beneficial insects move from flowering plants to nearby crops. California Agriculture 1998;52: 23-26
118. Nalepa C A, Bambara S B and Burroughs A M. Pollen and nectar feeding by Chilocorus kuwanae (Silvestri) (Coleoptera: Coccinellidae). Proceedings of the Entomological Society of Washington 1992;94: 596-597.
119. Pemberton R W, Vandenberg N J. Extrafloral nectar feeding by ladybird beetles (Coleoptera: Coccinellidae). Proceedings of the Entomological Society of Washington 1993; 95: 139-151.
120. Jervis M A. and Kidd N A C. Host-feeding strategies in hymenopteran parasitoids. Biological Reviews 1986;61: 395-434.
121. Langenberger M W and Davis A R. Temporal changes in floral nectar production, reabsorption, and composition associated with dichogamy in annual caraway (Carum carvi; Apiacea). American Journal of Botany 2002;1588-1598.
122. Quarles W and Grossman J. Insectary plants, intercropping and biological control. The IPM Practitioner 2002;24 (3): 1-11.
123. Fiedler A K and Landis D A. Plant characteristics associated with natural enemy abundance at Michigan native plants. Environmental Entomology 2007;36: 878-886.

124. Cowgill S E, Wratten S D and Sotherton N W. The effect of weeds on the numbers of hoverfly (Diptera: Syrphidae) adults and the distribution and composition of their eggs in winter wheat. Annals of Applied Biology 1993;123: 499-514.

125. Maredia K M, Gage S H, Landis D A and Scriber J M. Habitat use patterns by the seven-spotted lady beetle (Coleoptera: Coccinellidae) in a diverse agricultural landscape. Biological Control 1992;2: 159-165.

126. MacLeod A. Alternative crops as floral resources for beneficial hoverflies (Diptera: Syphidae). Proceedings of the Brighton Crop Protection Conference, pp. 997-1002. Brighton: British Crop Protection Council; 1992.

127. Lövei G L, Hodgson D J, MacLeod A and Wratten S D. Attractiveness of some novel crops for flower-visiting hover flies (Diptera: Syrphidae): comparisons from two continents. Pp. 368-370 in: Pest control and sustainable agriculture. Corey, S.; Dall, D.; Milne, W. ed. Australia, CSIRO; 1993. p514.

128. Patt J M, Hamilton G C and Lashomb J H. Foraging success of parasitoid wasps on flowers: interplay of insect morphology, floral architecture and searching behavior. Entomologia Experimentalis et Applicata 1997;83: 21-30.

129. Von Klinger K. Auswirkungen eingesater Randstreifen an einem Winterweizen-Feld auf die Raubarthropenfuana und den Getreideblattlausbefall. Journal of Applied Entomology 1987; 104: 47-58.

130. Ruppert V and Klingauf F. The attractiveness of some flowering plants for beneficial insects as exemplified by Syrphinae (Diptera: Syrphidae). Mitteilungen der Deutschen Gesellschaft für Allgemeine und Angewandte Entomologie 1988;6 (1-3): 255-261.

131. Gruys P. Hits and misses. The ecological approach to pest control in orchards. Entomolgia Experimentalis et Applicata 1982;31: 70-87.

132. Thomas M B, Wratten S D and Sotherton N W. Creation of island habitats in farmland to manipulate populations of beneficial arthropods: predator densities and species composition. Journal of Applied Ecology 1991;28: 906-917.

133. Dennis P and Fry G L A. Field margins: can they enhance natural enemy population densities and general arthropod diversity on farms? Agriculture Ecosystems and the Environment 1992;40: 95-115.

134. Wyss E. The effects of artificial weed strips on diversity and abundance of the arthropod fauna in a Swiss experimental apple orchard. Agriculture Ecosystems and Environment 1996;60: 47-59.

135. Letourneau D K and Altieri M A. Environmental management to enhance biological control in agroecosystems. In. Bellows, T.S., Fischer, T.W. (Eds.), Handbook of Biological Control. Academic Press, San Diego; 1999. p319-354.

136. Woodcock B A, Potts S G, Pilgrim E, Ramsay A J, Tscheulin T, Parkinson A, Smith R E N, Gundrey A L, Brown V K and Tallowin J R. The potential of grass field margin management for enhancing beetle diversity in intensive livestock farms. Journal of Applied Ecology 2007; 44: 60-69.

137. Jonsson M, Wratten S D, Landis D A and Gurr G M. Recent advances in conservation biological control of arthropods by arthropods. Biological Control 2008;45: 172-175.

138. Hossain Z, Gurr G M, Wratten S D and Raman A. Habitat manipulation in lucerne (Medicago sativa L.): arthropod population dynamics in harvested and 'refuge' crop strips. Journal of Applied Ecology 2002;39: 445-454.

139. Thomas M B, Wratten S D and Sotherton N W. Creation of island habitats in farmland to manipulate populations of beneficial arthropods: predator densities and species composition. Journal of Applied Ecology 1992;29: 524-531.

140. Perdikis D, Fantinou A and Lykouressis D. Enhancing pest control in annual crops by conservation of predatory Heteroptera. Biological Control 2011;59: 13-21.

141. Lavandero B, Wratten S D, Shishehbor P and Worner S. Enhancing the effectiveness of the parasitoid Diadegma semiclausum (Helen): movement after use of nectar in the field. Biological Control 2005;34: 152-158.

142. Tylianakis J M, Didham R K and Wratten S D. Improved fitness of aphid parasitoids receiving resource subsidies. Ecology 2004;85: 658-666.
143. Barbosa P. ed. Conservation Biological Control. Sand Diego, CA: Academic. 396 pp; 1998.
144. Gurr G M, Wratten S, and Altieri M A. Ecological engineering for pest management: advances in habitat manipulation for arthropods. CABI. Wallington. CABI, Wallingford, UK; 2004.
145. Heimpel G E and Jervis M A. Does floral nectar improve biological control by parasitoids? Plant-provided Food and Plant-Carnivore Mutualism (eds F.L. Wackers, P.C.J. Van Rijn & J. Bruin) Cambridge University Press, Cambridge, UK; 2005. p267-304.
146. Wäckers F L, Romesis J and van Rijn P. Nectar and pollen-feeding by insect herbivores and implications for tri-trophic interactions. Annual Review of Entomology 2007;52: 301-323.
147. Bowie M H, Wratten S D and White A J. Agronomy and phenology of "companion plants" of potential for enhancement of biological control. New Zealand Journal of Crop Horticultural Science 1995;23: 432-427.
148. Bigger D S and Chaney W E. Effects of Iberis umbellate (Brassicaceae) on insect pests of cabbage and on potential biological control agents. Environmental Entomology 1998;27: 161-167.
149. Finke D L and Denno R F. Spatial refuge from intraguild predation: implications for prey suppression and trophic cascades. Oecologia 2006;149: 265-275.
150. Hughes A R and Grabowski J H. Habitat context influences predator interference interactions and the strength of resource partitioning. Oecologia 2006;149: 256-264.
151. Haslett J R. Interpreting patterns of resource utilization: randomness and selectivity in pollen feeding by adult hoverflies. Oecologia 1989;78: 433-432.
152. Straub C S, Finke D L and Snyder W E. Are the conservation of natural enemy biodiversity and biological control compatible goals? Biological Control 2008;45: 225-237.
153. Risch S H. Insect herbivore abundance in tropical monocultures and polycultures: an experimental test of two hypotheses. Ecology 1981;62: 1325-1340.
154. Prokopy R C. Integration in orchard pest and habitat management: a review. Agriculture Ecosystems and Enviroment 1994;50: 1-10.
155. Baggen L R and Gurr G M. The influence of food on Copidosoma koehleri (Hymenoptera: encyrtidae), and the use of flowering plants as a habitat management tool to enhance biological control of potato moth, Phthorimaea operculella (Lepidoptera: Gelechiidae). Biological Control 1998;11: 9-17.
156. Hansen L S. Introduction of Aphidoletes aphidimyza (Rond.) (Diptera: Cecidomyiidae) from an open rearing unit for the control of aphids in glasshouses. Bulletin SROP 1983; 6: 146-150.
157. Bennison J A and Corless S P. Biological control of aphids on cucumbers: further development of open rearing units or 'bank plants' to aid establishment of aphid natural enemies. Bulletin OILB/SROP 1993;16: 5-8
158. Van Driesche R G, Lyon S, Sanderson J P, Bennett K C, Stanek E J, Zhang R T. Greenhouse trials of Aphidius colemani (Hymenoptera: Braconidae) banker plants for control of aphids (Hemiptera: Aphididae) in greenhouse spring floral crops. Florida Entomologist 2008;91: 583-591.
159. Stacey D L. Banker plant production of Encarsia formosa Gahan and its use in control of glasshouse whitefly on tomatoes. Plant Pathology 1997;26: 63-66.
160. Pickett C H, Simmons G S, Lozano E and Goolsby J A. Augmentative biological control of whiteflies using transplants. Biological Control 2004;49: 665-688.
161. Bennison J A. Biological control of aphids on cucumbers use of open rearing systems or 'banker plants' to aid establishment of Aphidius matricariae and Aphidoletes aphidimyza. Mededlingen van de Faculteit Landbouwwetenschappen. Universiteit Gent 1992;57: 457-466.

162. Frank S D. Biological control of arthropod pests using banker plant systems: Past progress and future directions. Biological Control 2010;52: 8-16.
163. Jacobson R J and Croft P. Strategies for the control of Aphis gossypii Glover (Hom: Aphididae) with Aphidius colemani Viereck (Hym: Braconidae) in protected cucumbers. Biocontrol Science and Technology 1998;8: 377-387.
164. Bribosia E, Bylemans D, Migon M and Van Impe G. In-field production of parasitoids of Dysaphis plantaginea by using the rowan aphid Dysaphis sorbi as substitute host. Biological Control 2005;50: 601-610.
165. Tooker J F and Frank S D. Genotypically diverse cultivar mixtures for insect pest management and increased crop yields. Journal of Applied Ecology 2012;49: 974-985.
166. Letourneau D K, Armbrecht I, Rivera B S, Lerma J M, Carmona E J, Daza M C, Escobar S, Gallindo V, Gutierrez C, Lopez S D, Mejia J L, Rangel A M A, Rangel J H, Rivera L, Saavedra C A, Torres A M and Trujillo A R. Does plant diversity benefit agroecosystems? A synthetic review. Ecological Applications 2011;21: 9-21.
167. Lin B B. Resilience in agriculture through crop diversification: adaptative management for environmental change. BioScience 2011;61: 183-193.
168. Beizhou S, Jie Z, Jinghui H, Hongying W, Yun K and Yuncong Y. Temporal dynamics of the arthropod community in pear orchards intercropped with aromatic plants. Pest Management Science 2011;67: 1107-1114.
169. Bone N J, Thomson L J, Ridland P M, Cole P and Hoffmann A A. Cover crops in Victorian apple orchards: effects on production, natural enemies and pests across a season. Crop Protection 2009;28: 675-683.
170. Bandara K A N P, Kumar V, Ninkovic V, Ahmed E, Pettersson J and Glinwood R. Can leek interfere with bean plant-bean fly interaction? Test of ecological pest management in mixed cropping. Journal of Economic Entomology 2009;102 (3): 999-1008.
171. Evans K A and Allen-Williams L J. Distant olfactory responses of the cabbage seed weevil, Ceuthorhynchus assimilis. To oilseed rape odour in the field. Physiological Entomology 1993;18: 251-256.
172. Judd G J R and Borden J H. Distant olfactory response of the onion fly, Delia antique, to host-plant odour in the field. Physiological Entomology 1989;14: 429-441.
173. Wigglesworth V B. The principles of insect physiology. Metheun, London, UK; 1939.
174. Kennedy J S. Olfactory responses to distant plants and other sources. Chemical Control of Insect Behaviour: Theory and Application (ed. by HH Shorey and JJ McKelvey, Jr), John Wiley and Sons, New York, NY, USA; 1977. p67 91.
175. Chapman R F. The Insects-Structure and Function. Hodder and Stoughton, London, UK; 1982.
176. Dethier V G, Barton Browne L and Smith S N. The designation of chemicals in terms of the responses they elicit from insects. Journal of Economic Entomology 1960;53: 134-136.
177. Hambäck P A, Pettersson J and Ericson L. Are associational refuges species specific? Functional Ecology 2003;17: 87-93.
178. Bukovinszky T, Potting R P J, Clough Y, van Lenteren J C and Vet L E M. The role of pre- and post-alighting detection mechanisms in the responses to patch size by specialist herbivores. Oikos 2005;109: 435-446.
179. Traw M R and Dawson T E. Differential induction of trichomes by three herbivores of black mustard. Oecologia 2002;131: 526-532.
180. Soroka J J, Holowachuk J M, Gruber M Y and Grenkow L F. Feeding by flea beetles (Coleoptera: Chrysomelidae; Phyllotreta spp.) is decreased on canola (Brassica napus) seedlings with increased trichome density. Journal of Economic Entomology 2011;104: 125-136.
181. Gao Z Z, Wu W J and Cui Z X. The Effect of Intercrop on the Densities of Phyllotreta Striolata (F.) Chinese Agricultural Science Bulletin 2004;20:214-216.

182. Mutiga S K, Gohole L S and Auma E O. Effects of integrating companion cropping and nitrogen application on the performance and infestation of collards by Brevicoryne brassicae. Entomologia Experimentalis et Applicata 2010;134: 234-244.

183. Bomford M K. Do tomatoes love basil but hate Brussels sprouts? Competition and land-use efficiency of popularly recommended and discouraged crop mixtures in biointensive agriculture systems. Journal of Sustainable Agriculture 2009;33: 396-417.

CHAPTER 4

Biochemical Insecticide Resistance in Tea Pests

Dhiraj Saha[1]

[1] Insect Biochemistry and Molecular Biology Laboratory, Department of Zoology, University of North Bengal, Raja Rammohunpur, P.O.- North Bengal University, Siliguri, District - Darjeeling, West Bengal, India

ABSTRACT

Polyphagous insect herbivores encounter numerous toxins (xenobiotics) as they pass through their life cycle; some toxins are produced naturally by the host plants (allelochemicals) and others by humans (insecticides) to manage these insects having pest status. The host plants have evolved defensive mechanisms for protection from herbivory, including chemical repellents and toxins (secondary metabolites). Many classes of insect repellents and toxic substances, such as isoflavonoids, furanocoumarins, terpenoids, alkaloids and cyanogenic glycosides are synthesized in plants. The biosynthetic pathways leading to these allelochemicals are continually evolving to generate new secondary metabolites. Similarly, to control the herbivorous insect pests, numerous chemicals of synthetic origin are used continuously against them. In response, the attacking organisms also evolve mechanisms that enable them to resist the defensive chemicals of their hosts and those toxins of synthetic origin applied for their control. A variety of defence mechanisms, including enzymatic detoxification systems, physiological tolerance and behavioural avoidance, protect insect herbivores from these xenobiotic compounds. Insect pests have evolved the mechanisms to degrade metabolically (enzymatically) or otherwise circumvent the toxic effect of many types of chemicals that we have synthesized as modern insecticides. The extent to which insects can metabolize and thereby degrade these antibiotics or toxins is of considerable importance for their survival in hostile chemical environment. These mechanisms continue to evolve as insects attempt to colonize new plant species or encounter newer molecules of synthetic insecticides. Generally, three main enzymes, general esterases (GEs), glutathione S-transferases (GSTs) and cytochrome P450-mediated monooxygenases (CYPs), are involved in the process of metabolic detoxification of insecticides. During the past 70 years, following the discovery and extensive use of synthetic insecticides, resistance of insects to insecticides has registered the greatest increase and strongest impact. The evolution of resistance to insecticides is an example of evolutionary process. An insecticide is the selection pressure, which results in a very strong but differential fitness of the individual in a population having susceptible and resistant genotypes. The survival and subsequent reproduction of resistant individuals lead to a change in the frequency of alleles conferring resistance in the population over time. While selection pressure acts to change allele frequencies within pest populations, the phenotype upon which selection operates is a function of both genotype and the environment. Recent studies in insect detoxifying enzymes have revealed further versatility in the adaptation of insects to their environment by the phenomenon of induction. This is the process in which a chemical stimulus enhances the activity of the detoxification enzyme systems by the production of additional enzymes that metabolize toxic chemical substances. Hence, the influence of environmental factors such as continuous usage of insecticides and the chemical constituents (allelochemicals) of host plants on phytophagous insects can have a great impact to induce the enzymatic detoxification systems of insects, thereby promoting the insecticide resistance mechanisms. While all insects do possess detoxification ability, its magnitude is expected to vary among the species with the nature of its recent environment and feeding ecology. The level and type of detoxifying mechanisms differ greatly, which therefore result in varying toxicity among different developmental stages, species and populations. Variation in detoxifying enzyme activity is responsible in part for the selective toxicity of different insecticides, the development of resistance to insecticides and selective adaptation to host plants. Over-expression of these detoxifying enzymes, capable of metabolizing insecticides, can result in a high level of metabolic tolerance/resistance to synthetic insecticides. Increased expressions of genes encoding the major xenobiotic metabolizing enzymes are the most common cause of insecticide resistance in insects.

Keywords: Insecticide resistance, tea, insect pests, detoxifying enzymes, cytochrome p450, carboxylesterases, glutathione S-transferases, monooxygenases, allelochemical, *Helopeltis theivora*, *Scirtothrips dorsalis*, *Empoasca flavescens*

1. INTRODUCTION

1.1. Tea, The Plantation Crop with Economic Value: Driving the Shrub to Cup

Tea is produced from young leaves and buds of tea plant, *Camellia sinensis* (L.) O. Kuntze is a native of China, and the Chinese are said to have discovered its use nearly 4,700 years ago. It is believed that Shen Nung, a Chinese emperor who lived some 4,700 years ago, discovered that tea leaves falling into boiling water make a refreshing drink. Tea then became a popular drink in China for both its flavour and medicinal qualities. Eventually, the habit of drinking tea spread throughout Asia and then throughout the world. Currently, tea is the second most preferred and popular drink after water. The word tea had its origin from t'e (pronounced as 'tay') in 'Amoy' dialect while in Cantonese it was called ch'a ('chah'). This is the name by which this wonderful beverage is known in Japan, Iran, Russia, Indonesia, Malaysia, Vietnam and India. Tea was introduced in Japan about AD 800 and was regarded as medicine for about 500 years. Tea was introduced in Europe in the early 17th century with the beginning of trade between Europe and the Southeast Asia.

Now tea is produced in almost every region of the globe. The tea plant is predominantly grown in Asia followed by in Africa and to a very small extent in Europe, South America, Australia and New Zealand. Now, tea is grown in 36 tropical and subtropical countries. Major tea producing countries are India, China, Sri Lanka, Kenya, Japan, Indonesia, Thailand, Bangladesh, Nepal, Vietnam, Turkey and Argentina.

Tea plants are native to East and South Asia and probably originated around the point of confluence of the lands of northeast India, north Burma, southwest China and Tibet. The commercially cultivated tea plants are derived from small-leaved China plants, *C. sinensis*, the Assam plants, *C. assamica* (Masters), and the Cambod plants, *C. assamica* ssp. *lasiocalyx* (Planchon ex Watt) Wight and numerous hybrids among them. Tea plants require warm humid climate, well-distributed rainfall and long sunshine hours. Shoots, comprising two or three tender leaves and a bud, are harvested and processed in factories to manufacture different types of tea.

2. INSECT PEST OCCURRENCES IN TEA

Being a widespread perennial monoculture crop, tea plantations provide the most congenial microclimate as well as continuous food supply to a number of arthropods. Every part of the tea plant, i.e. leaf, stem, root, flower and seed is subjected to attack by at least one arthropod pest species. According to recent estimates, globally, more than a thousand (1,034 species) arthropods and 82 species of nematodes are associated with tea plantations [1]. Among the insect pests, 32% are from order Lepidoptera, followed by 27% pest species from Hemiptera [2]. In India, only 300 species of arthropods are recorded as pest and about 167 species are from tea-growing Northeast India (Table 1) [3]. The dynamic adaptations of insects have facilitated them to exploit every part of the tea plant and the maximum numbers of pests occur on the foliage.

Among the arthropods that attack tea plant, insect and mite pests are the most damaging [2], causing on average a 5%–55% yield loss [4–6]. Insect pests alone can cause on an average 11%–55% yield loss, if left unrestrained [1].

Tea mosquito bug, thrips, jassids, tea caterpillars (looper, redslug and bunch), aphids, termites, cockchafers and red spider mite are among the various insects and mite pests that cause severe loss in tea production. The damage caused by the tea pests frequently leads to a significant impact on quality and yield, while the degree of pest infestation differs in different tea-growing areas depending on altitude, climate, forest cover and local cultural practices. Tea pests can be broadly classified into the different categories based on their feeding nature, time of occurrence and severity. The major pests are mite pests: red spider mite (*Oligonychus coffeae*), scarlet mite (*Brevipalpus phoenicis*), pink mite (*Acaphylla theae*), purple mite (*Calacarus carinatus*); sucking pests: tea mosquito bug (*Helopeltis theivora*), tea thrips (*Scirtothrips dorsalis* and *Myctrothrips setiventris*), tea jassids (*Empoasca flavescens*), aphids, scale insects; leaf eaters: looper caterpillars (*Hyposidra talaca, H. infixaria, Buzura suppressaria*), red slug caterpillar (*Eterusia magnifica*), bunch caterpillar (*Andraca bipunctata*); and soil-borne pests: termites (*Microcerotermes* sp. *Odontotermes* sp.), cockchafers, nematodes, weevils, cricket [8].

Table 1. Major insect and mite pests that occur on tea in India.

Sl. no.	Name of the pests		Nature of damage	Family: Order
1.	*Andraca bipunctata* Walk.	Bunch caterpillar	Folivores	Bombycidae: Lep.
2.	*Eterusia magnifica* Butl.	Red slug caterpillar	Folivores	Zygaenidae: Lep.
3.	*Buzura suppressaria* Guen.	Looper caterpillar	Folivores	Geometridae: Lep.
4.	*Hyposidra talaca. H. infixaria* Walk.	Black inch worm	Folivores	Geometridae: Lep.
6.	*Lymantria albulunata* Mre.	Sungma caterpillar	Folivores	Lymantridae: Lep.
7.	*Gracillaria theivora* Walsh.	Tea leaf roller	Folivores	Gracilariidae: Lep.
8	*Homona coffearia* Nietner	Tea tortrix	Folivores	Torticidae:Lep.
9.	*Laspeyresia leucostoma* Mayer.	Flushworm	Folivores	Eucosmidae: Lep.
10.	*Holotrichia impressa* Burm.	Cockchafer grubs	Root of young tea	Scarabaeidae: Col.
11.	*Serica assamensis* Brenske	Leaf-eating cockchafer	Leaf eater	Scarabaeidae: Col.
12.	*Asticus chrysochlorus* Wied.	Large green weevil	Leaf eater	Curculionidae: Col.
13.	*Agromyzidae* (Bigot) Meij	Tea leaf miner	Leaf eater	Agromyzidae: Dip.
14.	*Microtermes* spp.	Live wood-eating termite	Stem and root eater	Termitidae: Iso.
15.	*Odontotermes* sp.	Scavenging termite	Stem and root eater	Termitidae: Iso.
16.	*Helopeltis theivora* Waterhouse	Tea mosquito bug	Leaf sucker	Miridae: Hem.
17.	*Empoasca flavescens* Fabr.	Tea greenfly/tea jassid	Leaf sucker	Jassidae: Hem.
18.	*Toxoptera aurantii* Boyer de Fons.	Tea aphid	Leaf sucker	Aphididae: Hem.
19.	*Scirtothrips dorsalis* Hood	Yellow tea thrips	Leaf and bud sucker	Thripidae: Thy.
20.	*Mycterothrips (Teaniothrips) setiventris* Bagnall	Common thrips	Leaf and bud sucker	Thripidae: Thy.
21	*Oligonychus coffeae* Nietner	Red spider mite	Leaf sucker	Tetranychidae: Aca.
22.	*Brevipalpus phoenicis* Geijskes	Scarlet mite	Leaf sucker	Tenuipalpidae: Aca.

3. INSECT PESTS OF TEA AND MANAGEMENT PROBLEMS

Tea is produced from the young foliage, i.e. young leaves and a bud, and foliage production is increased by seasonal pruning which enhances the leaf cover. The major pests of the crop are those associated with the young foliage. The most important insect pest groups are the folivores (chewing) and sap suckers of the young tender leaves, buds and stems (sucking pest), which damage the most economic part of tea plant that is processed in tea industry for making the tea. These pests cause substantial loss in yield to the tea industry.

In India, different management practices are followed to protect the tea crop against different insect pest groups. Most of the plantations are managed conventionally i.e. using different organosynthetic insecticides, whereas some organic plantations use plant- and animal-based herbal and microbial insecticides. In conventional tea plantations, organo-synthetic insecticides of different functional groups such as organochlorines, organophosphates, synthetic pyrethroids (SPs) and neonicotinoids (NNs) are regularly used throughout the year to control the invasion of different insect pest groups (sucking, folivores and others) [8]. The use of insecticide is cost-effective to planters and a major concern for the environmental degradation due to contamination as well as in resurgence of primary pests [6], outbreak of secondary pests [9], development of insecticide resistance [10, 11], including undesirable residues in made tea [12]. Regular spraying of insecticides leads to the development of higher level of tolerance or resistance to insecticides in many insects [11, 13].

From the early forties onwards, dichlorodiphenyltrichloroethane (DDT) (organochlorine) was regularly used to manage the infestation of *H. theivora,* the major sucking pest in Northeast India [14]. In 1968, endosulfan (cyclodiene: organochlorine) was introduced in the tea plantations of the Dooars region of West Bengal, India, in the form of thiodan 35 EC [15]. Currently, in different conventional tea plantations of tea-growing regions of India, cypermethrin, deltamethrin, quinalphos, monocrotophos, chlorpyriphos, imidacloprid, etc. are extensively used during cropping season to control insect pests [16–18].

Recently, a number of insecticides have been found to be ineffective in controlling the insect and mite pests in different tea-growing regions of India [8]. The development of resistance to different classes of insecticides is one of the causes for persistence and resurgence of insect pests on tea crop [8, 19–21]. A major concern in managing the major insect pests of tea is its high potential to develop resistance rapidly to regularly used insecticides [11]. Continuous and repeated exposure to different classes of insecticides for many years, in addition to their high reproductive potential, short life cycle and numerous annual generations, has limited the management of major pests of tea [11]. Recently, there are reports on the development of resistance to many commonly used synthetic insecticides and consequent failure in controlling many tea pests [10, 22–26]. Such failures are already known in case of organochlorines (OCs), organophosphorus (OP) and synthetic pyrethroid (SP) insecticides and more recently for the newer compound such as neonicotinoids [19–21, 26]. The development of resistance in *H. theivora* populations to different classes of insecticides has been in the range of 1.47–62.99-fold for males and 1.25–62.82-fold for females in Northeast India [19]. Relative toxicity to commonly used insecticides has been observed to vary in *H.*

theivora populations from Jorhat, Assam [20], Darjeeling [27], and from sub-Himalayan Dooars region of Northeast India [28]. LC_{50} values of insecticides, when compared with the field dose against *H. theivora* recommended by TRA (Tea Research Association, Tocklai, Assam, India), revealed a pronounced shift in the level of susceptibility of *H. theivora* to all insecticides except acephate [20–21].

For the management of other sucking insects such as yellow tea thrips, *S. dorsalis,* insecticides are also used in conventional tea plantations. In tea ecosystem, control failure and the development of biochemical resistance have been reported in *S.* dorsalis [11, 24]. In India, *S. dorsalis* populations have developed a high degree of resistance to a range of organochlorine (DDT, BHC and endosulfan), organophosphate (acephate, dimethoate, phosalone, methyl-O-demeton and triazophos) and carbamate insecticide (carbaryl) in chili ecosystem [31]. *S. dorsalis* has also developed a high degree of resistance to various insecticides, viz. monocrotophos, acephate, dimethoate, phosalone, carbaryl and triazophos [32]. Recently, several insecticides have been tested on *S. dorsalis* in chili ecosystem in USA and found limited success with chlorfenpyr, spinosad and imidacloprid [33, 34]. The performance of novaluron, abamectin, spiromesifen, cyfluthrin, methiocarb and azadirachtin failed to provide effective control of this pest [35].

Similarly, in another emerging sucking insect pest of tea, tea greenfly, *E. flavescens,* repeated management failure and biochemical insecticide resistance in tea ecosystem from Northeast India have been reported [11, 24]. In China, chemical insecticides including fenvalerate, cyfluthin, cypermethrin and imidacloprid are sprayed to control the leafhoppers as frequently as seven times annually or even more frequently [36–37]. A high level of resistance against many insecticides has been reported in related species, *E. vitis* [38]. The resistance to thiamethoxam was highest and to cyperemethrin was lowest in *E. vitis.* Recently, in Fujian province of China, a regional diversity of resistance to eight insecticides in *E. vitis* has been reported in tea ecosystem with higher resistance level to bifenthrin, acetamiprid, imidacloprid, cartap and chlorfenapyr [39].

A high level of insecticide resistance in folivores, such as black hairy caterpillar, bunch caterpillar, looper pest complex (*Hyposidra talaca, H. infixaria, Buzura suppressaria, Eturesia magnifica*) and in termite of tea ecosystem, has been reported with reduced susceptibility against different insecticides.

Detoxification of insecticides is an important toxicokinetic mechanism for insect pests to tolerate regularly applied insecticides [8, 40–42]. Susceptibility levels against insecticides change mainly due to metabolic detoxification of the insecticides through the induction of some detoxifying enzymes under the stress of different management practices [43–45].

Generally, three principal enzymes, general esterases (GEs), glutathione S-transferases (GSTs) and cytochrome P450-mediated monooxygenases (CYP450s), are involved in the process of metabolic detoxification of insecticides [41]. Estimation of the activities of these metabolic defence-related detoxifying enzymes gives information on the level of tolerance/resistance of the insect pest population to insecticides and is a useful tool in monitoring the tolerance/resistance to insecticides at population level of the pest. The early detection of metabolic threats related to tolerance/resistance to insecticides in pest specimens is of crucial importance for devising pest control techniques that would minimize the development of tolerant/resistant forms and prevent any undesirable wastage of insecticide, money and manpower.

4. INSECTICIDE RESISTANCE MECHANISMS IN INSECT PESTS OF TEA

Insects come across with numerous toxins as they go through their life cycle. Some of these toxins are naturally produced by plants (plant allelochemical) and others by humans (synthetic insecticides). To protect themselves against the natural toxins, insects have evolved various detoxification mechanisms [41] These mechanisms also cross-protect insect pests when they are exposed to synthetic insecticides [25]. Herbivorous insect groups (agricultural pests) are significantly more diverse than their non-herbivorous sister groups [46]. The role of plant in promoting diversification in insects has occurred through co-evolutionary 'defence strategies' among them [47]. This diversification could also have been a result of insects 'tracking' plant phylogenies, with minor chemical changes in plants allowing the evolving populations of insects to change and speciate accordingly, which probably has occurred long after chemical changes in plants [48]. Evolution to herbivory preceded via mixed feeding on reproductive parts or spores, dead tissues of plants and animals and fungi. This progression implies that omnivory preceded generalized herbivory and the evolution of specialization on specific plant taxa was a later accomplishment [49].

Among sucking insect herbivores, the actual food used, i.e. digested whole tissue particularly parenchyma (as in *Helopeltis* sp.), cell content (thrips) and phloem flow (*Empoasca* sp.), influences both the feeding mechanism and feeding behaviour [48]. While the chewing insects (looper caterpillar complex) cause extensive damage, the sucking insects cause modest to barely perceptible damage. However, sucking insects, particularly phloem and digested tissue feeder,impose an additional challenge to the plants as they deplete photosynthates, act as vector of viruses and introduce chemical and protein effectors that alter plant defence mechanisms (signalling) and development [50]. When these attributes are combined with a broad host range, breeding strategies that promote invasiveness, highly evolved feeding strategies, the ability to adapt to a wide range plant habitats and the emergence of insecticide resistance, it is not surprising that sucking insects cause heavy losses in agriculture and horticulture [51].

Insecticide resistance is a genetic change in response to selection pressure of toxicants that impair pest control in the field [52]. Insecticide resistance does not occur unless a structural genetic change occurs that is heritable. Therefore, insecticide resistance is an evolutionary phenomenon that results under the selection pressure of a new toxicant in the environment [8].Thus insecticide resistance is different from insecticide tolerance. Insecticide tolerance is the natural ability of a population to withstand the toxic effect of a particular insecticide. It can develop within one generation as a result of physiological adaptation, i.e. induction of xenobiotic detoxifying enzymes. Hence, variation within a population may include individuals with genetic traits that make them better adapted to survive in exposure to an insecticide. If these individuals survive the insecticide exposure, then the tolerance traits can be passed on to the next generation, thereby enriching the gene pool with those genes. The mechanisms of development of insecticide tolerance can be divided into four levels:

Altered behaviour – avoidance of contact with the insecticide

Development of barrier tissues – reduced penetration of the insecticide though the integument

Enhanced detoxification – higher metabolism of the insecticide

Alteration of receptor – at the target site for the insecticide.

The first level, at which insecticide tolerance can develop, is when the insect encounters an insecticide. An altered behaviour helps the insect to avoid coming into contact with the insecticide. Once the insect comes in contact with the insecticide, a reduced and delayed penetration through the cuticle will reduce the effect of the insecticide at the target site; this is yet another level of resistance. Within the insect's body, the insecticide may be enzymatically metabolized and thereby inactivated. At the third level of resistance mechanism, three systems of xenobiotic detoxification enzymes operate: esterases, glutathione S-transferases and cytochrome P450-dependent monooxygenases. The increased activity of one of these enzyme systems in metabolizing insecticides will result in insecticide tolerance.

Alterations at the target site for the insecticide are the last level of insecticide resistance mechanisms. Different classes of insecticides bind to specific target sites and reduced binding at the target site, or increased number of target site molecules may confer insecticide resistance.

4.1. Behavioural Resistance

Behavioural resistance mechanisms are the least studied resistance mechanisms in insects, but this is not to say that behavioural resistance is the least significant. Behavioural resistance can be defined as '*evolved behaviours that will reduce an insect's exposure to toxic compounds or that allows an insect to survive in what would otherwise be a toxic and fatal environment*' [53]. Behavioural resistance has been observed in more than 30 species of insects [53]. Avoidance is the first step in the evolution of behavioural resistance [54]. In *H. theivora*, this kind of resistance has been seen [55]. *H. theivora* shows a different egg-laying strategy to avoid insecticide exposure. Even *E. flavescens* avoids exposure to direct sunlight and therefore prefers to stay on the underside of the tea leaves. This behaviour cross-protects it from the direct insecticide exposure in conventional tea plantations during spraying [8]. The same has been found in *S. dorsalis* which resides inside the leaf bud during development and underside of the leaf during adult stage, thereby avoiding direct exposure to insecticides [8].

4.2. Reduced Penetration

Reduced penetration of insecticides through barrier tissues of insects is another way in which an insect can modify the effective dose of insecticide at the target site. The mechanism may not prevent the insecticide from eventually entering the insect, but it can reduce the rate at which the insecticide reaches the target site. Reduced penetration has been shown to function as a resistance mechanism to many different insecticides, and, by the nature of this mechanism, cross-resistance is often found [56]. The rate of penetration of insecticides through the insect cuticle or other barriers (peritrophic membrane) depends on the physicochemical properties of the insecticide and the barrier. A reduced penetration contributes to DDT resistance in the tobacco budworm, *Heliothis virescens* F. (Lepidoptera: Noctuidae). DDT-resistant larvae had an altered composition of the cuticle. The protein and lipid contents are greater in the cuticle of resistant larvae and, furthermore, the cuticle of the resistant larvae probably had a higher degree of sclerotization [57]. In *M. domestica*, two resistant strains, with reduced penetration as one of the resistance mechanisms, also showed increased cuticular lipid content; more total lipids, mono-glycerides, fatty acids, sterols and phospholipids were present in the resistant strains compared to a susceptible strain [58]. Reduced penetration has been documented as a resistance mechanism only at the level of the insect cuticle, but any biological membrane may serve as a barrier and thereby give resistance [59]. As a single resistance mechanism it usually only confers low levels (less than threefold) of resistance [59]. Reduced penetration has been shown to function as a resistance mechanism to many different insecticides, including insecticides of the three major classes, OPs, carbamates and pyrethroids. Reduced penetration of OPs through the cuticular barrier has been reported, for example, diazinon in *M. domestica* [60], azinphos-methyl in the pear psylla, *Psylla pyricola* Foester (Hemiptera: Psyllidae) [61] and profenofos in *H. virescens* [62]. However, by slowing the penetration rate of insecticides, this mechanism reduces the risk that the insects' detoxification systems become overloaded, and the dose of insecticide reaches to a lethal level at the target site. In female *H. theivora*, a higher level of body lipid has been found which effectively reduces the penetration of insecticide to the target site [63]. No studies on resistance due to reduced penetration in *E. flavescens* and *S. dorsalis* or any other tea pests have been reported to date. The studies had shown that when different resistance mechanisms are combined in the same individuals, a synergistic effect, resulting in a high level of resistance, may arise [64, 65]. Therefore, even a small degree of reduced penetration can contribute significantly to the overall insecticide resistance of the insect pests.

4.3. Metabolic Detoxification

Metabolic detoxification of insecticides is an important toxicokinetic mechanism for insects to tolerate the toxic effects of insecticides. Generally, lipophilic (hydrophobic) insecticides are rapidly detoxified. Organophosphates, organochlorines, carbamates and pyrethroids are lipophilic compounds, and detoxification enzymes transform these insecticides to more hydrophilic and less biologically active compounds so that can be eliminated more easily by excretion. Increased detoxification of insecticides has often been reported in many resistant populations [40]. Three enzyme systems are generally recognized as the major detoxification systems involved in insecticide resistance in insects. These are carboxylesterases, cytochrome P450-dependent monooxygenases and glutathione S-transferases [40, 41].

4.4. Alteration at The Target Site for Insecticide (Target Site Insensitivity)

The biochemical sites for insecticide action differ for different insecticides and are a potential field of research for developing insecticides, which can act specifically or more efficiently on insect biochemical sites compared to mammals. The target site receptor for action of organophosphates, carbamates, organochlorines and pyrethroids is in the nervous system. The enzyme acetylcholinesterases (AChEs) (EC 3.1.1.7) are the target sites for organophosphates and carbamates, and voltage-gated sodium channel of the nerve membrane is the target of pyrethroids and DDT. Neurotoxic insecticides such as cyclodienes (e.g. dieldrin and endosulfan), y-HCH (lindane) and fipronil target gamma-aminobutyric acid (GABA)-receptor [68, 69] and nicotinyl insecticides (imidacloprid and nicotine) target the nicotinic acetylcholine receptor (nACHR) [70]. Alteration at the target site, to less a sensitive target for neurotoxic insecticides, is an important toxicodynamic resistance mechanism in insects [71].

In insects, the potent inhibitors of AChE are organophosphates and carbamates. These compounds inhibit the activity of AChE by forming a stable covalent intermediate, preventing the enzyme to hydrolyse acetylcholine. An accumulation of acetylcholine keeps the ion channel of the receptor permanently open, which eventually kills insect. OPs and carbamates are quasi-irreversible inhibitors of AChE. The organophosphates and carbamates phosphorylate and carbamylate the active site serine of AChE, respectively [72]. Generally, the reactivation time of phosphorylated or carbamylated AChE is long. However, the half-lives of reactivation vary considerably, from minutes to several days, depending on the compound interacting with AChE [73]. Carbamylated AChE generally reactivates faster than phosphorylated AChE. Reduced sensitivity of AChE to inhibition by OPs and carbamates is an important resistance mechanism in insects and is often referred to as altered or insensitive AChE [74]. The presence of insensitive AChE conferring resistance was first noticed in OP-resistant mites, *Tetranychus urticae* Koch (Acari: Tetranycidae) [74] and also found in several insect populations resistant to these compounds [75–77]. Insecticide susceptible and resistant insect pest populations differ in the level of AChE activity [78–80]. A higher level of AChE activity has been reported in *H. theivora* sampled from conventional tea plantations than from organic garden indicating the presence of resistance to insecticides in conventional tea ecosystems [81, 82]. There is no such report on *S. dorsalis* and *E. flavescens* in conventional tea ecosystems to date.

5. MAJOR METABOLIC DETOXIFYING ENZYMES IN INSECTS

Carboxylesterases, glutathione S-transferases and cytochrome P450-mediated monooxygenases are the three principal enzymes that facilitate the insects to metabolize different kind of toxins. These large enzyme families contain multiple forms with overlapping substrate specificities. Knowledge of insecticide detoxification helps in understanding the mechanism of insecticide resistance, hence the development of a sound resistance management strategy. Detoxification can be divided into phase I (primary) and phase II (secondary) processes (Figure 1).

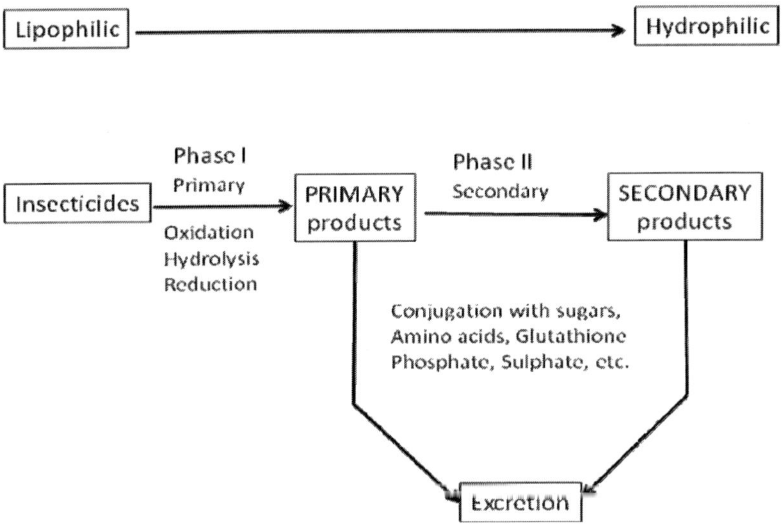

Figure 1. Insecticide detoxification pathways.

Phase I reactions consist of oxidation, hydrolysis and reduction. The phase I metabolites are sometimes polar enough to be excreted but are usually further converted by phase II reactions. In phase II reactions, the polar products are conjugated with a variety of endogenous compounds such as sugars, sulphate, phosphate, amino acids or glutathione and subsequently excreted. Phase I reactions are usually responsible for decreasing the biological activity of toxins, and therefore the enzymes involved are rate limiting with respect to toxicity. The most important function of biotransformation is to decrease the lipophilicity of insecticides, so that they can be excreted quickly [83].

5.1. Phase I Reactions

5.1.1. Cytochrome p450 monooxygenases (e.c. 1.14.-.-)

Oxidation is considered the most important among phase I reactions. The oxidative reactions are carried out mainly by a group of enzymes called cytochrome P450 monooxygenases [also known as mixed function oxidases (MFO) or polysubstrate monooxygenases (PSMO), microsomal oxidase, P450 enzymes]. Cytochrome P450, or *CYP* genes, constitutes one of the largest family of genes, with representatives in virtually all living organisms, from bacteria to protists, plants, fungi and animals [84].

In insects, P450 monooxygenases are involved in many processes including roles in the metabolism of plant allelochemicals by herbivores and in detoxification of insecticides. The human genome carries about 57 CYP genes, and insect genomes can carry from 36 CYP genes in the body louse *Pediculus humanus* [85] to 170 in a mosquito [86]. Each P450 protein is the product of a distinct CYP gene, and P450 diversity is the result of successive gene (or genome) duplications followed by sequence divergence [84].

The typically 45–55-kDa P450 proteins are heme-thiolate enzymes. Their essential common feature is the absorbance peak near 450 nm of their Fe^{II}–CO complex for which they are named [87]. P450 enzymes are best known for their monooxygenase role, catalysing the transfer of one atom of molecular oxygen to a substrate and reducing the other to water. The simple stoichiometry commonly describes the monooxygenase or mixed function oxidase reaction of P450:

$$RH + O_2 + NADPH + H^+ \rightarrow ROH + H_2O + NADP^+$$

However, oxygen atom transfer is not the only catalytic function of P450 enzymes. They also show activities such as oxidases, reductases, desaturases, isomerases, etc. and collectively are known to catalyse at least 60 chemically distinct reactions (Table 2).

The first insect P450 cloned and sequenced was CYP6A1 from *Musca domestica* in 1989 [88]. The P450 gene complement (CYPome) size of an insect genome is not a definite number [89]. Insects can survive with small CYPomes even in toxic environments. The human body louse *Pediculus humanus*, with 36 CYP genes, is known to become highly resistant to many classes of insecticides [85], and the honeybee, with just 46 CYP genes [90], is not more sensitive than other species in a comparison to the toxicity of 62 insecticides [91]. The main driver of CYPome evolution is of course gene duplication, followed by divergence (by neofunctionalization or subfunctionalization) or death (pseudogenization or deletion) [84].

Table 2. Enzymatic reactions catalysed by insect P450 enzymes (adapted from Feyereisen, 2005).

Reaction catalysed	P450
Oxidase activity O_2 to H_2O, H_2O_2, O_2^-	CYP6A1 (and probably most P450 enzymes)
Aliphatic hydroxylation C–H hydroxylation	CYP4C7, CYP6A1, CYP6A2, CYP6A8, CYP6G1, CYP6M2, CYP6CM1vQ, CYP9T2, CYP12A1, CYP18A1, CYP302A1 CYP306A1, CYP312A1, CYP314A1, CYP315A1
O-dealkylation	CYP6A1, CYP6D1, CYP6A5, CYP6B4, CYP6B17, CYP6B21, CYP6G1, CYP6Z2, CYP6CM1vQ, CYP9A12, CYP9A14, CYP12A1, CYP321A1,
Dehalogenation	CYP6G1
Epoxidation	CYP6A1, CYP6A2, CYP6B8, CYP6B27, CYP6AB3, CYP6 CYP6AB11, CYP9E1, CYP12A1, CYP15A1, CYP321A1
Aromatic hydroxylation	CYP6D1, CYP6G1, CYP6M2
Heteroatom oxidation and dealkylation Phosphorothioate ester oxidation	CYP6A1, CYP6A2, CYP6D1, CYP12A1
N-dealkylation	CYP6A5, CYP12A1
N-oxidation	+(Nicotine)
S-oxidation	+(Phorate)
Aldehyde oxidation	CYP18A1
Complex and atypical reactions	
Cyanogenic glucoside biosynthesis:	CYP405A2
Val/Ile to oximes	CYP332A3
Oximes to cyanohydrins	CYP6M2
Aryl ether cleavage	+?(Sterols, ecdysteroid)
Carbon–carbon cleavage	CYP4G1
Decarbonylation with C–C cleavage	+(Defensive steroids)
Aromatization	–
Reduction	–
Endoperoxide isomerisation	

5.1.2. Carboxylesterases (Ec 3.1.1.1)

Carboxylesterase or esterase is a collective term for the enzymes that hydrolyse carboxylic esters [92]. Classification of these enzymes is difficult because of their overlapping substrate specificity [93]. However, the esterase classification of Aldridge [94] is generally recognized. According to that classification, esterases inhibited by paraoxon in a progressive and temperature-dependent manner are called B-esterases and those which are not inhibited are A-esterases [94]. Some A-esterases can hydrolyse OPs, through an acylated cysteine in their active site, and are termed phosphoric triester hydrolases (EC 3.1.8.) [95, 96]. The term carboxylesterase is now mainly attributed to B-esterases [95, 96]. These enzymes have an active site serine residue, hence the terms B-esterase and serine hydrolase are synonymous. Insecticides such as organophosphates, carbamates, pyrethroids and some juvenoids, which contain ester linkages, are susceptible to hydrolysis. Esterases are hydrolases that split ester compounds by the addition of water to yield an acid and alcohol.

Esterases that metabolize organophosphates can be divided into three groups: A-esterases which are not inhibited by organophosphates but hydrolyse them; B-esterases, which are susceptible to organophosphate inhibition; and C-esterases which are uninhibited by organophosphates and do not degrade them [41].

There are two types of esterases that are important in metabolizing insecticides, namely, carboxylesterases and phosphatases (also called phosphorotriester hydrolases or phosphor-triesterases). Carboxyl esterases, which are B-esterases, play a significant role in degrading organophosphates, carbamates, pyrethroids, and some juvenoids in insects. The best example is malathion hydrolysis, which yields both α- and β-monoacids and ethanol [41]. Phosphatases are A-esterases that detoxify many organophosphorous insecticides especially phosphates in insects. In houseflies, paraoxon can be hydrolysed to diethyl phosphoric acid and *p*-nitrophenol. Phosphatases also hydrolyse the alkyl groups of organophosphates. Paraoxon is hydrolysed by the enzyme in houseflies. Several amides containing organophosphorous insecticides such as dimethoate and acephate have been shown to be hydrolysed by carboxylamidases to their corresponding carboxylic acid derivatives [41].

5.2. Phase Ii Reactions

Phase I reactions with xenobiotics result in the addition of functional groups such as hydroxyl, carboxyl and epoxide. These phase I products can further undergo conjugation reactions with endogenous molecules. These conjugations are called phase II reactions. The endogenous molecules include sugars, amino acids, glutathione, phosphate and sulphate. Conjugation products are usually more polar, less toxic and more readily excreted than their parent compounds. Thus, the process with only a few exceptions results in detoxifications.

Three types of conjugation reactions occur in insects. Type I requires an activated conjugating agent that then combines with the substrate to form the conjugated product. Type II involves the activation of the substrate to form an activated donor that then combines with an endogenous molecule to yield a conjugated product. In Type III, conjugation can proceed directly between the substrate and the conjugating agent without involving activation. Thus, Type I and II require information of high-energy intermediates before the conjugation reactions proceed. The chemical groups required for Type I are –OH, NH_2, COOH and SH (glucose conjugation, sulphate conjugation and phosphate conjugation); for Type II COOH (amino acid conjugation); and for Type III, halogens, alkenes, NO2, epoxides, ethers and esters (glutathione conjugation).

5.2.1. Glutathione s-transferases (ec 2.5.1.18.)

Glutathione conjugations are performed by a group of multifunctional enzymes known as glutathione S-transferases and are involved in detoxification mechanisms of many molecules. GSTs are involved in the transport of physiologically important lipophilic compounds. These enzymes catalyse reactions in which the sulphur atom of glutathione provides electron for nucleophilic attack on a second electrophilic substrate; the latter can be endogenous natural substrates such as epoxides, organic hydroperoxides, or activated alkenals resulting from oxidative metabolism. These enzymes catalyse the conjugation of reduced glutathione (GSH) with electrophilic substrates. Glutathione S-transferases perform a variety of reactions including:

The S-alkylation of GSH by alkyl halides and related compounds.

The replacement of labile aryl halogen or nitro groups by GSH.

The replacement of labile arakyl halogen and ester groups by GSH.

The addition of GSH to various epoxides.

The addition of GSH to α-, β-unsaturated compounds including aldehydes, ketones, lactones, nitriles and nitro compounds.

The *O*-alkyl and *O*-aryl conjugation of phosphorothiotes and phosphates with GSH.

The glutathione conjugate is subsequently transformed to mercapturic acid through the stepwise loss of glutamic acid and glycine to a cysteine conjugate, which is finally acetylated before excretion. Because of their broad substrate specificities, glutathione S-transferases are responsible for the detoxification of numerous xenobiotics [97]. More than 40 GST genes have been identified in insects [98, 99]. Mammalian GSTs have been classified into eight cytosolic classes (alpha, mu, pi, theta, sigma, zeta, kappa and omega) and a microsomal class on the basis of their amino acid sequence, immunological properties and substrate specificity. Each class shares 40% or higher amino acid in common. The classification of insect GST is not clear. The majority of insect GSTs do not belong to mammalian classes. Insect glutathione S-transferases consist of two subunits (homodimers and heterodimers) of molecular weight between 19 and 35 kD. Two classes of insect GSTs (Class I and Class II) were reported [100], which have been referred to as the Delta and Sigma class, respectively. Recently, a new class of insect GSTs, referred to as Epsilon has been described in several species of insects including *Anopheles gambiae* [101]. Purified cytosolic and microsomal glutathione S-transferase isozymes from fall armyworm larvae possessed cumenehydrperoxide peroxidase [102]. A Delta class GST purified from German cockroaches also showed high peroxidase activity [103]. The name of each GST is composed of the initials of the species scientific name, followed by the acronym GST, a capital letter to designate the class name and an Arabic number for the individual protein, such as AgGSTD2.

Glutathione S-transferases are important in the metabolism of organophosphorous insecticides resulting in detoxification [99, 104]. For example, methyl parathion is dealkylated by glutathione S-tranferases to form desmethyl parathion and methyl glutathione [41]. On the contrary, parathion can be de-arylated by glutathione S-transferases to produce diethyl phosphorothioic acid and S-(*p*-nitrophenyl) glutathione [41]. Interestingly, a glutathione S-transferase isozyme from the housefly exhibits DDT-dehydrochlorinase activity, showing that DDT-dehydrochlorinase (DDTase) is one of the glutathione S-tranferases [104]. DDT-dehydrochlorinase converts DDT to DDE, resulting in detoxification [41].

6. DETOXIFYING ENZYMES AND INSECTICIDE METABOLISM

The metabolism of insecticides by P450 enzymes is very often a key factor in determining toxicity to insects and to non-target species. The importance of monooxygenases in insecticide resistance became evident in the early 1960s, when it was shown that resistance to carbaryl could be abolished by the P450 inhibitor sesame [106]. Additional evidence of monooxygenase-based resistance quickly amassed [107, 108]. Monooxygenase-mediated detoxification is frequently found as a major mechanism of resistance, and unlike target site resistance, detoxification has the potential to confer cross-resistance to toxins independent of their target sites [109, 110]. Most cases of monooxygenase-mediated resistance result from an increase in detoxification (Table 3). However, in cases where the parent insecticide must undergo monooxygenase-mediated bioactivation, as is the case for many organophosphates, it is also possible that resistance could be achieved through decreased activation [111]. Although this has been reported once, it does not appear to be a common mechanism of resistance. This may explain why esterases are relatively more common than monooxygenases in resistance to some organophosphates [110, 112]. The classical example is probably the metabolism of phosphorothioate insecticides. In many cases, the active ingredients of organophosphorus insecticides are phosphorothioate (P=S) compounds (also known as phosphorothionates), whereas the molecule active at the acetylcholinesterase target site is the corresponding phosphate (P=O) (Figure 2).

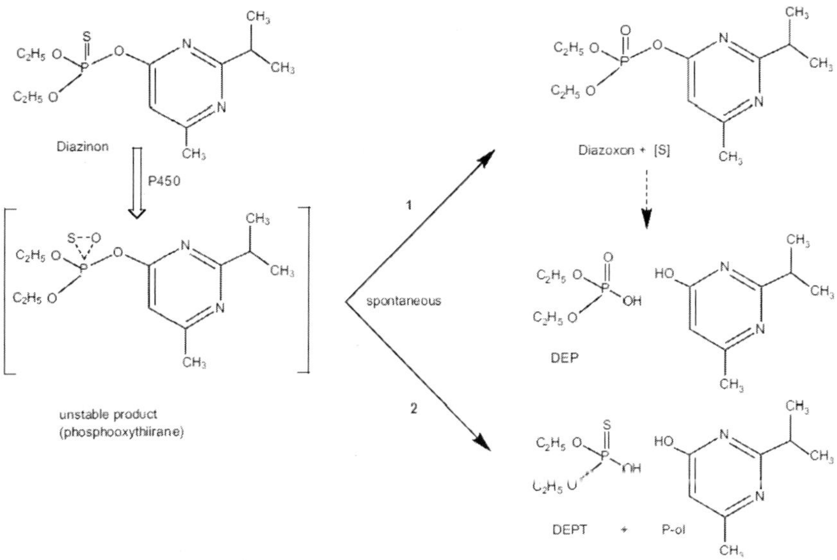

Figure 2. Metabolism of diazinon by cytochrome P450. Following an insertion of oxygen into the substrate, a reactive intermediate collapses (1) by desulphuration or (2) by cleavage of the ester linkage. DEP, diethylphosphate; DEPT, diethylphosphorothioate; P-ol, 2-isopropoxy-4-methyl-6-hydroxypyrimidine; [S], reactive form of sulphur released during the reaction. Adopted from Feyereisen, 2012.

P450 enzymes that metabolize OPs can metabolize other insecticides as well, and this sometimes leads to potentially useful interactions. Thus, enhanced detoxification of dicofol in spider mites can lead to enhanced chlorpyriphos activation, and hence negative cross-resistance [113]. Similarly, permethrin resistance in horn flies is suppressible by piperonyl butoxide, and negatively related to diazinon toxicity [114]. In *H. armigera* populations from West Africa, triazophos shows negative cross-resistance with pyrethroids, and in this case, the synergism shown by the OP towards the pyrethroid appears due to an enhanced activation to the oxon form [115]. Organophosphorus compounds (disulfoton and fenthion) are also activated by thioether oxidation (formation of sulphoxide and sulfone). The metabolism of pyrethroid by P450 enzymes is well studied in insects. Hydroxylations and further metabolism make pyrethroid metabolism and has been noticed for the single enantiomer of deltamethrin (Figure 3).

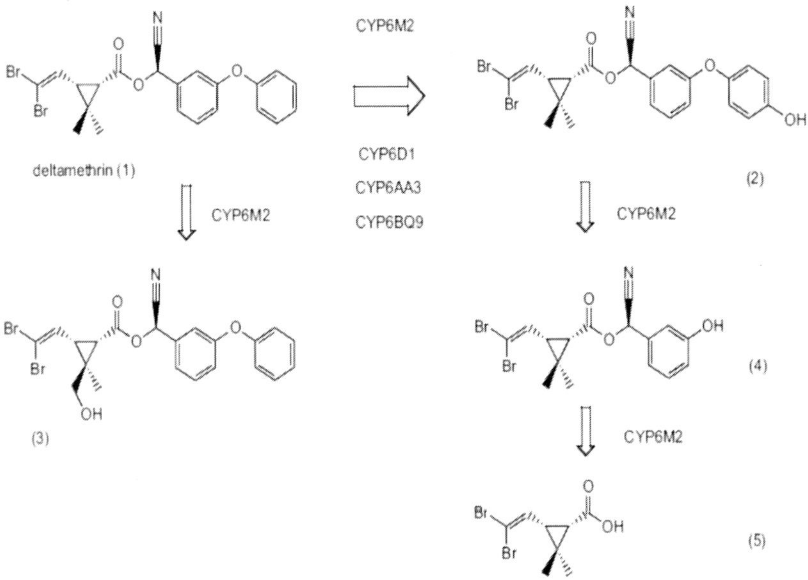

Figure 3. Metabolism of deltamethrin by insect P450 enzymes: (1) deltamethrin; (2) 4′ hydroxydeltamethrin; (3) *trans*-hydroxymethyl-deltamethrin; (4) cyano (3-hydroxyphenyl) methyl deltamethrate; (5) deltamethric acid. Adopted from Feyereisen, 2012.

The currently banned cyclodiene insecticides aldrin, heptachlor and isodrin are epoxidized by P450 enzymes to the environmentally stable toxic epoxides dieldrin, heptachlor epoxide and endrin, respectively [116]. Recombinant CYP6A1, -A2, -A8, -B8 and -B27; CYP12A1; and CYP321A1 can catalyse these epoxidations. Examples of pro-insecticide metabolism include the activation of chlorfenapyr by N-dealkylation [117] and diafenthiuron by S-oxidation [118]. In each case, the insect P450-dependent activation is a key in the selective toxicity of these pro-insecticides that target mitochondrial respiration. Recombinant housefly CYP6A1 catalyses the activation of chlorfenapyr (Figure 4).

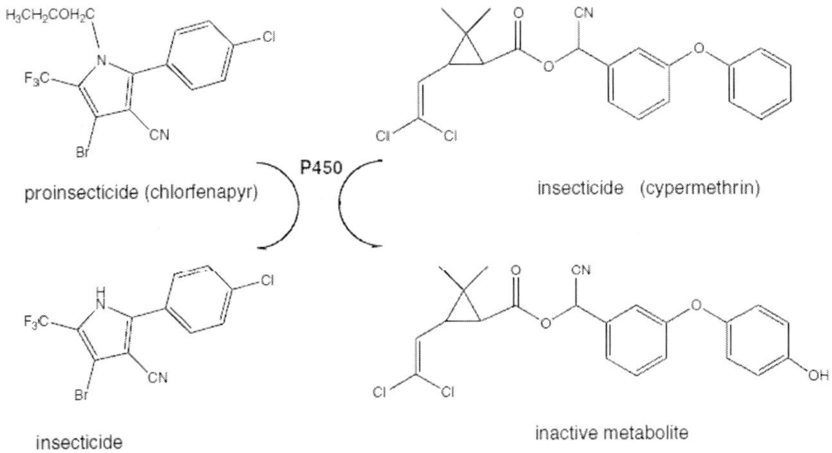

Figure 4. Chlorfenapyr and cypermethrin metabolism. The same P450 in *Heliothis virescens* probably activates the pyrrole and inactivates the pyrethroid, resulting in negative cross-resistance. Adopted from Feyereisen, 2012.

In *H. virescens*, the toxicity of chlorfenapyr is negatively correlated with cypermethrin toxicity [119]. The metabolism of imidacloprid is also of interest, particularly in relation to resistance. Piperonylbutoxide can synergize the toxicity of imidacloprid, and two P450 enzymes, CYP6G1 of *D. melanogaster* and CYP-6CM1vQ of *Bemisia tabaci*, have been shown to metabolize this neonicotinoid [120, 121] (Figure 5). Hydroxylations at the 4 and 5 positions can lead to the olefinic metabolite or to the dihydroxylated metabolite. In the whitefly, the 5-hydroxy metabolite is not toxic, but the 4-hydroxy metabolite is as toxic as the parent compound, so region selectivity may be of importance.

Despite the continuous use of insecticides, there are repeated failures in controlling the sucking insect pest species in recent years [11, 21] in different conventional tea plantations of Terai, the Dooars and Darjeeling foothill regions. Such a failure occurs due to changes in the susceptibility level of the pest species to the applied insecticides. Susceptibility level changes mainly due to metabolic detoxification of the insecticides through higher level of activity of some insecticide detoxifying enzymes under the stress of different management practices [8, 10, 11, 22]. In another mirid pest, *Lygus lineolaris,* metabolic resistance to insecticides due to enhanced level of activity has been reported by many authors [177, 178].

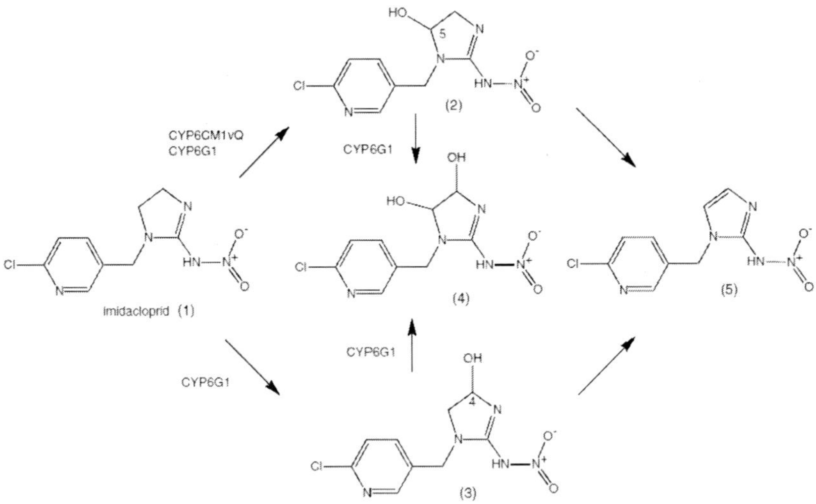

Figure 5. Metabolism of imidacloprid by insect P450 enzymes: (1) imidacloprid; (2) 5-hydroxyimidacloprid; (3) 4-hydroxyimidacloprid; (4) dihydroxyimidacloprid; (5) Non-enzymatically derived dehydroimidacloprid. Adopted from Feyereisen, 2005.

Table 3. Over-expressed CYP genes in insecticide-resistant strains.

Species	P450 Over-expressed	Resistance pattern
Musca domestica	CYP6A1	OP, carbamates [122]
		IGR [123, 124]
	CYP6A5v2, CYP6A36	Pyrethroids [125–127]
	CYP6D1, CYP6D3	Pyrethroids [128–130]
	CYP6D1	Pyrethroids [131]
	CYP6D1, CYP6D3v2	Pyrethroids [132]
	CYP6A24	Pyrethroids [133]
	CYP12A1	Pyrethroids [134]
Drosophila melanogaster	CYP6A2	DDT, malathion [135–137]
	CYP6A8	Malathion [137, 138]
		Malathion [137]
		DDT [139]
	CYP6G1	DDT [140, 141]
		DDT [142]
		Lufenuron, propoxur [141]
		Imidacloprid [141]
		Imidacloprid [142]
		DDT [143]

		Diazinon [144]
	CYP12D1/2	DDT [143]
		DDT [139]
	CYP12A4	Lufenuron [145]
Drosophila simultans	CYP6G1	DDT, imidacloprid, Malathion [139]
Anopheles gambiae	CYP6Z1	Pyrethroids [146]
	CYP325A3	Pyrethroids [147]
	CYP6M2, CYP6P3	Pyrethroids [148]
	CYP6P3	Permethrin [149]
	CYP6M2, CYP6Z2	Permethrin [150]
	CYP4C27, CYP4H15	DDT [151]
	CYP6Z1,2, CYP12F1, CYP314A1	DDT [147]
A. stephensi	CYP325C1	Pyrethroids [152]
A. funetus	CYP6P4, CYP6P9	Pyrethroids [153, 154]
Aedesaegypti	CYP9J10,27,32	Pyrethroids [155]
Culex pipien squinquefasciatus	CYP9M10	Permethrin [156]
	CYP6F1	Permethrin [157]
	CYP4H34, CYP6Z10, CYP9M10	Permethrin [158]
	CYP4H21, H22, H23, CYP4J4, CYP4J6	Deltamethrin [159]
Heliothis virescens	CYP9A1	Thiodicarb [160]
Helicoverpa zea	CYP6B8,B9	Cypermethrin [161]
H. armigera	CYP4G8	Pyrethroids [162]
	CYP6B7	Pyrethroids [163]
	CYP6B7, CYP9A12, CYP9A14	Pyrethroids [164]
	CYP4S1, CYP337B1	Fenvalerate [165]
	CYP4L5,11, CYP4M6,7, CYP6AE11, CYP9A14, CYP332A1, CYP337B1	Deltamethrin [166]
Plutella xylostella	CYP6BG1	Cypermethrin [167]
	CYP4M20	Cypermethrin [168]
Lygus lineolaris	CYP6X1	Permethrin [169]
Bemisia tabaci	CYP6CM1vQ	Imidacloprid [170]
Nilaparvata lugens	CYP6ER1	Imidacloprid [171]
Myzus persicae	CYP6CY3	Neonicotinoids [172]
Diabrotica virgifera	CYP4	Me-parathion [173]
Tribolium castaneum	CYP6BQ8,9,10, CYP436B1, B2	Deltamethrin [174]
Blattella germanica	P450MA,	Chlorpyriphos [175]
	CYP4G19	Pyrethroids [176]

In Western Flower Thrips, *Frankliniella occidentalis,* metabolic detoxification of insecticides has been reported by many authors [179]. In *Bemisia tabaci,* metabolic resistance due to enhanced activity of insecticide resistance-related enzymes has also been reported [180].

7. HOST ALLELOCHEMICALS, INDUCTION OF DETOXIFYING ENZYMES AND INSECTICIDE RESISTANCE

Understanding the diversity of insect responses to chemical pressures (plant allelochemicals and insecticides) in their local ecological context represents a key challenge in developing sustainable pest control strategies. Plants and insects have had co-existing relationships for a long time. Insects were suppressed either by other insects or toxins or by plant defence mechanisms in order to create a balance between the insect pest population and host. Each plant species has a unique set of defence traits ranging from morphological to phytochemical parameters that have behavioural and physiological ramifications for a potential herbivore consumer [181, 182]. Therefore, the resistance mechanisms evolved by insects to deal with the chemical defences of plants are similar to those mechanisms that have evolved to resist synthetic insecticides. The chemical structure of some synthetic insecticides is comparable to that of some plant-produced compounds (e.g. pyrethroids and nicotinoids). Insect resistance to plant allelochemicals interferes with their resistance to synthetic insecticides [183]. From the evolutionary perspective, despite the key role of the chemical 'arms race' in driving the co-evolution of plants and insects, much research has focused so far on describing the diversity of plant chemicals and their effects on herbivores. Hence, the understanding of insecticide resistance mechanisms as well as taking into account other ecological parameters is important in predicting the spread of insecticide resistance in natural pest populations and in choosing the optimum strategy for managing insect pest populations. Less is known about the multiple mechanisms evolved by insects to overcome these chemical defences (Table 4). These mechanisms include contact and ingestion avoidance, excretion, sequestration, degradation of the toxin and target site mutation.

Biotransformation of plant toxins is one of the major weapons that insects have evolved in their co-evolutionary arms race with plants [204]. To date, metabolic resistance to plant chemicals has been identified not only in herbivorous insects [194] but also in detrivorous insects such as mosquito larvae feeding plant debris [205]. Metabolic resistance often results from the overproduction of 'detoxification enzymes' that can metabolize plant xenobiotics (allelochemicals). This mechanism is often associated with phenotypic plasticity, as the production of detoxification enzymes is usually induced by the presence of plant xenobiotics in the diet of the insect.

Table 4. Plant allelochemicals and associated resistance mechanisms in insects.

Plant allelochemicals	Target (mechanism of effect)	Resistance mechanisms	Species
Alkaloids	Neuroreceptors (inhibition), ion channels (antagonists), nucleic acids (disruption of DNA synthesis), feeding (deterrent owing to bitterness), enzymes (inhibition)	Modification of nicotine synthesis by salivary glucose oxidase	*Helicoverpa zea* (Lep.) [184]
Cardenolides	Nervous system (depressing activity); Na^+, K^+-ATPase (specific inhibitor)	Canal trenching behaviour, target site mutation	*Danaus plexippus* (Lep.) [185] *Chrysochus* sp. (Col.) [186]
Cyanogenic glycosides	Electron transport (inhibition of mitochondrial cytochrome oxidase)	Ingestion avoidance, sequestration and detoxification	*Schistocerca americana* (Ort.), *Hypera brunneipennis* (Col), *Zygaena* sp. (Lep.), *Clossiana euphrosyne* (Lep.), *Heliconius sara* (Lep.) [187]
Glucosinolates	Respiration (inhibition)	Detoxification by GSTs, detoxification by glucosinolate sulphatase, formation of nitriles instead of isothiocyanate detoxification by P450s, detoxification by N-oxidation and sequestration	*Myzus persicae* (Hem.) [188] *Plutella xylostella* (Lep.) [189] *Pieris rapae* (Lep.) [190] *Drosophila melanogaster* [191] *Estigmene acrea* (Lep.) [192] *Tyria jacobaeae* (Lep.) [193]
Flavonoids and phenolic acids	Respiration (inhibition), growth (inhibition)	Ingestion avoidance, decrease of toxin levels in gall tissue, glycosylation by UDP-glycosyl transferase, sequestration and/or excretion	*Manduca sexta* (Lep.) [194] *Pontania* sp. (Hym) [195] *Bombyx mori* (Lep.) [196]
Iridoid glycosides	Feeding (deterrent owing to bitterness), nucleic acids (inhibition of DNA polymerase), proteins (denaturant and cross-linking activities)	Sequestration	*Longitarsus* sp. (Col.) [197]
Coumarins and furanocoumarins	Nucleic acids (photoactivie DNA binding) pro-oxidant activity	Detoxification by P450s, detoxification by GSTs	*Papilio polyxenes* (Lep.) [198] *Depressaria pastinacella* [199] *Spodoptera frugiperda* (Lep.) [200]
Protease inhibitors	Digestive system (inhibition of protease)	Over-expression of insensitive protease	*Callosobruchus maculatus* (Col.) [201]
Terpinoids	Nervous system (inhibition of acetylcholine esterases); feeding (deterrent owing to physical barrier and bitterness); growth and development inhibitor (pheromone analogue)	Repression of genes involved in biosynthetic pathways	*Spodoptera exigua* (Lep.) [202]
Tannins	Feeding (complexation of salivary and gut proteins); pro-oxidant activity	Synthesis of anti-oxidant compounds	*Orgyia leucostigma* (Lep.) [203]

Lep., Lepidoptera; Col., Coleoptera; Ort., Orthoptera; Hym., Hymenoptera; Dip., Diptera; Hem., Hemiptera.

Induction of insect detoxifying enzyme activities by plant allelochemicals is a clear manifestation of biochemical phenotypic plasticity and has been documented in several instances [206, 207]. Many of the theories and some of the experiments implicitly or explicitly deal with the insect's ability to metabolize plant secondary substances by P450 and other enzymes. In those studies, a 'higher activity of midgut microsomal oxidase enzymes in polyphagous than in monophagous species indicates that the natural function of these enzymes is to detoxify natural insecticides present in the larval food plants'. The estimation of aldrin epoxidation in gut homogenates of last instar larvae from 35 species of Lepidoptera showed that polyphagous species had on average a 15 times higher activity than monophagous species. This trend was seen in sucking insects as well. A 20-fold lower aldrin epoxidase activity was found in the oleander aphid *Aphis nerii* (specialist feeder on two plant families, Asclepiadaceae and Apocynaceae) when compared to the potato aphid *Myzus euphorbiae* or to the green peach aphid *Myzus persicae* (both are generalists found on 30–72 plant families) [208]. A similar type of observation was made for other detoxification enzymes. In mites, predatory mite has a five times lower aldrin epoxidase activity than its herbivorous prey [209]. The toxicity of the natural phototoxin α-terthienyl is inversely proportional to the level of its metabolism in Lepidoptera and is related to diet breadth. Metabolism is highest in *Ostrinia nubilalis,* which feeds on numerous phototoxic Asteraceae; lower in *Helicoverpa virescens*, which has a broad diet, including some Asteraceae that are non-phototoxic; and lowest in *Manduca sexta*, a specialist of Solanaceae [210].

In addition to insecticides, insect carboxylesterases also metabolize many glycosides. B-glucosidase enzyme is active towards a variety of glucosides in fall armyworms, corn earworms, cabbage loopers and valvet bean caterpillars. The *p*-nitrophenyl β-D-glucoside, 4-methyl umbelliferyl β-D-glucoside, D (+)-cellobiose, D-amygdalin and helicon were preferred substrates whereas sinigrin, phloridzin, α-solanine, tomatine and linamarin were poor substrates for these insects and many other insects reported to date [211]. β-Glycosidases have been shown to play important roles in the survival of certain phytophagous insects [211]. The ability of peach tree borer, *Synanthedon exitiosa,* larvae to survive well on prunasin-containing peach tree is because they can metabolize cyanogenic glycosides through β-glucosidase and detoxify the released cyanide by β-cyanoalanine synthase, thereby allowing them to utilize peach trees [212]. Another example is the larvae of the tiger swallowtail, *Papilio glaucus,* which feeds on quacking aspen, which contains various phenolics glycosides (e.g. salicortin). These larvae hydrolyse the glycosides by β-glucosidase and detoxify the released phenolics aglycone by a highly active esterase, thereby allowing them to survive on aspen [213, 214].

Glutathione S-transferases are also involved in the metabolism of many toxic plant allelochemicals. These plant allelochemicals may be of many diverse groups including α, β-unsaturated carbonyl compounds (e.g. *trans-*

cinnamaldehyde, trans-2-hexanal), isothiocyanates (e.g. allylisothiocyanate, benzyl isothiocyanate) and organothiocyanates (e.g. benzyl thiocyanate) as have been documented in some instances [41]. The glutathione S-transferase activities are lower in the specialist insects than in the generalists. In the crucifer-adapted cabbage looper for the metabolism of isothiocyanates (plant allelochemical), the activity of this enzyme wasfound to be two- to sixfold higher than that in the fall armyworm [215, 216]. These findings strongly advocate that glutathione S-transferases play an important role in developing resistance towards plant allelochemicals in phytophagous lepidopteran insects [215–217]. Many plant allelochemicals are potent inhibitors of glutathione S-transferases in many insects [218]. Many flavonoids, other phenols and α-, β-unsaturated carbonyl compounds are also found to be potent inhibitors of the enzymes.

H. theivora, E. flavescens and *S. dorsalis* all are polyphagous in nature. *H. theivora* known to feed on at least sixteen different plant families reported till date [11]. Similarly, *E flavescens* is also polyphagous [11]. *S. dorsalis* has been documented to attack more than 150 hosts from at least 40 different plant families [219]. Hence, these pests are exposed to a wide variety of plant allelochemicals of diverse groups having the potential to induce the activity of these resistance-related enzymes. A higher level of detoxifying enzyme activity in *H. theivora* has been reported when reared on two alternative hosts, i.e. *Mikania micrantha* (Asteraceae) and *Psidium guajava* (Myrtaceae), than on tea [220]. Over the four hundred million years of co-evolution with plants, phytophagous insects have developed diverse resistance mechanisms to cope with plant chemical defences. Because insects face a geographical mosaic of chemical environments, from non-toxic to highly toxic plants, the costs associated with resistance traits vary with the probability of encountering a toxin. Moreover, other selection pressures, such as the presence or absence of competitors and predators, can also influence the costs and selection of particular resistance traits. Thus, the complexity of the local community composition is a key factor in maintaining the diversity of adaptive mechanisms to plant xenobiotics. These mechanisms are more plastic and complex compared with those involved in resistance to insecticides, perhaps because environments in which insecticides are heavily used also tend to have communities of low diversity and complexity. However, because some detoxification enzymes are involved in plant toxins and insecticides metabolism, cross-resistance mechanisms can be predicted to be observed under specific environmental conditions. Deciphering the impact of allelochemicals in cross-resistance mechanisms with insecticides at a local scale, and comparing the molecular and evolutionary mechanisms of resistance to phytotoxins and synthetic insecticides, represent promising areas of research for developing long-term sustainable insect control strategies for the effective management of pest concern [220].

8. GENETICS AND INSECTICIDE RESISTANCE IN TEA PESTS

Earlier, common visible markers including mophometrices, eye colour, body spots or bands and hairs or spines, wing venation were used as phenotypic markers in studying the pattern of dispersal, mating behaviour, population variability and inheritance of genetic traits in insects [221, 222]. Although the phenotypic markers are found at all times of life span of the organism and can be readily used for studies in field conditions, they suffer from many practical limitations. The major drawback is that these visible phenotypes are relatively infrequent and often hard to score. Because the phenotype markers are rare, use of these markers in mapping a trait is difficult. For all such difficulties and with the concurrent advancement in biochemical methodologies, protein markers then became more popular. Protein markers made a significant contribution in the early periods when DNA technologies were not so much advanced, as it is now [223]. A diverse range of novel molecular (DNA) markers are now available for entomological investigations. Currently, both DNA and protein markers have revolutionized the biological sciences and have enhanced many fields of insect study, especially agricultural entomology [224].

Insecticide resistance is the result of an increase in the ability of individuals of an insect species to survive insecticide application and is an important example of man-driven evolution [225]. Alleles conferring resistance may arise and spread in populations and to other populations with variable success, depending on factors such as selective forces, genetic variability, gene flow, population size and environmental conditions [220]. Studies that map the population structure of pest insects, as well as the potential for gene flow between populations, are needed to understand the development of resistance and prevention of its spread [226, 227]. Development of resistance is often rapid in isolated populations that have been treated by insecticides [228]. The rate of development of insecticide resistance may, however, be influenced by gene flow between treated and untreated populations by maintaining the frequency of resistance alleles at a low level [229]. Contaminant exposure was a poor predictor of population structure and the level of gene flow was a better predictor of relatedness [230]. Gene flow may balance divergence by opposing the effect of selection pressures [229]. Population genetic patterns should therefore be investigated with reference to geographical variability, as well as selection pressure. Detoxification resistance occurs when enhanced levels or modified activities of biotransformation enzymes prevent the insecticide from acting on its site of action because the metabolites produced have little or no activity compared with the original substance [231]. These changes may be due to mutations resulting in a protein with slightly different properties or altered expression.

As chemical control is frequently used to avoid economic damage, the sucking insects have been subjected to major selection pressure. Insecticides will probably continue to be the main control method in the near future and therefore it is important to study the structure of sucking insect population and change in insecticide susceptibility. There are several techniques for estimating the genetic diversity such as randomly amplified polymorphic DNA analysis, microsatellites, minisatellites, restriction fragment length polymorphism analysis and amplified fragment length polymorphism (AFLP) analysis. DNA markers are also suitable for use with small amounts of insect material and can be used with stored, dry or old samples. Some have complex multi-locus banding patterns, which may be of a non-Mendelian nature (e.g. randomly amplified polymorphic DNAs (RAPDs)). They have an expanding range of applications, many involving intra- and interspecific discriminations.

9. ACKNOWLEDGEMENTS

The author expresses his sincere thanks to the Head, Department of Zoology, University of North Bengal, for providing laboratory space. The author also expresses sincere appreciation to those scientists and authors based on whose concept, hypothesis and work this chapter has been developed. The author also expresses his thanks to the University of North Bengal for providing uninterrupted local area network that has immensely helped in searching and collecting information. The author also expresses his sincere thanks to InTech Open Access Publishing Group, editor of the book '*Insecticide Resistance*' Prof. Stanislav Trdan (University of Ljubljana, Biotechnical Faculty, Dept. of Agronomy, Chair for Phytomedicine, Agricultural Engineering, Crop Production, Pasture and Grassland Management, Slovenia) and Ms Sndra Bakic (InTech Europe, Rijeka, Croatia).

REFERENCES

1. Chen Z. M. and Chen X. F. (1989) An analysis of world tea fauna. *Journal of Tea Science* 9: 73–88.
2. Muraleedharan N. (1992) Pest control in Asia. In Tea: Cultivation to Consumption. Wilson, K.C. and Clifford, M.N. (edited by N. Muraleedharan). Chapmann & Hall, London, pp. 375–412.

3. Das G. M. (1965) Pests of tea in North East India and their control. Memorandom No. 27, pp. 169–173, Tocklai Experimental Station, Tea Research Association, Jorhat, Assam.
4. Muraleedharan N. (2007) Tea insects: ecology and control, pp. 672–674. In Encyclopedia of Pest Management. CRC Press, London.
5. Rattan P. S. (1992) Pest and disease control in Africa, pp. 331–352. In Tea: Cultivation to Consumption (edited by K. C. Wilson and M. N. Clifford). Chapman & Hall, London.
6. Sivapalan P. (1999) Pest management in tea, pp. 625–646. In Global Advances in Tea Science (edited by N. K. Jain). Aravali Books, New Delhi.
7. Hazarika L. K., Bhuyan M. and Hazarika B. N. (2009) Insect pests of tea and their management. *Annual Review of Entomology* 54: 267–284
8. Saha D. and Mukhopadhyay A. (2013) Insecticide resistance mechanisms in three sucking insect pests of tea with reference to North East India - an appraisal. *International Journal of Tropical Insect Science* 33(1): 46–70.
9. Cranham J. E. (1966) Tea pests and their control. *Annual Review of Entomology* 11: 491–514.
10. Saha D., Roy S. and Mukhopadhyay A. (2012) Insecticide susceptibility and activity of major detoxifying enzymes in female Helopeltis theivora Waterhouse (Heteroptera: Miridae) from sub-Himalayan tea plantations of North Bengal, India. *International Journal of Tropical Insect Science* 32: 85–93.
11. Saha, D. (2014) Assessment of Population Variability at Subcellular Level of Some Common Sucking Tea Pests from Darjeeling Hill and its Adjoining Plain. A Ph.D. Thesis. University of North Bengal, Sliguri-734013, District - Darjeeling, West Bengal, India.
12. Chaudhuri T. C. (1999) Pesticide residues in tea, pp. 369–378. In Global Advances in Tea Science (edited by N. K. Jain). Aravali Books, New Delhi.
13. Komagata O., Kasai S. and Tomita T. (2010) Overexpression of cytochrome P450 genes in pyrethroid- resistant Culex quinquefasciatus. *Insect Biochemistry and Molecular Biology* 40: 146–152.
14. Das G. M. (1963) Some important pests of tea. *Two and a Bud* 10: 4–8.
15. Mukerjea T. D. (1968) Thiodan: a broad spectrum new insecticide. *Two and a Bud* 15: 6–10.
16. Barbora B. C. and Biswas A. K. (1996) Use pattern of pesticides in tea estates of North East India. *Two and a Bud* 47: 19–21.
17. Sannigrahi S. and Talukdar T. (2003) Pesticide use patterns in Dooars tea industry. *Two and a Bud* 50: 35–38.
18. Gurusubramanian G., Rahman A., Sarmah M., Roy S. and Bora S. (2008) Pesticide usage pattern in tea ecosystem, their retrospects and alternative measures. *Journal of Environmental Biology* 29: 813–826.
19. Gurusubramanian G. and Bora S. (2007) Relative toxicity of some commonly used insecticides against adults of Helopeltis theivora Waterhouse (Miridae: Hemiptera) collected from Jorhat area tea plantations, South Assam, India. *Resistant Pest Management Newsletter* 17: 8–12.
20. Gurusubramanian G. and Bora S. (2008) Insecticidal resistance to tea mosquito bug, Helopeltis theivora Waterhouse (Miridae: Heteroptera) in North East India. *Journal of Environmental Research Development* 2: 560–567.
21. Gurusubramanian G., Senthilkumar N., Bora S., Roy S. and Mukhopadhyay A. (2008) Change in susceptibility in male Helopeltis theivora Waterhouse (Jorhat population, Assam, India) to different classes of insecticides. *Resistant Pest Management Newsletter* 18: 36–39.
22. Saha D., Mukhopadhyay A. and Mahadur M. (2010) Variation in detoxifying enzymes of Assam thrips, Scirtothrips dorsalis Hood (Thysanoptera: Thripidae) from organically and insecticide managed tea plantations. North Bengal University. *Journal of Animal Science* 4: 45–52.

23. Saha D., Mukhopadhyay A. and Mahadur M. (2012) Detoxifying enzymes activity in tea greenfly, Empoasca flavescens Fabricius (Homoptera: Jassidae) from sub Himalayan tea plantations, pp. 257–266. In Proceedings of National Symposium on Biodiversity Status and Conservation Strategies with special Reference to North East India (edited by R. Varatharajan). Centre for Advanced Studies, Department of Life Sciences, Manipur University, Manipur, India.
24. Saha D., Roy S. and Mukhopadhyay A. (2012) Seasonal incidence and enzyme-based susceptibility to synthetic insecticides in two upcoming sucking insect pests of tea. *Phytoparasitica* 40: 105–115.
25. Saha D., Mukhopadhyay A. and Bahadur M. (2012) Variation in the activity of three principal detoxifying enzymes in major sucking pest of tea, Helopeltis theivora Waterhouse (Heteroptera: Miridae) from sub- Himalayan tea plantations of West Bengal, India. *Proceedings of Zoological Society*. (July-Dec 2013) 66(2): 92–99.
26. Kumar, B., Saha, D. and Mukhopadhyay, A. (2014) Enhancement of resistant ratio vis-a-vis defence-enzymes activity in tea mosquito bug, *Helopeltis theivora* Waterhouse (Hemiptera: Miridae) selected through exposure to sub-lethal dose of monocrotophos. *Proceedings of Zoological Society*.
27. Bora S., Rahman A., Sarmah M. and Gurusubramanian G. (2007) Relative toxicity of pyrethroid and nonpyrethroid insecticides against male and female tea mosquito bug (Darjeeling strain). *Journal of Entomological Research* 37: 37–41.
28. Roy S., Gurusubramanian G. and Mukhopadhyay A. (2008) Insecticide persistence and residual toxicity monitoring in tea mosquito bug, *Helopeltis theivora* Waterhouse (Hemiptera: Heteroptera: Miridae). *Resistant Pest Management Newsletter* 17: 9–14.
29. Rahman A., Sarmah M., Phukan A. K., Roy S., Sannigrahi S., Borthakur M. and Gurusubramanian G. (2005) Approaches for the management of tea mosquito bug, Helopeltis theivora Waterhouse (Miridae: Heteroptera), pp. 146–161. In Proceedings of the 34th Tocklai Conference 'Strategies for Quality in the Digital Era' (edited by A. K. Barooah, M. Borthakur and J. N. Kalita). Tocklai Experimental Station, TRA, Jorhat, Assam.
30. Roy S. (2010) Evaluation of the levels of insecticide susceptibility of *Helopeltis theivora* Waterhouse (Heteroptera: Miridae) and development of an efficacious strategy for management of the pest in Dooars tea plantation of North Bengal, PhD thesis, University of North Bengal, India.
31. Reddy G. P. V., Prasad V. D. and Rao R. S. (1992) Relative resistance in chilli thrips, Scirtothrips dorsalis (Hood) populations in Andhra Pradesh to some conventional insecticides. *Indian Journal of Plant Protection* 20: 218–222.
32. Vanisree K., Upendhar S. and Rajasekhar P. (2011) Toxicity of certain novel insecticides against chilli thrips, Scirtothrips Dorsalis (Hood). *Resistant Pest Management Newsletter* 21: 17–21.
33. Seal D. R., Ciomperlik M., Richards M. L. and Klassen W. (2006) Comparative effectiveness of chemical insecticides against the chilli thrips, Scirtothrips dorsalis Hood (Thysanoptera: Thripidae), on pepper and their compatibility with natural enemies. *Crop Protection* 25: 949–955.
34. Seal D. R., Klassen W. and Sabines C. (2007) Management of chilli thrips, Scirtothrips dorsalis (Thysanoptera: Thripidae): effectiveness of neonicotinoids and spinosyns and ineffectiveness of pyrethroids. *Proceedings of the Caribbean Food Crops Society* 43: 39–48.
35. Doğramaci M., Arthurs S. P., Chen J., McKenzie C., Irrizary F. and Osborne L. (2011) Management of chilli thrips Scirtothrips dorsalis (Thysanoptera: Thripidae) on peppers by *Amblyseius swirskii* (Acari: Phytoseiidae) and *Orius insidiosus* (Hemiptera: Anthocoridae). *Biological Control* 59: 340–347.
36. Shi C. H., Shang J. N. and Chen Y. F. (2001) Dynamic residues of imidacloprid in tea products and its application to control of tea insect pests, pp. 165–166. In Integrated Pest Management in Relation to Safe Agricultural Products (edited by Y. J. Yu and Q. H. Zhang). China Agriculture Press, Beijing.

37. Tao T., Hu K. M., She Y. P., Zhu Q. Z. and Luo Z. (1996) Studies on the integrated management techniques of smaller green leafhopper in tea plantation of south Yunnan. *Acta Phytophysiologica Sinica* 23: 310–314.
38. Nian-wu W., Jin-han X. U., Zheng C., Hui W., Ling-ling Z. and Xiong G. (2004) Resistance level of Empoasca vitis (Gothe) in different tea plantations [J]. *Journal of Tea Science*, 2-.
39. Jia-Xiang Z., Jian-Wei F. U., Qing-Quan S. U., Jian-Yu L. I. and Zhi-Xiong Z. (2009) The regional diversity of resistance of Tea Green Leafhopper (Gothe), to insecticides in Fujian Province. *Journal of Tea Science* 2-.
40. Soderland D. M. and Bloomquist J. R. (1990) Molecular mechanisms of insecticide resistance, pp. 58–96. In Pesticide Resistance in Arthropods (edited by R. T. Roush and B. E. Tabashnik). Chapman and Hall, New York.
41. Yu S. J. (2008) The Toxicology and Biochemistry of Insecticides. CRC Press, Florida. 296 p.
42. Wu S., Yang Y., Yuan G., Campbell P. M., Teese M. G., Russell R. J., Oakeshott J. G. and Wu Y. (2011) Overexpressed esterases in a fenvalerate resistant strain of the cotton bollworm, *Helicoverpa armigera*. *Insect Biochemistry and Molecular Biology* 41: 14–21.
43. Buès R., Bouvier J. C. and Boudinhon L. (2005) Insecticide resistance and mechanisms of resistance to selected strains of *Helicoverpa armigera* (Lepidoptera: Noctuidae) in the south of France. *Crop Protection* 24: 814–820.
44. Cao C.W., Zhang J. Gao X.W. Liang P. and Guo H.L. (2008) Overexpression of carboxylesterase gene associated with organophosphorous insecticide resistance in cotton aphids, *Aphis gossypii* (Glover). *Pesticide Biochemistry and Physiology* 90: 175–180.
45. Cao C.W., Zhang J., Gao X.W., Liang P. and Guo H.L. (2008) Differential mRNA expression levels and gene sequences of carboxylesterase in both deltamethrin resistant and susceptible strains of the cotton aphid, *Aphis gossypii*. *Insect Science* 15: 209–216.
46. Mitter C., Farrell B. D. and Futuyma D. J. (1991) Phylogenetic studies of insect/plant interactions: Insights into the genesis of diversity. *Trends in Ecology and Evolution* 6: 290–293.
47. Ehrlich P. R. and Raven P. H. (1964) Butterflies and plants: a study in coevolution. *Evolution* 18: 586–608.
48. Bernays E. A. (1998) Evolution of feeding behavior in insect herbivores: success seen as different ways to eat without being eaten. *Bioscience* 48: 35–44.
49. Dethier V. G. (1954) Evolution of feeding preferences in phytophagous insects. *Evolution* 8: 33–54.
50. Kaloshian I. and Walling L. L. (2005) Hemipterans as plant pathogens. *Annual Review of Plant Biology* 43: 491–521.
51. Goggin F. L. (2007) Plant–aphid interactions: molecular and ecological perspectives. *Current Opinion in Plant Biology* 10: 399–408.
52. Sawicki, R. M. (1987) Definition, detection and documentation of insecticide resistance, pp.105-112. In Combating resistance to xenobiotics: Biological and chemical approaches (edited by E. Hoorwood). Chichester, England,
53. Sparks T. C., Lockwood J. A., Byford R. L., Graves J. B. and Leonard B. R. (1989) The role of behavior in insecticide resistance. *Journal of Pesticide Science* 26: 383–399.
54. Brattsten L. B. (1988) Potential role of plant allelochemicals in the development of insecticide resistance, pp. 313–348. In Novel Aspects of Insect–Plant Interactions (edited by P. Barbosa and D. K. Letourneau). John Willey & Sons, New York.
55. Roy S. and Mukhopadhyay A. (2011) Insecticide-Induced Change in Egg-Laying Strategy of *Helopeltis theivora* (Hemiptera: Miridae) on Tea Shoot (*Camellia sinensis*). *Proceedings of Zoological Society* 64 (1): 54–56.
56. Price N. R. (1991) Insect resistance to insecticides: mechanisms and diagnosis. *Comparative Biochemistry and Physiology* 100C, 319–326.

57. Vinson S. B. and Law P. K. (1971) Cuticular composition and DDT resistance in the tobacco budworm. *Journal of Economic Entomology* 64: 1387–1390.
58. Patil V. L. and Guthrie F. E. (1979) Cuticular lipids of two resistant and a susceptible strain of houseflies. *Pesticide Science* 10: 399–406.
59. Scott J. G. (1990) Investigating mechanisms of insecticide resistance: methods, strategies, and pitfalls, pp. 39–57. In Pesticide Resistance in Arthropods (edited by R. T. Roush and B. E. Tabashnik). Chapman and Hall, New York.
60. Forgash A. J., Cook B. J. and Riley R. C. (1962) Mechanisms of resistance in diazinon-selected multi resistant *Musca domestica*. *Journal of Mechanisms of Economic Entomology* 55: 544–551.
61. Van de Baan H. E. and Croft B. A. (1991) Resistance to insecticides in winter- and summer-forms of pear psylla, *Psylla pyricola*. *Pesticide Science* 32: 225–233.
62. Kanga L. H. B. and Plapp F. W. J. (1995) Target-site insensitivity as the mechanism of resistance to organophosphorus, carbamate, and cyclodiene insecticides in tobacco budworm adults. *Journal of Economic Entomology* 88: 1150–1157.
63. Roy S., Gurusubramanian G. and Mukhopadhyay A. (2008) Variation in endosulfan susceptibility and body lipid content of *Helopeltis theivora* Waterhouse (Heteroptera: Miridae) in relation to the use pattern of the insecticide, in sub-Himalayan Dooars tea plantations. *Journal of Plantation Crops* 36: 388–392.
64. Martinez-Torres D., Foster S. P., Field L. M., Devonshire A. L. and Williamson M. S. (1999) A sodium channel point mutation is associated with resistance to DDT and pyrethroid insecticides in the peach-potato aphid, Myzus persicae (Sulzer) (Hemiptera: Aphididae). *Insect Molecular Biology* 8: 339–346.
65. Mutero A., Pralavorio M., Bride J. M. and Fournier D. (1994) Resistance-associated point mutations in insecticide-insensitive acetylcholinesterase. *Proceedings of the National Academy of Sciences USA* 91: 5922–5926.
66. Brown T. M. and Brogdon W. G. (1987) Improved detection of insecticide resistance through conventional and molecular techniques. *Annual Review of Entomology* 32: 145–162.
67. Scharf M. E., Neal J. J. and Bennett W. G. (1998) Changes of insecticide resistance levels and detoxication enzymes, following insecticide selection in the German cockroach, *Blattella germanica* (L.). *Pesticide Biochemistry and Physiology* 59: 67–79.
68. Casida J. E. (1993) Insecticide action at the GABA-gated chloride channel: recognition, progress, and prospects. *Archives of Insect Biochemistry and Physiology* 22: 13–23.
69. Narahashi, T. (1996) Neuronal ion channels as the target sites of insecticides. *Pharmacology and Toxicology* 79: 1–14.
70. Sattelle D. B. (1985) Acetylcholine receptors, pp. 395–434. In Comprehensive Insect Physiology, Biochemistry, and Pharmacology (edited by G. A. Kerkut and L. I. Gilbert). Vol. 11. Pergamon, Oxford.
71. Byrne F.J. and Toscano N. C. (2001) An Insensitive Acetylcholinesterase Confers Resistance to Methomyl in the Beet Armyworm *Spodoptera exigua* (Lepidoptera: Noctuidae) *Journal of Economic Entomology* 94(2): 524–528.
72. Eldefrawi A. T. (1985) Acetylcholinesterases and anticholinesterases, pp. 115–130. In Comprehensive Insect Physiology, Biochemistry, and Pharmacology (edited by G. A. Kerkut and L. I. Gilbert). Vol. 12. Pergamon Press, Oxford.
73. Aldridge W. N. and Reiner E. (1972). Enzyme inhibitors as substrates. North-Holland Publishing Co., Amsterdam.
74. Smissaert, H. R. (1964) Cholinesterase inhibition in spider mite susceptible and resistant to organophosphate. *Science* 143: 129–131.
75. Fournier D. and Mutero A. (1994) Modification of acetylcholinesterase as a mechanism of resistance to insecticides. *Comparative Biochemistry and Physiology* 108C, 19–31.
76. Hama H. (1983) Resistance to insecticides due to reduced sensitivity of acetylcholinesterase, pp. 299–331. In Pest Resistance to Pesticides (edited by G. P. Georghiou and T. Saito). Plenum Press, New York.

77. Oppenoorth F. J. (1985) Biochemistry and genetics of insecticide resistance, pp. 731–773. In Comparative Insect Physiology, Biochemistry and Pharmacology (edited by G. A. Kerkut and L. I. Gilbert). Vol. 12. Pergamon Press, Oxford.
78. Guedes R. N. C., Zhu K. Y., Kambhampati S. and Dover B. A. (1997) An altered acetylcholinesterase conferring negative cross-insensitivity to different insecticidal inhibitors in organophosphate-resistant lesser grain borer, *Rhyzopertha dominica. Pesticide Biochemistry and Physiology* 58: 55–62.
79. Harold J. A. and Ottea J. A. (1997) Toxicological significance of enzyme activities in profenofos resistant tobacco budworms. *Heliothis virescens* (F.). *Pesticide Biochemistry and Physiology* 58: 23–33.
80. Levitin E. and Cohen E. (1998) The involvement of ace-tylcholinesterase in resistance of the California red scale, *Aonidiella aurantii* to organophosphorus pesticides. *Entomologia Experimentalis et Appllicata* 88: 115–121.
81. Sarker M. and Mukhopadhyay A. (2006a) Studies on salivary and midgut enzymes of a major sucking pest of tea, Helopeltis theivora (Heteroptera: Miridae) from Darjeeling plains, India. *Journal of the Entomological Research Society* 8: 27–36.
82. Sarker M. and Mukhopadhyay A. (2006b) Studies on some enzymes related to insecticide resistance in Helopeltis theivora Waterhouse (Insecta: Heteroptera: Miridae) from Darjeeling foothills and plains. *Journal of Plantation Crops* 34: 423–428.
83. Li X., Schuler M. A. and Berenbaum M. R. (2007) Molecular mechanisms of metabolic resistance to synthetic and natural xenobiotics. *Annual Review of Entomology* 52: 231–253.
84. Feyereisen R. (2012) Insect CYP genes and P450 enzymes, pp. 236–316. In Insect Molecular Biology and Biochemistry (edited by L. I. Gilbert). Academic Press, London.
85. Lee S. H., Kang J. S., Min J. S., Yoon K. S., Strycharz J. P., Johnson, R., Mittapalli, O., Margam, V.M., Sun, W., Li, H.M., Xie, J., Wu, J,, Kirkness, E.F., Berenbaum, M.R., Pittendrigh, B.R. and Clark, J.M. (2010) Decreased detoxification genes and genome size make the human body louse an efficient model to study xenobiotic metabolism. *Insect Molecular Biology* 19: 599–615.
86. Arensburger P., Megy K., Waterhouse R. M., Abrudan J., Amedeo P., Antelo, B., Bartholomay L., Bidwell S., Caler E., Camara F., Campbell C.L., Campbell K.S., Casola C., Castro M.T., Chandramouliswaran I., Chapman S.B., Christley S., Costas J., Eisenstadt E., Feschotte C., Frasher-Liggett C., Guigo R., Haas B., Hammond M., Hansson B.S., Hemingway J. Hill S.R., Howarth C., Ignell R., Kennedy R.C., Kodira C.D., Lobo N.F., Mao C., Mayhew G., Michel K., Mori A., Liu N., Naveira H., Nene V., Nguyen N., Pearson M.D., Pritham E.J., Puiu D., Qi Y., Ranson H., Ribeiro J.M. Roberston H.M., Severson D.W., Shumway M., Stanke M., Strausberg R.L., Sun C., Sutton G., Tu Z.J., Tubio J.M., Unger M.F., Van Lan Dingham D.L., Vilella A.J., White O., White J.R., Wondji C.S., Wortman J., Zdobnov E.M., Birren B., Christensen B.M., Collins F.H., Cornel A., Dimopoulos G., Hannick L.I., Higgs S., Lanzaro G.C., Lawson D., Lee N.H., Muskavitch M.A., Raikhel A.S., Atkinson P.W., (2010) Sequencing of *Culex quinquefasciatus* establishes a platform for mosquito comparative genomics. *Science* 330: 86–88.
87. Omura T. and Sato R. (1964) The carbon monoxide-binding pigment of liver microsomes I. Evidence for its hemoprotein nature. *Journal of Biological Chemistry* 239: 2370–2378.
88. Feyereisen R., Koener J. F., Farnsworth D. E. and Nebert D. W. (1989) Isolation and sequence of cDNA encoding a cytochrome P-450 from an insecticide-resistant strain of the house fly, *Musca domestica. Proceedings of the National Academy of Science USA* 86: 1465–1469.
89. Feyereisen R. (2011) Arthropod CYPomes illustrate the tempo and mode in P450 evolution. *Biochimica et Biophysica Acta* 1814: 19–28.
90. Claudianos C., Ranson H., Johnson R. M., Biswas S., Schuler M. A., Berenbaum M. A., Feyereisen R. and Oakeshott J. G. (2005) A deficit of detoxification enzymes:

pesticide sensitivity and environmental response in the honeybee. Insect Molecular Biology 15, 615–636.

91. Hardstone M. C. and Scott J. G. (2010) Is *Apis mellifera* more sensitive to insecticides than other insects? *Pest Management Science* 66: 1171–1180.

92. Hemingway J. and Karunaratne S. H. P. P. (1998) Mosquito carboxylesterases: a review of the molecular biology and biochemistry of a major insecticide resistance mechanism. *Medical and Veterinary Entomology* 12: 1–12.

93. Heymann E. and Jakoby W.B. (1980) Carboxylesterases and amidases, pp. 291–323. In Enzymatic basis of detoxication (edited by E. Heymann and W. B. Jokoby). Academic Press, New York,.

94. Aldridge W.N. (1993) The esterases: perspectives and problems. *Chemico-Biological Interactions* 87: 5–13.

95. Reiner E. (1993) Recommendations of the IUBMB nomenclature committee – comments concerning classification and nomenclature of esterases hydrolysing organophosphorus compounds. *Chemico-Biological Interactions* 87: 15, 16.

96. Walker C.H. (1993) The classification of esterases which hydrolyse organophosphates–recent developments. *Chemico-Biological Interactions* 87: 17–24.

97. Yu S.J. (1996) Insect glutathione *S*-transferases. *Zoological Studies* 35: 9–19.

98. Ranson H., Claudianos C., Ortelli F., Abgrall C., Hemingway J., Sharakhova M.V., Unger M.F., Collins F.H. and Feyereisen R. (2002) Evolution of supergene families associated with insecticide resistance. *Science* 298: 179–181.

99. Ranson, H. and Hemingway, J. (2005) Mosquito glutathione transferases. Review. *Methods in Enzymology* 401: 226–241.

100. Fournier D., Bride J.B., Poirie M., Berge J.P. and Plapp F.W. Jr. (1992) Insect Glutathione S-transferases: Biochemical characteristics of the major forms from house flies susceptible and resistant to insecticides. *Journal of Biological Chemistry* 267: 1840–1845.

101. Ranson H., Rossiter L., Ortelli F., Jensen B., Wang X., Roth C.W., Collins F.H. and Hemingway J. (2001) Identification of a novel class of insect glutathione S-transferases involved in DDT resistance in the malaria vector, *Anopheles gambiae. Biochemical Journal* 359: 295–304.

102. Yu S.J. (2002) Biochemical characteristics of microsomal and cytosolic glutathione S-transferases in larvae of the fall armyworm, *Spodoptera frugiperda* (J. E. Smith). *Pesticide Biochemistry and Physiology* 72: 100–110.

103. Ma B. and Chang F.N. (2007) Purification and cloning of a delta class glutathione s-transferase displaying high peroxidase activity isolated from German cockroach, *Blattella germanica. FEBS Journal* 274: 1793–1803.

104. Che-Mendoza A., Penilla R.P. and Rodríguez D.A. (2009) Insecticide resistance and glutathione S-transferases in mosquitoes: A review. *African Journal of Biotechnology* 8 (8): 1386–1397

105. Clark A.G. and Shamaan N.A. (1984) Evidence that DDT-dehydrochlorinase from the house fly is a glutathione S-transferase. *Pesticide Biochemistry and Physiology* 22: 249–261.

106. Eldefrawi M.E., Miskus R. and Sutcher V. (1960) Methylenedioxyphenyl derivatives as synergists for carbamate insecticides on susceptible, DDT-and parathion-resistant house flies. *Journal of Economic Entomology* 53: 231–234.

107. Georghiou G. P. and Metcalf R. L. (1961) The adsorption and metabolism of 3-isopropylphenyl-n-methylcarbamate by susceptible and carbamate selected strains of house flies. *Journal of Economic Entomology* 54(2): 231–233.

108. Schonbrod R.D., Phillco W.W. and Terriere L.C. (1965) Hydroxylation as a factor in resistance in house flies and blow flies. *Journal of Economic Entomology* 58(1): 74–77.

109. Agosin M. (1985) Role of microsomal oxidations in insecticide degradation, pp. 647–712. In Comparative Insect Physiology, Biochemistry and Pharmacology (edited by L. I. Gilbert and G. A. Kerkut). Vol. 12. Pergamon, Oxford.

110. Scott J. G. (1991) Insecticide resistance in insects, pp. 663–677. In Handbook of Pest Management (edited by D. Pimental). CRC Press, Boca Raton.

111. Konno T. E. H. and Dauterman W. C. (1989) Studies on methyl parathion resistance in Heliothis virescens. *Pesticide Biochemistry and Physiology* 33: 189–199.

112. Oppenoorth F. J. (1985) Biochemistry and genetics of insecticide resistance, pp. 731–773. In Comparative insect physiology, biochemistry and pharmacology (edited by G. A. Kerkut and L. I. Gilbert). Vol. 12. Pergamon Press Oxford, UK.

113. Hatano R., Scott J. G. and Dennehy T. J. (1992) Enhanced activation is the mechanism of negative crossresistance to chlorpyrifos in the dicofol-IR strain of *Tetranychus urticae* (Acari: Tetranychidae). *Journal of Economic Entomology* 85: 1088–1091.

114. Cilek J. E., Dahlman D. L. and Knapp F.W. (1995) Possible mechanism of diazinon negative cross-resistance in pyrethroid-resistant horn flies (Diptera: Muscidae). *Journal of Economic Entomology* 88: 520–524.

115. Martin T., Chandre F., Ochou O. G., Vaissayre M. and Fournier D. (2003) Oxidases responsible for resistance to pyrethroids sensitize *Helicoverpa armigera* (Hubner) to triazophos in West Africa. *Insect Biochemistry and Molecular Biology* 33: 883–887.

116. Drabek J. and Neumann R. (1985) Proinsecticides, pp. 35–86. In Insecticides (edited by D. H. Huston and T. R. Roberts). Wiley, London.

117. Black B. C., Hollingworth R. M., Ahammadsahib K. I., Kukel C. D. and Donovan S. (1994) Insecticidal action and mitochondrial uncoupling activity of AC303, 630 and related halogenated pyrroles. *Pesticide Biochemistry and Physiology* 50: 115–128.

118. Kayser H. and Eilinger P. (2001) Metabolism of diafenthiuron by microsomal oxidation: Procide activation and inactivation as mechanisms contributing to selectivity. *Pest Management Science* 57: 975–980.

119. Pimprale S. S., Besco C. L., Bryson P. K. and Brown T. M. (1997) Increased susceptibility of pyrethroid-resistant tobacco budworm (Lepidoptera: Noctuidae) to chlorfenapyr. *Journal of Economic Entomology* 90: 49–54.

120. Joussen N., Heckel D. G., Haas M., Schuphan I. and Schmidt B. (2008) Metabolism of imidacloprid and DDT by P450 CYP6G1 expressed in cell cultures of *Nicotiana tabacum* suggests detoxification of these insecticides in Cyp6g1-overexpressing strains of *Drosophila melanogaster*, leading to resistance. *Pest Management Science* 64: 65–73.

121. Karunker I., Morou E., Nikou D., Nauen R., Sertchook R., Stevenson B. J., Paine M. J. I., Morin S. and Vontas J. (2009) Structural model and functional characterization of the *Bemisia tabaci* CYP6CM1vQ, a cytochrome P450 associated with high levels of imidacloprid resistance. *Insect Biochemistry and Molecular Biology* 39: 697–706.

122. Feyereisen R., Koener J. F., Farnsworth D. E. and Nebert D. W. (1989) Isolation and sequence of cDNA encoding a cytochrome P-450 from an insecticide-resistant strain of the house fly, *Musca domestica*. *Proceedings of the National Academy of Science USA* 86: 1465–1469.

123. Cariño F., Koener J. F., Plapp F. W., Jr. and Feyereisen R. (1992) Expression of the cytochrome P450 gene CYP6A1 in the housefly, *Musca domestica*. *ACS Symposium Series* 505: 31–40.

124. Sabourault C., Guzov V.M., Koener J.F., Claudianos C., Plapp F.W. and Feyereisen R., (2001) Overproduction of a P450 that metabolizes diazinon is linked to a loss-of-function in the chromosome 2 ali-esterase (Mda E7) gene in resistant house flies. Insect Molecular Biology 10: 609–618

125. Zhu F., Liu N. (2008a) Differential expression of CYP6A5 and CYP6A5v2 in pyrethroid resistant house flies, *Musca domestica*. Archives of Insect Biochemistry and Physiology 67: 107–119.

126. Zhu F., Feng J., Zhang L., Liu N. (2008b) Characterization of two novel cytochrome P450 genes in insecticide-resistant house-flies. Insect Molecular Biology 17: 27–37.

127. Zhu F., Li T., Zhang L., Liu N. (2008c) Co-up-regulation of three P450 genes in response to permethrin exposure in permethrin resistant house flies, Musca domestica. BMC Physiology 8: 18.

128. Liu N. and Scott J. G. (1996) Genetic analysis of factors controlling high-level expression of cytochrome P450, CYP6D1, cytochrome b5, P450 reductase, and monooxygenase activities in LPR house flies, *Musca domestica*. *Biochemical Genetics* 34: 133–148.

129. Kasai S. and Scott J. G. (2001a) Cytochrome P450s CYP6D3 and CYP6D1 are part of a P450 gene cluster on autosome 1 in the house fly. *Insect Molecular Biology* 10: 191–196.

130. Kasai S. and Scott J. G. (2001b) Expression and regulation of CYP6D3 in the house fly, *Musca domestica* (L.). *Insect Biochemistry and Molecular Biology* 32: 1–8.

131. Kasai S. and Scott J. G. (2000) Overexpression of cytochrome P450 CYP6D1 is associated with monooxygenase-mediated pyrethroid resistance in house flies from Georgia. *Pesticide Biochemistry and Physiology* 68: 34–41.

132. Kamiya E., Yamakawa M., Shono T. and Kono Y. (2001) Molecular cloning, nucleotide sequences and gene expression of new cytochrome P450s (CYP6A24, CYP6D3v2) from the pyrethroid resistant housefly, *Musca domestica* L. (Diptera: Muscidae). *Applied Entomology and Zoology* 36: 225–229.

133. Shono T., Kasai S., Kamiya E., Kono Y. and Scott J. G. (2002) Genetics and mechanisms of permethrin resistance in the YPER strain of house fly. *Pesticide Biochemistry and Physiology* 73: 27–36.

134. Guzov V. M., Unnithan G. C., Chernogolov A. A. and Feyereisen R. (1998) CYP12A1, a mitochondrial cytochrome P450 from the house fly. *Archives of Biochemistry and Biophysics* 359: 231–240.

135. Waters L. C., Zelhof A. C., Shaw B. J. and Ch'ang L. Y. (1992) Possible involvement of the long terminal repeat of transposable element 17.6 in regulating expression of an insecticide resistance-associated P450 gene in *Drosophila*. *Proceedings of the National Academy of Science USA* 89: 4855–4859.

136. Maitra S., Dombrowski S. M., Waters L. C. and Ganguly R. (1996) Three second chromosome-linked clustered Cyp6 genes show differential constitutive and barbital-induced expression in DDT-resistant and susceptible strains of *Drosophila melanogaster*. *Gene* 180: 165–171.

137. Maitra S., Dombrowski S. M., Basu M., Raustol O., Waters L. C. and Ganguly R. (2000) Factors on the third chromosome affect the level of cyp6a2 and cyp6a8 expression in *Drosophila melanogaster*. *Gene* 248: 147–156.

138. Maitra S., Price C. and Ganguly R. (2002) *Cyp6a8* of *Drosophila melanogaster*: Gene structure, and sequence and functional analysis of the upstream DNA. *Insect Biochemistry and Molecular Biology* 32: 859–870.

139. Le Goff G., Boundy S., Daborn P. J., Yen J. L., Sofer L. Lind R., Sabourault C., Madi-Ravazzi L. and ffrench-Constant R.H. (2003) Microarray analysis of cytochrome P450 mediated insecticide resistance in *Drosophila*. *Insect Biochemistry and Molecular Biology* 33: 701–708.

140. Daborn P., Boundy S., Yen J., Pittendrigh B. and ffrench-Constant R. (2001) DDT resistance in *Drosophila* correlates with *Cyp6g1* over-expression and confers crossresistance to the neonicotinoid imidacloprid. *Molecular Genetics and Genomics* 266: 556–563.

141. Daborn P. J., Yen J. L., Bogwitz M. R., Le Goff G., Feil E., Jeffers S., Tijet N., Perry T., Heckel D., Batterham P., Feyereisen R., Wilson T. G. and ffrench-Constant R. H. (2002) A single P450 allele associated with insecticide resistance in *Drosophila*. *Science* 297: 2253–2256.

142. Catania F., Kauer M. O., Daborn P. J., Yen J. L., ffrench-Constant R. H. and Schlotterer C. (2004) World-wide survey of an Accord insertion and its association with DDT resistance in *Drosophila melanogaster*. *Molecular Ecology* 13: 2491– 2504.

143. Brandt A., Scharf M., Pedra J. H., Holmes G., Dean A., Kreitman M. and Pittendrigh B. R. (2002) Differential expression and induction of two *Drosophila* cytochrome P450 genes near the Rst (2) DDT locus. *Insect Molecular Biology* 11: 337–341.

144. Pyke F. M., Bogwitz M. R., Perry T., Monk A., Batterham P. and McKenzie J. A. (2004) The genetic basis of resistance to diazinon in natural populations of *Drosophila melanogaster*. *Genetica* 121: 13–24.
145. Bogwitz M. R., Chung H., Magoc L., Rigby S., Wong W., O'Keefe M., McKenzie J.A., Batterham P. and Daborn P. J. (2005) *Cyp12a4* confers lufenuron resistance in a natural population of *Drosophila melanogaster*. *Proceedings of the National Academy of Science USA* 102: 12807–12812.
146. Nikou D., Ranson H. and Hemingway J. (2003) An adult specific CYP6 P450 gene is overexpressed in a pyrethroids resistant strain of the malaria vector, *Anopheles gambiae*. *Gene* 318: 91–102.
147. David J. P., Strode C., Vontas J., Nikou D., Vaughan A., Pignatelli P.M., Louis C., Hemingway J. and Ranson H. (2005) The *Anopheles gambiae* detoxification chip: A highly specific microarray to study metabolic-based insecticide resistance in malaria vectors. *Proceedings of the National Academy of Science USA* 102: 4080–4084.
148. Djouaka R. F., Bakare A. A., Coulibaly O. N., Akogbeto M. C., Ranson H., Hemingway J. and Strode C. (2008) Expression of the cytochrome P450s, CYP6P3 and CYP6M2 are significantly elevated in multiple pyrethroid resistant populations of *Anopheles gambiae* s.s. from Southern Benin and Nigeria. *BMC Genomics* 9: 538.
149. Muller P., Warr E., Stevenson B. J., Pignatelli P. M., Morgan J. C., Steven A., Yawson A. E., Mitchel S. N., Ranson H., Hemingway J. Paine M. J. I. Donnelly M. J. (2008) Field-caught Permithrin-Resistant *Anopheles gambiae* overexpress CYP6P3, a P450 that metabolises Pyrethroids. *PLoS Genetics* 4 (11): e1000286.
150. Muller P., Donnelly M. J. and Ranson H. (2007) Transcription profiling of a recently colonised pyrethroid resistant *Anopheles gambiae* strain from Ghana. *BMC Genomics* 8: 36.
151. Vontas J., Blass C., Koutsos A. C., David J. P., Kafatos F. C., Louis C., Hemingway J. Christophides G. K. and Ranson H. (2005) Gene expression in insecticide resistant and susceptible *Anopheles gambiae* strains constitutively or after insecticide exposure. *Insect Molecular Biology* 14: 509–521.
152. Vontas J., David J. P., Nikou D., Hemingway J., Christophides G. K., Louis C. and Ranson H. (2007) Transcriptional analysis of insecticide resistance in *Anopheles stephensi* using cross-species microarray hybridization. *Insect Molecular Biology* 16: 315–324.
153. Amenya D. A., Naguran R., Lo T. C., Ranson H., Spillings B. L., Wood O.R., Brooke B. D., Coetzee M. and Koekemoer L. L. (2008) Over expression of a cytochrome P450 (CYP6P9) in major African malaria vector, *Anopheles funestus*, resistant to pyrethroids. *Insect Molecular Biology* 17: 19–25.
154. Wondji C. S., Irving H., Morgan, J., Lobo N. F., Collins F. H., Hunt R.H., Coetzee M., Hemingway J. and Ranson H. (2009) Two duplicated P450 genes are associated with pyrethroid resistance in *Anopheles funestus*, a major malaria vector. *Genome Research* 19: 452–459.
155. Strode C., Steen K., Ortelli F. and Ranson H. (2006) Differential expression of the detoxification genes in the different life stages of the malaria vector *Anopheles gambiae*. *Insect Molecular Biology* 15: 523–530.
156. Hardstone M. C., Komagata O., Kasai S., Tomita T. and Scott J. G. (2010) Use of isogenic strains indicates CYP9M10 is linked to permethrin resistance in *Culex pipiens quinquefasciatus*. *Insect Molecular Biology* 19: 717–726.
157. Kasai S., Weerashinghe I. S., Shono T. and Yamakawa M. (2000) Molecular cloning, nucleotide sequence and gene expression of a cytochrome P450 (CYP6F1) from the pyrethroid-resistant mosquito, *Culex quinquefasciatus* Say. *Insect Biochemistry and Molecular Biology* 30: 163–171.
158. Komagata O., Kasai S. and Tomita T. (2010) Overexpression of cytochrome P450 genes in pyrethroid-resistant *Culex quinquefasciatus*. *Insect Biochemistry and Molecular Biology* 40: 146–152.

159. Shen B., Dong H. Q., Tian H. S., Ma L., Li X. L., Wu G. L. and Zhu C. L. (2003) Cytochrome P450 genes expressed in the deltamethrin- susceptible and -resistant strains of *Culex pipiens pallens*. *Pesticide Biochemistry and Physiology* 75: 19–26.

160. Rose R. L., Goh D., Thompson D. M., Verma K. D., Heckel D. G., Gahan L. J., Roe R. M. and Hodgson E. (1997) Cytochrome P450 (CYP) 9A1 in *Heliothis virescens*: The first member of a new CYP family. *Insect Biochemistry and Molecular Biology* 27: 605–615.

161. Hopkins B. W., Longnecker M. T. and Pietrantonio P. V. (2010) Transcriptional overexpression of CYP6B8/CYP6B28 and CYP6B9 is a mechanism associated with cypermethrin survivorship in field-collected *Helicoverpa zea* (Lepidoptera: Noctuidae) moths. *Pest Management Science* 67: 21–25.

162. Pittendrigh B., Aronstein K., Zinkovsky E., Andreev O., Campbell B., Daly J., Trowell S. and ffrench-Constant R. H. (1997) Cytochrome P450 genes from *Helicoverpa armigera*: Expression in a pyrethroid-susceptible and -resistant strain. *Insect Biochemistry and Molecular Biology* 27: 507–512.

163. Ranasinghe C. and Hobbs A. A. (1998) Isolation and characterization of two cytochrome P450 cDNA clones for CYP6B6 and CYP6B7 from *Helicoverpa armigera* (Hubner): Possible involvement of CYP6B7 in pyrethroid resistance. *Insect Biochemistry and Molecular Biology* 28: 571–580.

164. Yang Y., Chen S., Wu S., Yue L. and Wu Y. (2006) Constitutive overexpression of multiple cytochrome P450 genes associated with pyrethroid resistance in *Helicoverpa armigera*. *Journal of Economic Entomology* 99: 1784–1789.

165. Wee C. W., Lee S. F., Robin C. and Heckel D. G. (2008) Identification of candidate genes for fenvalerate resistance in *Helicoverpa armigera* using cDNA-AFLP. *Insect Molecular Biology* 17: 351–360.

166. Brun-Barale A., Hema O., Martin T., Suraporn S., Audant P., Sezutsu H. and Feyereisen R. (2010) Multiple P450 genes overexpressed in deltamethrin- resistant strains of *Helicoverpa armigera*. *Pest Management Science* 66: 900–909.

167. Bautista M. A., Tanaka T. and Miyata T. (2007) Identification of permethrin-inducible cytochrome P450s from the diamondback moth, *Plutella xylostella* (L.), and the possibility of involvement in permethrin resistance. *Pesticide Biochemistry and Physiology* 87: 85–93.

168. Baek J. H., Clark J. M. and Lee S. H. (2010) Cross-strain comparison of cypermethrin-induced cytochrome P450 transcription under different induction conditions in diamondback moth. Pesticide Biochemistry and Physiology 96: 43–50.

169. Zhu Y. C. and Snodgrass G. L. (2003) Cytochrome P450 CYP6X1 cDNAs and mRNA expression levels in three strains of the tarnished plant bug *Lygus lineolaris* (Heteroptera: Miridae) having different susceptibilities to pyrethroids insecticide. *Insect Molecular Biology* 12: 39–49.

170. Karunker I., Benting J., Lueke B., Ponge T., Nauen R., Roditakis E., Vontas J., Gorman K., Denholm I. and Morin S. (2008) Over-expression of cytochrome P450 CYP6CM1 is associated with high resistance to imidacloprid in the B and Q biotypes of *Bemisia tabaci* (Hemiptera: Aleyrodidae). *Insect Biochemistry and Molecular Biology* 38: 634–644.

171. Bass C., Carvalho R. A., Oliphant L., Puinean A. M., Field L. M., Nauen R., Williamson M. S., Moores, G. and Gorman, K. (2011) Overexpression of a cytoichrome P450 monooxygenase, CYP6ER1, is associated with resistance to imidacloprid in the brown plant hopper, *Nilapatvata lugens*. *Insect Molecular Biology* 20(6): 763–773.

172. Puinean A. M., Foster S. P., Oliphant L., Denholm I., Field L. M. Millar N. S., Williamson M. S. and Bass C. (2010) Amplification of a cytochrome P450 gene is associated with resistance to neonicotinoid insecticides in the aphid *Myzus persicae*. *PLoS Genetics* 6(6): e1000999.

173. Scharf M. E., Parimi S., Meinke L. J., Chandler L. D. and Siegfried B. D. (2001) Expression and induction of three family 4 cytochrome P450 (CYP4)* genes identified from insecticide-resistant and -susceptible western corn rootworms, *Diabrotica virgifera virgifera*. *Insect Molecular Biology* 10: 139–146.

174. Zhu F., Parthasarathy R., Bai H., Woithe K., Kaussmann M., Nauen R., Harrison D. A. and Palli S. R. (2010) A brain-specific cytochrome P450 responsible for the majority of deltamethrin resistance in the QTC279 strain of *Tribolium castaneum. Proceedings of the National Academy of Sciences USA* 107: 8557–8562.

175. Scharf M. E., Lee C. Y., Neal J. J. and Bennett G. W. (1999) Cytochrome P450 MA expression in insecticide-resistant German cockroaches (Dictyoptera: Blattellidae). *Journal of Economic Entomology* 92: 788–793.

176. Pridgeon J. W., Zhang L. and Liu N. (2003) Overexpression of CYP4G19 associated with a pyrethroid-resistant strain of the German cockroach, *Blattella germanica* (L.). *Gene* 314: 157–163.

177. Zhu Y. C., Snodgrass G. L. and Ming Shun Chen M. S. (2004) Enhanced esterase gene expression and activity in a malathion-resistant strain of the tarnished plant bug, Lygus lineolaris. *Insect Biochemistry and Molecular Biology* 34: 1175–1186.

178. Zhu Y. C., West S., Snodgrass G. and Luttrell R. (2011) Variability in Resistance-related Enzyme Activities in Field Populations of the Tarnished Plant Bug, *Lygus lineolaris. Pesticide Biochemistry and Physiology* 99: 265–273.

179. Ferrari J. A., Morse J. G., Georghiou G. P., Sun Y (1993) Elevated esterase activity and acetylcholinesterase insensitivity in citrus thrips (Thysanoptera: Thripsidae) populations from the San Joaquin Valley of California. *Journal of Economic Entomology* 86: 1645–1650.

180. Vassiliou V., Emmanouilidou M., Perrakis A., Morou E., Vontas J., Tsagkarakou A. and Roditakis E. (2011) Insecticide resistance mechanisms in tea pests 69 Insecticide resistance in *Bemisia tabaci* from Cyprus. *Insect Science* 18 30–39.

181. Wink M. and Waterman P. (1999) Chemotaxonomy in relation to molecular phylogeny of plants. *Annual Plant Reviews* 2, 300–341

182. Wink M. (2003) Evolution of secondary metabolites from an ecological and molecular phylogenetic perspective. *Phytochemistry* 64: 3–19.

183. Yu S. J. and Ing R. T. (1984) Microsomal biphenyl hydroxylase of fall armyworm larvae and its induction by allelochemicals and host plants. *Comparative Biochemistry and Physiology* 78C: 145–152.

184. Musser R. O., Hum-Musser S. M., Eichenseer H., Peiffer M., Ervin G., Murphy J. B. and Felton G. W. (2002) Herbivory: caterpillar saliva beats plant defences – a new weapon emerges in the evolutionary arms race between plants and herbivores. *Nature* 416: 599–600.

185. Helmus M. R. and Dussourd D. E. (2005) Glues or poisons: which triggers vein cutting by monarch caterpillars? *Chemoecology* 15: 45–49

186. Labeyrie E. and Dobler S. (2004) Molecular adaptation of *Chrysochus* leaf beetles to toxic compounds in their food plants. *Molecular Biology and Evolution* 21: 218–221.

187. Zagrobelny M., Bak S., Rasmussen A. V., Jørgensen B., Naumann C. M. and Møller B. L. (2004) Cyanogenic glucosides and plant–insect interactions. *Phytochemistry* 65: 293–306.

188. Francis F., Vanhaelen N. and Haubruge E. (2005) Glutathione S-transferases in the adaptation to plant secondary metabolites in the Myzus persicae aphid. *Archives of Insect Biochemistry and Physiology* 58: 166–174.

189. Ratzka A., Vogel H., Kliebenstein D. J. Mitchell-Olds T. and Kroymann J. (2002) Disarming the mustard oil bomb. *Proceedings of the National Academy of Science USA* 99: 11223–11228.

190. Wittstock U., Agerbirk N., Stauber E. J., Olsen C. E., Hippler M., Mitchell-Olds T., Gershenzon J. and Vogel H. (2004) Successful herbivore attack due to metabolic diversion of a plant chemical defense. *Proceedings of the National Academy of Science USA* 101: 4859–4864.

191. Fogleman J.C. (2000) Response of Drosophila melanogaster to selection for P450-mediated resistance to isoquinoline alkaloids. *Chemico-Biological Interactions* 125: 93–105.

192. Hartmann T., Theuring C., Beuerle T., Klewer N., Schulz S., Singer M. S. and Bernays E. A. (2005) Specific recognition, detoxification and metabolism of pyrrolizidine

alkaloids by the polyphagous arctiid Estigmene acrea. *Insect Biochemistry and Molecular Biology* 35: 391–411.

193. Naumann C., Hartmann T. and Dietrich Ober D. (2002) Evolutionary recruitment of a flavin dependent monooxygenase for the detoxification of host plant acquired pyrrolizidine alkaloids in the alkaloid-defended arctiid moth Tyria jacobaeae. *Proceedings of the National Academy of Science USA* 99: 6085–6090.

194. Glendinning J. I. (2002) How do herbivorous insects cope with noxious secondary plant compounds in their diet? *Entomologia Experimentalis et Applicata* 104: 15–25.

195. Nyman T. and Julkunen-Tiitto R. (2000) Manipulation of the phenolic chemistry of willows by gall-inducing sawflies. *Proceedings of the National Academy of Science USA* 97: 13184–13187.

196. Luque T., Okano K. and O'Reilly D. R. (2002) Characterization of a novel silkworm (*Bombyx mori*) phenol UDP-glucosyltransferase. *European Journal of Biochemistry* 269: 819–825.

197. Willinger G. and Dobler S. (2001) Selective sequestration of iridoid glycosides from their host plants in Longitarsus flea beetles. *Biochemical Systematics and Ecology* 29: 335–346.

198. Petersen R. A., Zangerl A. R., Berenbaum, M. R. and Schuler M. A. (2001) Expression of CYP6B1 and CYP6B3 cytochrome P450 monooxygenases and furanocoumarin metabolism in different tissues of *Papilio polyxenes* (Lepidoptera: Papilionidae). *Insect Biochemistry and Molecular Biology* 31: 679–690.

199. Nitao J. K., Berhow M., Duval S. M., Weisleder, D., Vaughn S. F., Zangerl A. and Berenbaum M. R. (2003) Characterization of furanocoumarin metabolites in parsnip webworm, *Depressaria pastinacella*. *Journal of Chemical Ecology* 29: 671–682.

200. Yu S. J. (2002) Biochemical characteristics of microsomal and cytosolic glutathione S-transferases in larvae of the fall armyworm, *Spodoptera frugiperda* (J. E. Smith). *Pesticide Biochemistry and Physiology* 72: 100–110.

201. Moon J., J., Salzman R. A., Ahn J. E., Koiwa H. and Zhu-Salzman K. (2004) Transcriptional regulation in cowpea bruchid guts during adaptation to a plant defence protease inhibitor. *Insect Molecular Biology* 13: 283–291.

202. Bede J. C., Musser R. O., Gary W. Felton G. W. and Korth K. L. (2006) Caterpillar herbivory and salivary enzymes decrease transcript levels of *Medicago truncatula* genes encoding early enzymes in terpenoid biosynthesis. *Plant Molecular Biology* 60: 519–531.

203. Barbehenn R. V., Poopat U. and Spencer B. (2003) Semiquinone and ascorbyl radicals in the gut fluids of caterpillars measured with EPR spectrometry. Insect Biochemistry and Molecular Biology 33: 125–130

204. Berenbaum M. R. (2002) Postgenomic chemical ecology: from genetic code to ecological interactions. Journal of Chemical Ecology 28(5): 873–896.

205. Meyran J. C., David J. C., Rey J. P., Cuany D. Bride A. and Amichot J. M. (2002). The biochemical basis of dietary polyphenols detoxification by aquatic detritivorous Arthropoda. *Recent Research Developments in Analytical Biochemistry* 2: 185–199.

206. Fraenkel G. S. (1959) The raison d'etre of secondary plant substances. *Science* 129: 1466–1470.

207. Krieger R. I., Feeny P. P. and Wilkinson C. F. (1971) Detoxication enzymes in the guts of caterpillars: an evolutionary answer to plant defenses? *Science* 172: 579–581.

208. Mullin C. A. (1986) Adaptive divergence of chewing and sucking arthropods to plant allelochemicals, pp. 175–209. In Molecular Aspects of Insect-Plant Associations (edited by L. B. Brattsten and S. Ahmad). Plenum Press, New York.

209. Mullin C. A., Croft B. A., Strickler K., Matsumura F. and Miller J. R. (1982) Detoxification enzyme differences between a herbivorous and predatory mite. *Science* 217: 1270, 1271.

210. Iyenger S., Arnason J. T., Philogene B. J. R., Werstiuk N. H. and Morand P. (1990) Comparative metabolism of the photoxic allelochemical α-terthienyl in three species of lepidopterans. *Pesticide Biochemistry and Physiology* 37(2): 154–164.

211. Yu S. J. (1983) Induction of detoxifying enzymes by allelochemicals and host plants in the fall armyworm. *Pesticide Biochemistry and Physiology* 19: 330–336.

212. Reilly C. C., Gentry C. R. and McKay J. R. (1987) Biochemical evidence for resistance of root stocks to the peach tree borer and species separation of peach tree borer and lesser peach tree borer (Lepidoptera: Sesiidae) on peach trees. *Journal of Economic Entomology* 80: 338–343.

213. Lindroth R. L. (1989) Host plant alteration of detoxication activity in Papilio glaucus glaucus. *Entomologia Experimentalis et Applicata* 50: 29–35.

214. Lindroth R. L., Scriber J. M. and Hsia M. T. S. (1988) Chemical ecology of the tiger swallowtail: mediation of host use by phenolic glycosides. *Ecology* 69, 814–822.

215. Wadleigh R. W. and Yu S. J. (1987) Glutathione transferase activity of fall armyworm larvae toward a, b-unsaturated carbonyl allelochemicals and its induction by allelochemicals. *Insect Biochemistry* 17: 759–764.

216. Wadleigh R. W. and Yu S. J. (1988) Detoxification of isocyanate allelochemicals by glutathione transferase in three lepidopterous species. *Journal of Chemical Ecology* 14: 1279–1288.

217. Yu S. J. (1989) Purification and characterization of glutathione transferases from five phytophagous Lepidoptera. *Pesticide Biochemistry and Physiology* 35: 97–105.

218. Yu S. J. and Abo-Elghar G. E. 2000. Allelochemicals as inhibitors of glutathione S-transferase in the fall armyworm. *Pesticide Biochemistry and Physiology* 68: 173–183.

219. Mound L. A. and Palmer G. M. (1981) Identification, distribution and host plants of the pest species of Scirtothrips (Thysanoptera: Thripidae). *Bulletin of Entomological Research* 71: 467–479.

220. Saha D., Mukhopadhyay A. and Bahadur M. (2012) Effect of host plants on fitness traits and detoxifying enzymes activity of major sucking insect pest of tea, *Helopeltis theivora* (Heteroptera: Miridae). *Phytoparasitica* 40: 433–444.

221. Bartlett A. C., Wilson N. M. and Mattix E. B. (1968) The fate of genetic markers in populations of boll weevils. *Journal of Economic Entomology* 61: 808–812.

222. Bartlett A. C. and Butler G. D. Jr. (1975) Genetic control of the cabbage looper by a recessive lethal mutation. *Journal of Economic Entomology* 68: 331–335.

223. Loxdale H. D. and Lushai G. (1998) Molecular Markers in Entomology. *Bulletin of Entomological Research* 88: 577–600.

224. Loxdale H. D., Brookes C. P. and De Barro, P. J. (1996) Application of novel molecular markers (DNA) in agricultural entomology. In The Ecology of Agricultural Pests (edited by W. O. C. Symondson and J. E. Liddell). Chapman and Hall, London.

225. Daly J. C. (1993) Ecology and genetics of insecticide resistance in *Helicoverpa armigera:* interactions between selection and gene flow. *Genetica* 90: 217–226.

226. Roush R. T. and Daly J. C. (1990) The role of population genetics in resistance research and management, pp. 97–152. In Pesticide Resistance in Arthropods (edited by R. T. Roush and B. E. Tabashnik). Chapman & Hall, New York.

227. Labbe P., Lenormand T. and Raymond M. (2005) On the world wide spread of an insect resistance gene: a role for local selection. *Journal of Evolutionary Biology* 18: 1471–1484.

228. Denholm I., Sawicki R. M. and Farnham A. W. (1985) Factors affecting the resistance to insecticides in houseflies, *Musca domestica.* IV: the population biology of the flies on animal farms in south-eastern England and its implications for the management of resistance. *Bulletin of Entomological Research* 75: 143–158.

229. Lenormand T. (2002) Gene flow and the limits to natural selection. Trends in Ecology and Evolution 17: 183–189.

230. Whitehead A., Anderson S. L., Kuivila K. M., Roach J. L. and May B. (2003) Genetic variation among interconnected populations of Catostomus occidentalis: implications for distinguishing impacts of contaminants from biogeographical structuring. Molecular Ecology 12: 2817–2833.

231. Brogdon W. G. and McAlliester J. C. (1998) Insecticide resistance and vector control. Emerging Infectious Diseases 4: 605–613.

CHAPTER 5

Semiochemicals and Their Potential Use in Pest Management

Hamadttu Abdel Farag El-Shafie[1*] and Jose Romeno Faleiro[2]

[1] *Date Palm Research Center of Excellence, King Faisal University, Al Ahsa, Kingdom of Saudi Arabia*

[2] *Food and Agriculture Organization of the UN, Date Palm Research Center, Ministry of Agriculture, Al-Ahsa, Kingdom of Saudi Arabia*

ABSTRACT

This chapter gives an account on the general concept of insect semiochemicals, their definitions, classification, formulation, utilization in integrated pest management programs, and the shortcomings of their application. The different semiochemically based insect management techniques, such as mass trapping, mating disruption, and attract-and-kill, are highlighted. The chapter also summarizes a case study from 7-year research on semiochemicals of the invasive red palm weevil, *Rhynchophorus ferrugineus*.

Keywords: olfactometer, mating disruption, chemical ecology, pheromones, kairomones

1. INTRODUCTION

Chemical communication plays an important and essential role in the survival of insects, which enable them to appraise immediate environment through modification of their behavior. Semiochemicals are organic compounds used by insects to convey specific chemical messages that modify behavior or physiology [1]. The term semiochemical is derived from the Greek word "semeon" which means sign or signal. Insects use semiochemicals to locate mate, host, or food source, avoid competition, escape natural enemies, and overcome natural defense systems of their hosts. Semiochemicals have the advantage of being used to communicate message over relatively long distances compared with other insect means of communication such as touch. Semiochemicals have different molecular weights depending on carbon chain. They are biologically active at very low concentration in the environment, thus their chemical characterization is complicated.

Expensive equipment items are needed for extraction and chemical characterization of semiochemicals. These equipment items include solid-phase microextraction (SPME), gas chromatography-electroantennography (GC-EAG), gas chromatography-mass spectrometry (GC-MS), and nuclear magnetic resonance (NMR) [2]. For the development of new synthetic pheromone blend, a lot of work on electrophysiological and behavioral bioassay is required [3].

Semiochemicals are species-specific and harmless to the environment. These advantages over conventional insect pest control agents make semiochemicals promising tools for the management of agricultural pests particularly under organic cropping systems.

2. CLASSIFICATION OF SEMIOCHEMICALS

Semiochemicals are classified based on their effect or function and this should be taken into account because the same molecule could act as a pheromone for one insect species and as a kairomone or allomone for another species. Semiochemicals are divided into two broad groups: pheromones that mediate interactions among individuals of the same species (intraspecific reactions) and allelochemicals that mediate interactions among individuals of different species (interspecific interactions). According to the behavioral response, pheromones are further subdivided into primer pheromones that have long-term physiological changes and releaser pheromones that elicit short-term or immediate behavioral response. Allelochemicals are divided into kairomones that mediate interactions favoring the recipient, allomones, on the other hand, favor the emitter. Synomones favoring both the emitter and the recipient, and apneumones, which are substances, produced by nonliving material that elicit behavioral response favorable to the receiving organism but harmful to a second organism found on the nonliving material. Schematic diagram showing the classification of semiochemicals is shown in **Figure 1** .

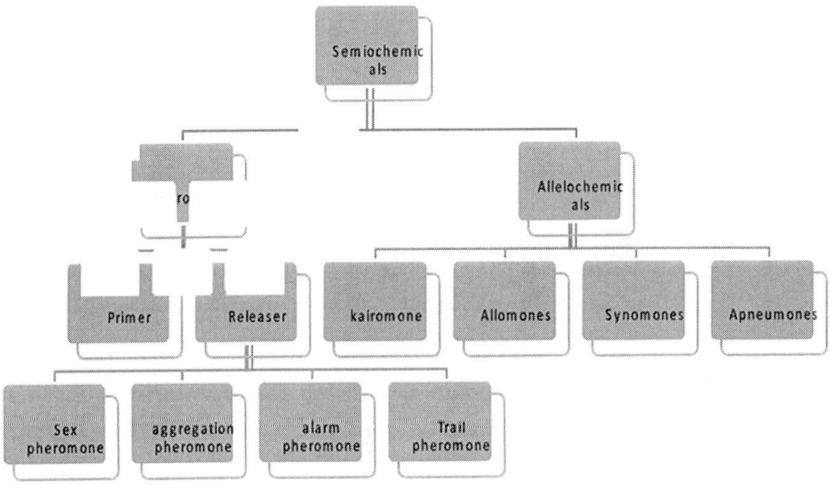

Figure 1. Classification of semiochemicals based on their effect and role in specific interactions.

2.1. Insect Pheromones

Karlson and Lüscher [4] first proposed the term pheromone to describe chemical signals that mediate intraspecific interactions. The sex pheromone of the silkworm moth, *Bombyx mori* was the first pheromone to be chemically identified in 1959 and is considered as the most important semiochemical used in pest management. Other pheromones include aggregation pheromone, which are produced by males and attract both sexes of conspecific individuals. The sex pheromone of moths is the most studied and widely used in insect pest management than other pheromones [3]. One-day-old female moth emits the sex pheromone usually at a rate of a few tens of pictogram per second at a certain time of the day or night. It has a characteristic behavior of raising the abdomen and exposing the pheromone glands at the end of the abdomen, this behavior is termed as calling posture [3]. Male moths, on the other hand, synchronize their daily activity to calling females for mating to be successful. The males respond to pheromone by flying upwind in the plume from 10 to 100 m downwind to locate the source [3]. Insect pheromones diffuse from their source in the form of strands of odor that drift downwind and become stretched, twisted, and ripped apart into substrands, which interspersed with pockets of clean air to form odor plume that produce the sustained upwind flight or what is termed as attraction [5, 6].

2.2. Insect Parasitoid Kairomones

Semiochemicals play an important role in host-parasitoid relationship, which was categorized by [7] into three stages: habitat-location, host-location and host-acceptance, and oviposition. These semiochemicals included but not limited to aldehydes, alcohols, sulfur-containing compounds, esters, terpenes, alkanes, heterocyclic aromatic compounds, proteins, amino acids, triglycerides, and salts. He stated that semiochemicals identified in the habitat-location step were likely to be from the host-plant of the host insect, while in host-location and host-acceptance and oviposition steps, semiochemicals are predominantly from the host. Kairomones used by parasitoids to locate their hosts can be divided into two groups, external to the host, which are long-chain hydrocarbons, ketones of fatty acids, esterified cholesterols or proteins found in either host frass or glue used to attach eggs to a substrate [7]. The internal kairomones represented by amino acids and salts in the hemolymph, which have normally been sensed the ovipositors and serve as indicators for suitability of the host for the parasitoid offspring, a kind of maternity care found in many insect species.

3. IDENTIFICATION AND BEHAVIORAL CHARACTERIZATION OF SEMIOCHEMICALS

Identifying the range of volatiles that insects can detect in their environment is an important step toward understanding the role of olfaction in modulating insect behavior [8]. The process of semiochemical identification involves extraction or headspace collection, identification of active compounds, characterization of chemical composition of the identified compound, and elucidation of behavioral response of the insect to the active product. If the organ that produces the semiochemical is known, for example, the exocrine gland or the gut of the insect, it can be extracted and identified. However, nonrelevant compounds can also be extracted, which complicate the identification process. Accordingly, headspace collection is preferred instead [2] where a charcoal-filtered air is passed over the insect or its organ in an isolated aeration chamber and the odor-laden air is withdrawn by vacuum for analysis [8]. The air containing a mixture of volatiles can be analyzed *per se* or after being absorbed by super Q using columns. In the 1950s, it has been discovered that a measurable voltage arised between the tip and base of the insect antenna when exposed to odors of biological significance for the insect. This antennal response to different odors is known as electroantennogram or EAG [9]. This voltage is thought to represent the summed potentials of multiple responding olfactory neurons within the antenna, and the amplitude of the voltage roughly corresponds to an insect's sensitivity to a particular compound. The EAG has been widely used in entomology for pheromones' identification. The electroantennogram was improved through time and the insect antenna has been used as a detector (EAD) for a capillary-column gas chromatography, which is coupled with the flame ionization detector (FID) that is sensitive to all organic molecules. The GC-EAD is gas chromatography coupled with electroantennographic detection of compounds present in complex mixtures. This analytical procedure allows for rapid and accurate identification of insect odors. It is widely used to discover and identify semiochemicals like insect pheromones and repellents [8, 10]. The apparatus consists of injector (heated chamber), the column (10–100 m long, 1 mm wide) lined with a semisolid wax or polymer, flame ionization detector, insect antennal preparation, and a monitor to display the voltage output of the detector and the insect antenna (voltage on the Y-axis against time on the X-axis). Retention time and peak of each molecule give the identity and amount of the compound in the mixture, respectively. The FID output can be used to confirm the presence, identity, and quantity of compounds exposed to the antenna while the antennal (EAD) output indicates the presence/absence of olfactory sensitivity to eluting compounds and provides a relative measure of the intensity of olfactory stimulation. The FID peaks of the test compounds can be compared with retention times of those of commercially available versions of the same compounds injected into the GC. Identifications can be confirmed by re-analyzing the extract with a coupled gas chromatography-mass spectrometer (GC-MS) using the same column and GC operating parameters as used in the GC-EAD analysis. Testing the behavioral activity of EAD-active volatiles is an important and complementary step in the identification of semiochemicals that modify insect behavior. This behavioral bioassay test can be carried out using wind tunnel or an olfactometer [11, 12].

4. INTERACTION OF INSECTS AND PLANTS SEMIOCHEMICALS

Insects live in an environment with many volatile compounds including insect herbivore, host plant, and insect carnivore semiochemicals. These volatile chemicals interact with each other and finally modify the behavior and the physiology of insect pest species. Some insects sequester or acquire host plant compounds and use them as sex pheromone or sex pheromone precursors [13]. Many butterflies, moths, beetles, grasshopper, and aphids used pyrrolizidine alkaloids from their host plants as strong feeding deterrents against their natural enemies or predators [14]. It has been shown that the oil palm *Elaeis guineensis*, the host plant of the African palm weevil *Rhynchophorus phoenicis* produce a mixture of volatile esters of which the ethyl acetate induces male weevils to release the aggregation pheromone E-6-methyl-2-hepten-4-ol or rhyncophorol [15]. The males of orchid bees collect a mixture of terpenoids from the orchids and use them as aggregation pheromone to induce the formation of leks or the sites where males compete for females [16].

5. POTENTIAL USE OF SEMIOCHEMICALS IN INSECT PEST MANAGEMENT

Semiochemicals have been used for insect pest management more than 100 years ago [2]. Insect sex pheromones are the semiochemicals that are widely used for the management of insect pest particularly members of the order Lepidoptera. Aggregation pheromones from the order Coleoptera are also used for the management of agricultural insect pests of economic importance. Several serious agricultural pests including the carob moth *Ectomyelois ceratoniae*, the armyworm *Spodoptera frugiperda*, tomato leaf miner *Tuta absoluta*, fruit flies *Bactrocera* sp., mountain pine beetle (MPB) *Dendroctonus ponderosae*, Asian citrus psyllid *Diaphorina citri*, and the red palm weevil (RPW) *Rhynchophorus ferrugineus* have been successfully managed by using semiochemicals.

Semiochemicals are considered safe and environmentally friendly molecules due to their natural origin, low persistency in the environment, and species specificity, which attribute much to their harmless effect on nontarget organisms [17]. However, there are some difficulties in the practical applications of semiochemicals in pest management, and due to these challenges Semiochemically-based pest methods are still at the beginning[2]. Baker [3] mentioned the reasons that promoted or hindered the adoption of pheromones in the management programs of insect species as follows:

a. The biological differences in the mate-finding behavior of different species.

b. The chemistries of the pheromones used.

c. The successful engineering of the controlled-release dispenser and the use of proper trap design

d. The different political, economic, and use-pattern in different countries particularly the regulation of pheromones' application.

6. SEMIOCHEMICAL FORMULATIONS

Strong plumes of the correct blend of pheromones that create the above-threshold plume strands for downwind of the pheromone source are the key points to the optimal use of pheromones in integrated insect pest management programs. To achieve this, a controlled-release system of pheromone dispenser that mimics the natural pheromone release by insect pest is required [3]. In addition, optimization of trap density, design, and trap position is essential to achieve trap efficiency.

Mafra-Neto et al. [18] listed the shortcomings in semiochemical formulations as follows:

a. High cost of semiochemicals in terms of method of deployment in the field compared to conventional synthetic pesticides. Most semiochemicals used in insect pest management are formulated in the form of devices which require manual application.

b. The physical limitations such as instability, volatility, and sensitivity of the active ingredient of the semiochemicals to environmental factors like temperature and light.

c. Inconsistency of semiochemical product to maintain release rate and short-field longevity.

d. Mechanism of behavior manipulation in some techniques in which the semiochemical is used (e.g., mating disruption formulation that acts by camouflage requires large quantity of pheromone to be deployed which means more application cost).

For the above-mentioned reason, ISCA technologies developed an innovative semiochemical application technique for agricultural and forest insect pests. The technique is called specialized pheromone and lure application technology (SPLAT®). It is an amorphous, flowable, and controlled-release emulsion, with chemical and physical properties that may be adjusted by small changes in composition in processing or application method. A shear-thinning thixotropic, non-Newtonian fluid enters a liquid state by agitation, but quickly solidifies when agitation is stopped. These physical characteristics give the formulation flexibility in application. SPLAT® can provide a continuous controlled-release of semiochemicals for a period ranging from 2 weeks up to 6 months [18].

Semiochemicals are utilized for the management of insect pests through the following tactics:

a. Detection of invasive species and in delimiting surveys.

b. Monitoring the populations of endemic species to synchronized the timing of insecticide treatments

c. Evaluation of the effectiveness of pest management tactics through post-application assessment.

d. Improvement of old method of insect counts used for decision-making.

e. Increasing the effectiveness of biological control by increasing the predation/parasitism rates of predators and parasitoids. Kairomones could be applied to plants to increase the rate of parasitization through increasing the search rate of *Trichogramma* sp. [19].

f. Reduction of pest population through mating disruption, attract and kill, mass trapping, and repellency techniques.

7. SEMIOCHEMICAL-BASED PEST CONTROL TECHNIQUES

7.1. Attract and Kill (A&K)

The technique as the name implies simply use an attractant or semiochemical to lure an insect to a point source that contains a killing agent (insecticide, pathogen, or sterilant), hence the technique is termed attract and kill, attract and infect, and attract and sterilize, respectively. The technique leads to the reduction of the insect population by killing the target insect or reducing its fitness and fecundity or disabling it by causing disease. For more information on this technique regarding the fundamental requirements for both the attractant and the killing agent, please see Mafra-Neto et al. [18].

7.2. Mating Disruption

The technique is most commonly used in semiochemical-based pest management. It manipulates insect behavior in such a way that leads to population reduction. The environment where specific insect pest needs to be controlled is saturated with synthetic sex pheromones so that the abilities of males to locate the natural pheromone plume emitted by females are disrupted [18]. Mating disruption using synthetic pheromones or parapheromones does not completely shut off mating, but the delay in females mating may reduce their fecundity by approximately 50% [3]. Insect females have a critical time to mate and reproduce and any delay of mating may affect their fitness and their abilities to select the suitable sites for oviposition [18]. The mating system of some insects involves the transfer of certain peptides that trigger the egg laying behavior in the females. Four mechanisms were proposed to explain how mating disruption occurs, and these are:

a. Competitive attraction or false trail following

This happens when males respond to synthetic pheromone plumes produced by semiochemical dispenser rather than the natural plume emitted by the calling female [18]. This mechanism is density-dependent and decrease in efficiency as the population of pest increases.

b. Camouflage

This mechanism requires complete saturation of the environment with the synthetic pheromone. In this case, the male cannot locate the positions of the females and it is density-independent.

c. Desensitization

Adaptation of the male olfactory receptor system or habituation of the central nervous system may occur due to the overexposure to synthetic pheromone.

d. Sensory imbalance

Adaptation of the male olfactory receptor system or habituation of the central nervous system may occur due to the overexposure to synthetic pheromone.

7.3. Mass Trapping

It is a pheromone technique commonly used for direct insect population suppression. The technique is defined as the deployment of sufficient high density of pheromone traps that eliminate enough adults from the population and thus reduce subsequent larval damage [3]. Pheromones for monitoring are usually used at low density and the trapped insects have no effect on reduction of the population [20]. Mass trapping is effective in the case of male-emitted pheromone system that attracts females such as weevils (red palm weevil) and snout beetle. In this system, females are trapped, thus mass trapping directly reduces egg laying. The technique is effective with insects having relatively low population, live a long time before egg-laying, lay small number of eggs, and the emerging larvae cause considerable damage [21].

7.4. Repellents

A repellent is defined as a substance that deters or inhibits insects from finding, feeding on, or ovipositing on an attractive host substrate [18]. Several semiochemicals with repellent effect are available for the management of agricultural and forest pests. However, their practical application is limited due to the availability of cheap and effective pest control alternatives, lack of adequate formulations for delivery, and regulatory obstacles including registration [22].The repellent "verbenone" is now commercially available for the management of mountain pine beetle (MPB), *Dendroctonus ponderosae*. The repellents can be used alone or in combination with attractants for the management of insect pests as part of a push-pull strategy. Cook et al. [23] defined the push-pull strategy as the use of semiochemicals to make a protected resource an attractive or unsuitable for the pests (push) while luring them to an attractive source (pull) where the pests can be removed. Compared to other semiochemically based pest management techniques such as mating disruption or attract-and-kill, push-pull strategy requires more understanding of the chemical ecology of the insect pests [23].

8. A CASE STUDY OF THE RED PALM WEEVIL IN DATE PALM PLANTATION

The red palm weevil (RPW), *Rhynchophorus ferrugineus* (Coleoptera: *Curculionidae*), is an invasive and destructive insect pest of date palm worldwide. RPW was first reported as a pest of coconut palm in India and since the 1980s, it has invaded many countries around the globe. The weevil is a relatively large insect exhibiting different color morphs and sexual dimorphism where the male can be easily distinguished by the presence of dense hairs on the rostrum which are absent in the female. The weevil is extremely difficult to manage because all the life stages are concealed inside the date palm tissues and are difficult to detect at an early stage of damage. Several tactics including preventive and curative measures are adopted to manage the weevil in date palm plantations [24]. However, RPW is currently managed in date palm groves through semiochemically based integrated management using male-produced aggregation pheromone or (4-methyl-5-nonanol) for both monitoring and mass trapping of adult weevils [25]. This pheromone was first identified and synthesized in the early 1990s and has since proven a valuable tool for the management of RPW [26]. The aggregation pheromone acts synergistically with 4-methyl-5-nonanone (ketone) and is more attractive to RPW when combined with kairomones or volatiles emitted from the host to increase the efficiency of the blend [25]. The RPW pheromone (Ferrolure +) is commercially available (a mixture of 4-methyl-5-nonanol and 4-methyl-5-nonanone with a ratio of 9:1) released at 3–10 mg/day. The addition of ethyl acetate (released at 200–400 mg/day), fermenting mixture of dates and water increases trapping efficiency. The pheromone/food-based trapping system is considered as an environmentally friendly approach compared to the use of insecticides, which is currently being applied for the control of RPW [27]. Ferrugineol is attractive to both sexes; however, several researchers reported that traps baited with this attractant tend to capture significantly more females than males, usually with a ratio of two females to one male [24, 28, 29]. In addition, the captured weevils were found to be young, gravid, and fertile, indicating significant impact of trapping on the population reduction of the weevil in a given locality [30, 31]. The advantages of semiochemicals over other methods of pest control are that they are naturally occurring substances with species-specific character. Semiochemicals, a major component of IPM strategy for the management of red palm weevil in date palm [32], are environmentally friendly and have no adverse effects on natural enemies and pollinators in the agro-ecosystem. The aggregated nature of RPW distribution in the field, the long-life cycle, adult longevity, reliance on aggregation pheromone and host kairomone, and the relatively low population make the use of semiochemicals ideal for the management of this notorious pest [33]. To have an efficient semiochemical-based management program for RPW, a highly optimized pheromone product is needed in terms of attraction to lure the weevil directly to the trap (point source). The weevil must successfully locate the trap, arrested and enter inside; otherwise, it will infest the palm, which should be avoided when using traps. The trap should be more attractive and arresting for the weevils than the natural kairomone emitted by the date palm in the field. Optimizing the RPW trapping system requires better understanding of semiochemical ecology of the weevil in date palm plantation. Accordingly, the authors conducted a series of laboratory and field experiments that expand for 7 years to understand the chemical ecology and semiochemicals of red palm weevil in the date palm plantations in order to optimize trapping efficiency for an effective semiochemically based integrated management of this notorious pest. A summary of the results of these experiments is given in the following paragraphs.

8.1. Research Methodology

All laboratory experiments were conducted in the Date Palm Research Center of Excellence (latitude 25.16′6.9780″N, longitude 49.42′27.2772″E, and altitude 153 m), King Faisal University, Kingdom of Saudi Arabia. The field experiments were conducted in highly infested date palm groves selected based on data of trap catches obtained from the Directorate of Agriculture, Al-Ahsa, Ministry of Agriculture, and Kingdom of Saudi Arabia. Weevils used in olfactometer assays were obtained from a colony of RPW, which was established in the laboratory on bolts of the popular date palm cultivar "Khalas" that represent more than 85% of the cultivated date palms in Al-Ahsa oasis where the study has been carried out. To obtain virgin weevils, pupae were collected from the reared colony, and each pupa was kept separately in 20-ml-plastic jar with perforated lid. The jars were then kept in an incubator at a temperature of 30°C and 70% RH until adult eclosion. Emerged adult weevils were fed ad libitum on sugar cane for at least 3 days before being used for the different experiments.

8.1.1. Olfactometer Bioassays

Olfactometers are used to gauge the odor detection threshold of substances. A four arm-choice olfactometer®, a custom-made by Analytical Research Systems, Inc., Florida (ARS Inc., Florida) (www.ars-fla.com/mainpages/Bio-Assay/4 & 6-choice.htm) was used to study the weevils' preference to different lures (**Figures 2 – 4**). This olfactometer was connected to a pump that maintained a constant flow of pure air through the four arms while at the same time vacuuming out through central suction the odors emitted by the treatments. Four-flow meters control airflow in the olfactometer. Each of the four arms of olfactometer is provided with an odor source to be bioassayed, thus providing four odor fields for the test weevils to make a choice. Weevil directional preference to one of the four arms is used as an indicator to assess its behavioral response to that odor. The apparatus is manipulated in such a way that other factors, which might affect the weevil response to the odor, are controlled. More details on the functioning of the olfactometer are given in Section 8.3 of this chapter. The olfactometer was calibrated before carrying out the experiments according to the specifications shown in **Table 1** .

Table 1. Calibration of the olfactometer® to study the response of the red palm weevil (RPW) to different RPW pheromone lures.

Inlet/outlet pressure and air flow rate	Test calibration
1. Olfactometer pressure (10–20 PSI)	15 PSI
2. Source inlet pressure (50–150 PSI)	60 PSI
3. Olfactometer vacuum: center suction (−5 to −22″Hg)	−10″Hg
4. Vacuum pump pressure (60 + PSI)	+60 PSI
5. Olfactometer air inlet flow (0–1.3 LPM)	0.9 LPM

ARS Inc, Florida, USA.
PSI = pounds/square inch; ″Hg = inch mercury; LPM = liters per minute.

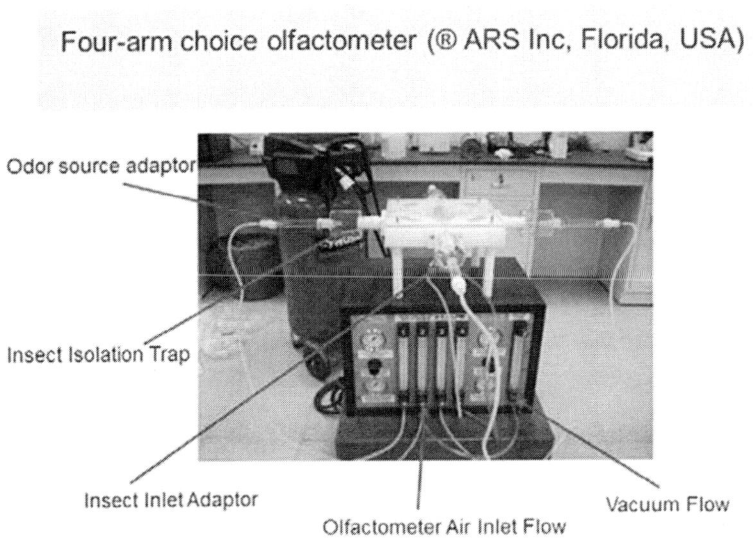

Figure 2. The layout of the four-arm choice olfactometer is shown. It consists of three units: a pump to provide purified air, clean air delivery system (vacuum), and odor exposure arena with polyethylene body and removable lid made of plexiglas.

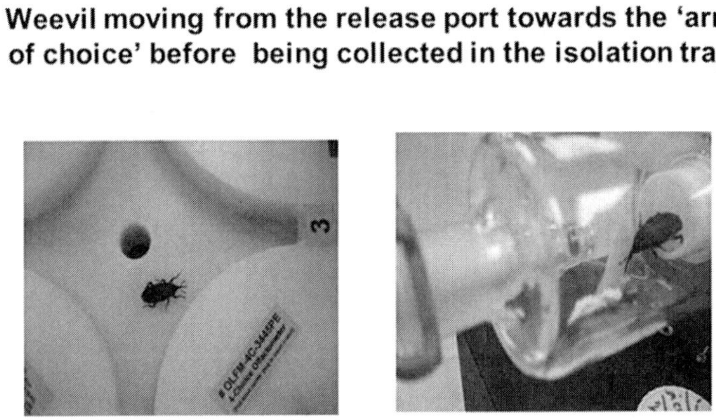

Figure 3. Dorsal view of the olfactometer showing insect inlet adaptor and isolation trap. An adult weevil moving upwind responding to odor source (*left*) before ending up in the insect isolation trap (*right*).

8.2. Optimizing Components Of Pheromone-Baited Traps

Several assays were carried out to evaluate the response of both male and female RPW adults to the aggregation pheromone (Ferrolure™) (El-Shafie and Faleiro, unpublished data). The test insects were (i) newly emerged unmated/mated insects and (ii) 1-month-old mated insects. Newly emerged insects were tested individually (20 replicates) while the 1-month-old adults were tested in batches of five weevils (10 replicates). The internal odor source (IOS) adapter of the olfactometer was moved from one arm to another sequentially, so that each arm of the olfactometer had the same number of replicates thereby eliminating bias of the instrument and environment if any. Besides the commercial pheromone, the internal odor source of the instrument containing the lure was charged with 1 ml fermented date solution dispensed in a perforated tube. A 5-min period was allowed for the test insects to move toward the arms from the insect release tray (IRT). Weevils collected in the insect isolation trap (IIT) at the end of 5 min were recorded.

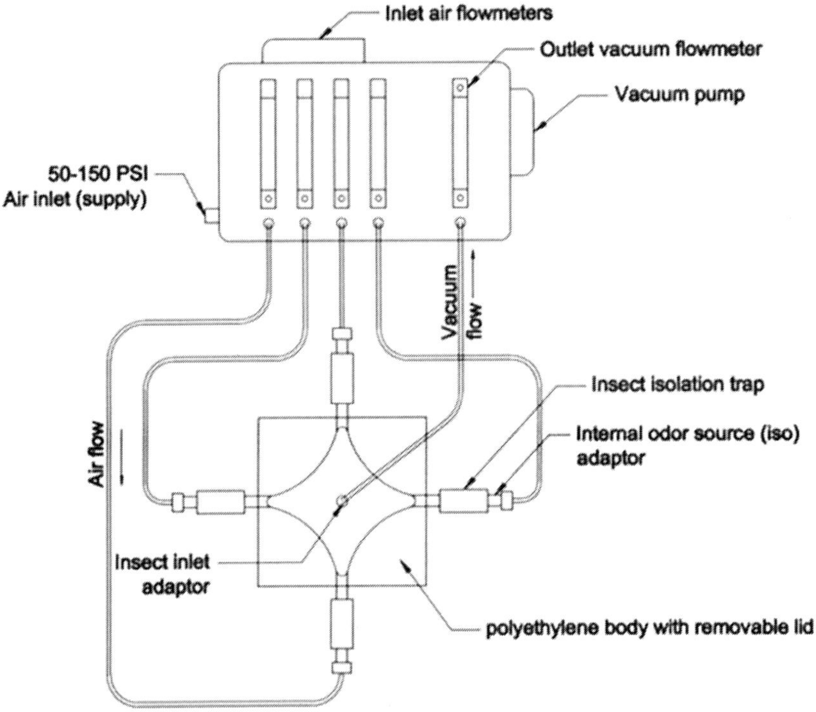

Figure 4. Schematic diagram of a four-arm olfactometer illustrating how it works. The olfactometer was connected to a pump that maintained a constant flow of pure air through the four arms while at the same time vacuuming out the odors emitted by the treatments through central suction. Four-flow meters controlled airflow in the olfactometer. Each of the four arms of olfactometer is provided with an odor source to be bioassayed, thus providing four odor fields for the test weevils to make a choice. Weevil directional preference to one of the four arms is used as an indicator to assess its behavioral response to that odor.

The results presented in **Figure 5** reveal significant and high attraction (>50%) to the aggregation pheromone in newly emerged unmated male and female weevils. Upon mating both newly emerged male and female weevils were significantly attracted to the pheromone but at a reduced (~30%) rate as compared to newly emerged unmated (virgins) individuals. With age (1 month old), the attraction to the pheromone reduced still further with only 14% of the female weevils responding significantly to the pheromone, whereas 1-month-old male weevil did not show significant attraction to the pheromone. In general, around 35% of the test insects (male and females) were attracted to the pheromone. The RPW uses the aggregation pheromone (ferrugineol) to recruit conspecific mates over long distance to colonize its host. This biological characteristic has been extensively utilized to manage this weevil through monitoring and mass trapping. Our results revealed that the response of RPW (males and females) to aggregation pheromone increased with mating as supported by the finding of Poorjavad et al. [34] who studied the effect of different doses of the aggregation pheromone on the diurnal response of virgin and mated males and females RPW under laboratory conditions using a two-choice pitfall static olfactometer. The authors concluded that in both sexes, the response to ferrugineol increased with mating. Mated females showed strong response to ferrugineol than unmated ones, due to the searching behavior of the former for egg laying sites.

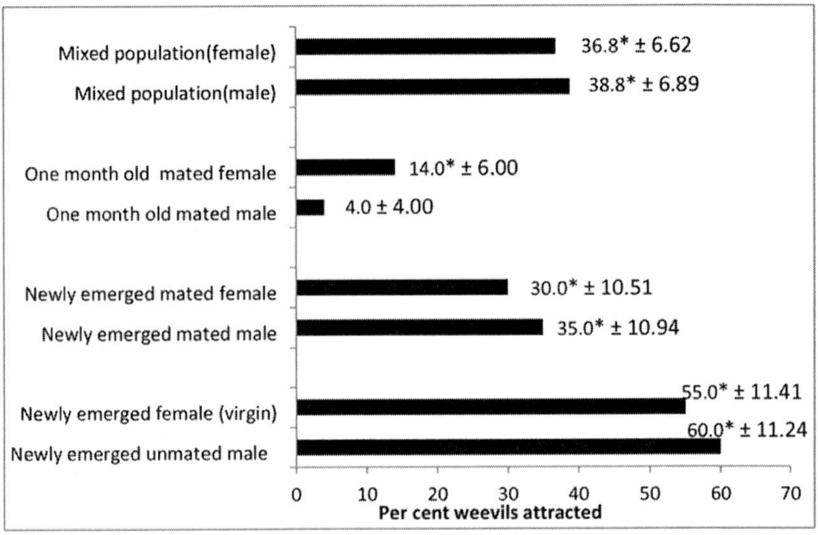

Figure 5 Response of *R. ferruginous* to the aggregation pheromone in choice olfactometer assays (*t*-test; *significant at p = 0.05).

Weissling et al. [35] stated that the optimal attraction of weevil to trap is affected by the proportional changes in volatile chemicals from fermentation overtime. These volatile chemicals or kairomones, as determined by GC-MS, included palm esters, ethyl acetate, ethyl propionate, ethyl butyrate, and ethyl isobutyrate [36]. Kumar et al. [37] tested different pheromone lures in India and reported differences in their efficiencies. This discrepancy could be attributed to difference in environmental conditions between India and Saudi Arabia where the present investigation was carried out. In this context, Faleiro and Chellapan [38] reported a difference in longevity of the lures in different seasons (winter and summer). They stated that the release of pheromone into the environment is faster in summer than in winter and that was attributed to higher temperature and day light. Thus, they recommended putting traps under shade conditions to sustain the efficiency of the trapping system. The fact that only 35% of the test weevil (males and females) responded to the aggregation pheromone, which is supported by El-Sayed et al. [27], require that the remaining population in the field have to be managed through other IPM tactics.

8.3. Determining The Extent of Attraction of Weevils To Date Palm Volatiles

Faleiro et al. [12] studied the mechanisms of resistance against RPW in seven major date palm cultivars of the Al-Ahsa oasis in Saudi Arabia viz. Khalas, Sheshi, Reziz, Khasab, Hatmi, Shahal and Gaar by determining the extent of attraction of female RPW adults to fresh palm volatiles emitted from date palm frond tissue through four-arm choice olfactometer assays. In each of the inlet odor source (IOS) adapters of the olfactometer, freshly cut palm petiole pieces (5 × 1 × 1 cm) of a single cultivar were placed. Two experiments were carried out wherein four cultivars were tested at a time with Khalas as the control treatment in both the experiments. Fifteen-day-old field collected gravid adult female weevils were used in the assays. Five female weevils were placed in the insect release device of the olfactometer. After 5 min, the number of adult female weevils collected in the insect isolation trap (IIT) was noted. Each experiment was replicated eight times. At the end of each assay (replication), palm tissue pieces and test insects used in the assay were discarded. New palm tissue pieces and test insects were used for every replication (assay). The IOS was moved sequentially to the next arm of the olfactometer at the end of each test replication so that every treatment was at the same arm of the olfactometer twice during each trial. This was done to eliminate bias if any in the instrument and environment. Results revealed that the popular date palm cultivar Khalas had the least antixenotic effect on female RPW adults where a high degree of attraction to palm tissue volatiles was recorded which was statistically similar to the cultivars Reziz, Sheshi, and Hatmi. The cultivars Khasab, Shahal, and Gaar exhibited high degree of nonpreference (antixenosis) (**Figure 6**). Identifying the chemical components of tissue volatiles that trigger antixenosis in date palm to RPW will pave the way for future studies on chemical ecology of RPW and its interactions with date palm as a main host.

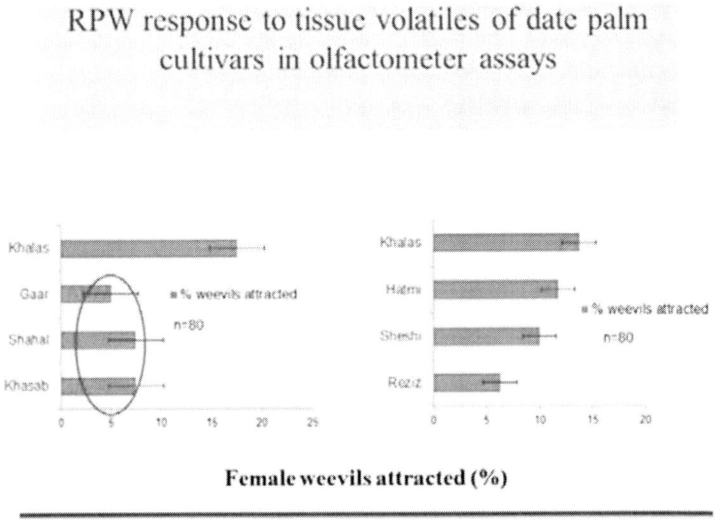

Figure 6. Red palm weevil response to tissue volatiles of date palm cultivars in olfactometer assays.

8.4. Testing The Efficacy Of Hook^{tm} Rpw Versus Food-Baited Pheromone Traps

Food-baited pheromone traps (FBPTs) have been used to manage the red palm weevil in date plantations through monitoring or mass trapping techniques for population reduction. The FBPTs are simply made of plastic buckets baited with ferrugineol and fermenting dates as sources of kairomone (**Figures 7** and **8**). The main problem with FBPTs is that they require frequent service (replacement of food bait and water) which makes their utilization over large area laborious and expensive. ISCA technologies developed Hook RPW is an attract-and-kill formulation that combines the aggregation pheromone and the insecticide cypermethrin. The formulation does not require service once deployed in the field, thus it can drastically reduce application cost particularly in area-wide RPW integrated management.

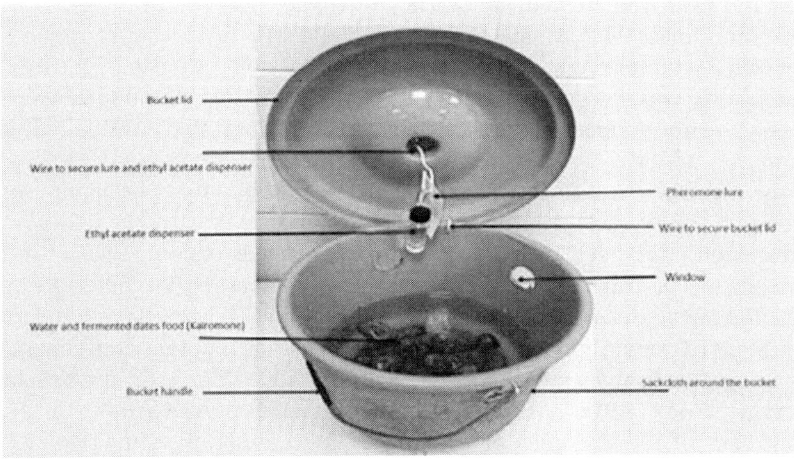

Figure 7. The standard RPW bucket trap showing the main components.

Figure 8. A marked RPW male walking toward the food-baited pheromone trap.

El-Shafie et al. [39] tested the efficacy of Hook™ RPW in an infested date plantation in Eastern province of Saudi Arabia. The product was applied in hundred 3 g dollops at an application rate of 250 dollops per ha. The area of treated plot was 0.4 ha containing ca. 60 date palms of the popular variety "Khalas." The number of weevils attracted to source point and killed was compared with the catches of the conventional FBPTs commonly deployed by the Directorate of Agriculture to manage RPW in date palm plantations. The data of weevils captured per week in the treatment plots was converted to number of RPW caught per ha per week to compensate for the discrepancy in trap density between the two trapping methods. Results revealed no significant difference between Hook RPW and FBPTs indicating that SPLAT formulation can be successfully used as a potent component in the integrated management of RPW. In a separate trial concerning bait-lure synergy, Hook RPW sustained the same level of attraction to PRW regardless of the presence or absence of food bait [39]. SPLAT for RPW proved to maintain field longevity and efficacy for 3 months under the dry summer conditions of Saudi Arabia.

A 10-fold increase in pheromone trap number in Al-Hasa date palm oasis during the period from 2007 to 2012 in Saudi Arabia decreased the RPW infestation in an area of 1104 ha by 86%. Likewise, the application of insecticides was reduced by 91% and the felling and eradication of infested date palm trees dropped by 89% [40]. Similar stories of successful management of RPW using semiochemicals in the Gulf and the Middle East are documented [28, 41]. The research on semiochemical repellents of RPW has been initiated and Guarino et al. [42] identified α-pinene and methyl salicylate for being potential repellents for RPW. This could open an avenue for the future use of push-pull strategy for integrated management of red palm weevil.

9. CONCLUSIONS

Semiochemicals have been exploited in several ways to manage insect pests. These include monitoring and detection, population suppression through mating disruption, mass trapping and attract-and-kill techniques. The male-produced aggregation pheromone is successfully used in food-baited traps for the area-wide integrated management of red palm weevil in date palm, coconut palm, and Canary Island palm plantations. Mated RPW females responded more to the aggregation pheromone than virgin females indicating that an egg-laying stimulus may be responsible for deriving these females to the aggregation pheromone. Large number of RPW adult weevils fly to traps located near to their colony to nearby traps and increasing distance to trap from infestation spot increase the probability weevil will not find the trap. This behavioral response of RPW to aggregation pheromone should be considered when specifying trap density per ha for mass trapping. Some date palm cultivars exhibited different kinds of resistances to RPW the mechanisms of which vary; however, antixenosis or egg-laying nonpreference is important. Future research on semiochemicals for insect pest management should focus on innovative formulation for field deployment as well as on optimization of controlled-release technologies and trapping efficiency. More research on the chemical ecology of target insect pest is of paramount importance for development of semiochemically based insect pest management programs.

REFERENCES

1. El-Sayed, AM. 2015. The pherobase: database of insect pheromones and semiochemicals, 2008. http://www.pherobase.com
2. Soroker, V, Harari, A and Faleiro, JR. 2015. The role of semiochemicals in date pest management. In "Sustainable Pest Management in Date Palm: Current Status and Emerging Challenges" (Editors): Wakil, W, Faleiro, JR and Miller. T. http://www.springer.com/us/book/9783319243955. ISBN 978-3-319-24397-9. Springer International Publishing. Switzerland. 445p.
3. Baker, TC. 2011. Insect pheromones: useful lessons for Crustacean pheromone programs? In: Chemical Communication in Crustacean (Editors): Breithaupt, T and Thiel, M. Springer Science + Business Media LLC. Springer-Verlag New York.
4. Karlson, P and Lüscher, M. 1959. Pheromones: a new term for a class of biologically active substances. Nature. 183: 55–56.
5. Baker, TC, Willis, MA, Haynes, KF, Phelan, PL. 1985. A pulse cloud of sex pheromone elicits upwind flight in male moths. Physiological Entomology. 10: 57–65.
6. Mafra-Neto, A, Carde, RT. 1994. Fine-scale structure of pheromone plumes modulates upwind orientation of flying moths. Nature. 369: 142–144.
7 Rutledge, CE. 1996. A survey of identified kairomones and synomones used by insect parasitoids to locate and accept their hosts. Chemoecology. 7: 121–131.
8. Fraser, AM, Mechaber, WL and Hildebrand, JG. 2003. Electroantennographic and behavioral responses of the sphinx moth *Manduca sexta* to host plant headspace volatiles. Journal of Chemical Ecology. 29: 1813–1833.
9. Schneider, D. 1957. Elektrophysiologische Untersuchungen von Chemo-und Mechanorezetoren der Antenne des Seidenspinners. *Bombyx mori* L. Zeitschrift fuer Vergleichende Physiologie. 40: 8–41.
10. Struble, DL and Arn, H. 1984. Combined gas chromatography and electroantennogram recording of insect olfactory responses, pp. 161–178, in (Editors): Hummel, HE and Miller, TA. Techniques in pheromone research. Springer-Verlag. New York.
11. Park, KC, Ochieng, SA, Zhu, J, and Baker, TC 2002. Odor discrimination using insect electroantennogram responses from an insect antennal array. Chemical Senses. 27: 343–352.
12. Faleiro, JR, El-Shafie, HAF, Ajlan, AM, and Sallam, AA 2014. Screening date palm cultivars for resistance to red palm weevil, *Rhynchophorus ferrugineus* (Coleoptera: Curculionidae). Florida Entomologist. 97(4): 1529–1536.
13. Reddy, GVP, Guerrero, A. 2004. Interactions of insect pheromones and plant semiochemicals. Trends in Plant Science. 9: 253–261.

14. Nishida, R, 2002. Sequestration of defensive substances from plants by Lepidoptera. Annual Review of Entomology. 47: 57–92.
15. Jaffé, K, Sánchez, P, Cerda, H, Hernández, JV, Jaffé, R, Urdaneta, N, Guerra, G, Martínez, R, Miras, R. 1993. Chemical ecology of the palm weevil *Rhynchophorus palmarum* (L.) (Coleoptera: Curculionidae): attraction to host plants and to male-produced aggregation pheromone. Journal of Chemical Ecology. 19: 1703–1720.
16. Dressler, RL. 1982. Biology of the orchid bees (Euglossini). Annual Review of Ecology, Evolution, and Systematics. 13: 373–394.
17. Horowitz, R, Ellsworth, PC, Ishaaya, I. 2009. Biorational pest control-an overview. In: Ishaaya, I. and Orowitz, AR (Editors): Biorational Control of Arthropod Pests. Springer Science + Business Media B. V.. Springer Netherlands.
18. Mafra-Neto, A, Fettig, CJ, Unson, AS, Rodriguez-Saona, C, Holdcraft, R, Faleiro, JR, El-Shafie, H, Reinke, M, Bernardi, C, Villagran, KM. 2014. Development of specialized pheromone and lure application technologies (SPLAT®) for management of Coleopteran pests in agricultural and forest systems. In (Editors): Gross, AD, Coats, JR, Duke, SO and Seiber, JN . Biopesticides: State of the Art and Future Opportunities. ACS Symposium Series: American Chemical Society. Washington, DC.
19. Lewis, WJ, Jones, RL, Nordlund, DA, Sparks, AN. 1975. Kairomones evaluation for increasing rate of parasitization by *Trichogramma* spp. in the field. Journal of Chemical Ecology. 1: 343–347.
20. Baker, TC. 2008. Use of pheromones in IPM. In (Editors): Radcliffe, T and Hutchinson, B . Integrated pest management, Cambridge University Press. Cambridge. pp. 271–285.
21. Oehlschlager, AC, Chinchilla, C, Castillo, D, Gonzalez, L. 2002. Control of red ring disease by mass trapping of *Rhynchophorus palmarum*(Coleoptera: Curculionidae). Florida Entomologist. 85: 507–513.
22. Isman, MB. 2006. Botanical insecticides, deterrents and repellents in modern agriculture and an increasingly regulated world. Annual Review of Entomology. 51: 45–66.
23. Cook, SM, Khan, ZR, Pickett, JA. 2007. The use of push-pull strategies in integrated pest management. Annual Review of Entomology. 52: 375–400.
24. Faleiro, JR. 2006. A review of the issues and management of the red palm weevil *Rhynchophorus ferrugineus* (Coleoptera: Rhynchophoridae) in coconut and date palm during the last one hundred years. International Journal of Tropical Insect Science. 26: 135–150.
25. Hallett, R, Oehlschlager, A, Borden, J. 1999. Pheromone trapping protocols for the Asian palm weevil, *Rhynchophorus ferrugineus* (Coleoptera: Curculionidae). International Journal of Pest Management. 45: 231–237.
26. Hallett, RH, Gries, G, Gries, R, Borden, JH, Czyzewska, E, Oehlschlager, AC, Pierce, HD, Angerilli, NPD, Rauf, JR. 1993. Aggregation pheromones of two Asian palm weevils, *Rhynchophorus ferrugineus* and *Rhynchophorus vulneratus.* Naturwissenschaften. 80: 328–331.
27. El-Sayed, AM, Suckling, D, Wearing, C, Byers, J. 2006. Potential of mass trapping for long-term pest management and eradication of invasive species. Journal of Economic Entomology. 99: 1550–1564.
28. Soroker, V, Blumberg, D, Haberman, A, Hamburger-Rishad, M, Reneh, S, Talebaev, S, Anshelevich, L, Harari, AR. 2005. Current status of red palm weevil infestation in date palm plantations in Israel. Phytoparasitica. 33: 97–106.
29. Ávalos, JA, Soto, A. 2015. Study of chromatic attraction of the red palm weevil, *Rhynchophorus ferrugineus* using bucket traps. Bulletin of Insectology. 68: 83–90.
30. Avand-Faghih, A. 1996. The biology of red palm weevil, *Rhynchophorus ferrugineus*, (Coleoptera: Curculionidae) in Saravan region (Sistan and Baluchistan province, Iran). Applied Entomology and Phytopathology. 63: 16–18.

31. Faleiro, JR, Abraham, VA, Al-Shuaibi, MA, Kumar, TP. 2000. Field evaluation of red palm weevil, *Rhynchophorus ferrugineus* Oliv. Pheromone (ferrugineol) lures. Indian Journal of Entomology. 62: 427–433.
32. Abbas, MS, Hanounik, SB, Sahdad, AS, Al-Bagham, SA. 2006. Aggregation pheromone traps, a major component of IPM strategy for the red palm weevil, *Rhynchophorus ferrugineus* (Coleoptera: Curculionidae) in date palms. Journal of Pest Science. 79 (2): 69–73.
33. Giblin-Davis, RM, Oehlschlager, AC, Perez, A, Gries, G, Gries, R, Weissling, TJ, Chinchilla, CM, Pena, JE, Hallett, RH, Pierce, HD, Gonzalez, LM. 1996. Chemical and behavioural ecology of palm weevils (Coleoptera: Curculionidae). The Florida Entomologist. 79: 153–166.
34. Poorjavad, N, Goldansaz, SH, Avand-Faghih, A. 2009. Response of the red palm weevil, *Rhynchophorus ferrugineus* to its aggregation pheromone under laboratory conditions. Bulletin of Insectology. 62: 257–260.
35. Weissling, TJ, Giblin-Davis, RM, Scheffrahn, RH, Marban-Mendoza, NM. 1992. Trap for capturing and retaining *Rhynchophorus cruentatus*(Coleoptera: Curculionidae) adults using Sabal palmetto as bait. Florida Entomologist. 75: 212–221.
36. Gries, G, Gries, R., Perez, AL, Gonzalez, LM, Pierce, HDR, Oehlschlager, AC, Hainds, M, Zebeyou, M and Kouame, B. 1994. Ethyl propionate: synergistic kairomone for African palm weevil, Rhynchophorus phoenicis L., (Coleoptera: Curculionidae). Journal of Chemical Ecology. 20: 889–897.
37. Kumar, KR, Maheswari, P and Dongre, TK. 2004. Study on comparative efficacy of different types of pheromones in trapping the red palm weevil, *Rhynchophorus ferrugineus* (Oliv) of coconut. Indian Coconut Journal. 34 (12): 3–4.
38. Faleiro, JR and Chellapan, M. 1999. Attraction of red palm weevil *Rhynchophorus ferrugineus* Oliv. To ferrugineol, based pheromone lures in coconut gardens. Journal of Tropical Agriculture. 1 (2): 60–63.
39. El-Shafie, HAF, Faleiro, JR, Al-Abbad, AH, Stoltman, L, Mafra-Neto, A. 2011. Bait-free attract and kill technology (Hooktm RPW) to suppress red palm weevil *Rhynchophorus ferrugineus* (Coleoptera: Curculionidae) in date palm. Florida Entomologist. 94: 774–778.
40. Hoddle, MS, Al-Abbad, AH, Elshafie, HAF, Faleiro, JR, Sallam, AA, Hoddle, CD. 2013. Assessing the impact of area wide pheromone trapping, pesticide applications, and eradication of infested date palms for *Rhynchophorus ferrugineus* (Coleoptera: Curculionidae) management in Al Ghowaybah, Saudi Arabia Crop Protection. 53: 152–160.
41. Al-Shawaf, AM, Al-Shagag, A, Al-Bagshi, M, Al-Saroj, S, Al-Bather, S, Al-Dandan, AM, Ben Abdallah, A, Faleiro, JR. 2013. A quarantine protocol against red palm weevil *Rhynchophorus ferrugineus* (Olivier) (Coleoptera: Curculionidae) in date palm. Journal of Plant Protection Research. 53: 409–415.
42. Guarino, S, Peri, E, Lo Bue, P, Pia Germaná, M, Colazza, S, Anshelevich, L, Ravid, U, Soroker, V. 2013. Assessment of synthetic chemicals for disruption of *Rhynchophorus ferrugineus* response to attractant-baited traps in an urban environment. Phytoparasitica. 41: 79–88.

CHAPTER 7

Use of Banker Plant System for Sustainable Management of the most Important Insect Pest in Rice Fields in China

Xusong Zheng[1], Yanhui Lu[1], Pingyang Zhu[2], Facheng Zhang[2], Junce Tian[1], Hongxing Xu[1], Guihua Chen[2], Christian Nansen[1,3] & Zhongxian Lu[1]

[1] *State Key Laboratory Breeding Base for Zhejiang Sustainable Pest and Disease Control, Institute of Plant Protection and Microbiology, Zhejiang Academy of Agricultural Sciences, Hangzhou 310021, China.*

[2] *Jinhua Plant Protection Station, Jinhua 321017, China.*

[3] *Department of Entomology and Nematology, UC Davis Briggs Hall, Room 367, Davis, CA 95616, USA.*

ABSTRACT

To meet the World's food demand, there is a growing need for sustainable pest management practices. This study describes the results from complementary laboratory and field studies of a "banker plant system" for sustainable management of the rice brown planthopper (BPH) (*Nilaparvata lugens* Stål) – the economically most important rice pest in Asian rice growing areas. The banker plant system consisted of planting a grass species, *Leersia sayanuka*, adjacent to rice fields. *L. sayanuka* is the host plant of a planthopper, *Nilaparvata muiri*. An egg parasitoid, *Anagrus nilaparvatae*, parasitizes eggs of both BPH and *N. muiri*, and its establishment and persistence are improved through plantings of *L.*

sayanuka and thereby attraction of *N. muiri*. Laboratory results showed that BPH was unable to complete its life cycle on *L. sayanuka*, and *N. muiri* could not complete its life cycle on rice. Thus, planting *L. sayanuka* did not increase the risk of planthopper damage to rice fields. Field studies showed that BPH densities were significantly lower in rice fields with banker plant system compared to control rice fields without banker plant system.

INTRODUCTION

Rice is the principal food of more than 50% of the world's population, and for 60% of the Chinese population1. It has been estimated that the world's rice production has to increase drastically over the next three decades to meet the growing food demand in Asia2. With expanding cities and declining value of arable land, this increase in rice production cannot be met by expansion of cultivation area but must come from higher productivity from existing agricultural land. Throughout Asia, there has been an intensification of rice production through higher fertilizer and pesticides inputs in the last decades3,4. Although this intensification has led to yield increases, there is growing concern about the sustainability of current practices5. Furthermore, the continent has faced a stark increase in insect pest outbreaks of planthoppers (Homoptera: Delphacidae) and other serious insect pests5,6 and these pest outbreaks have been directly linked to intensification of rice and reduction of biodiversity in rice-based ecosystem7.

The brown planthopper (BPH), *Nilaparvata lugens* (Homoptera: Delphacidae) is one of the key pests in Asian rice production systems, including China5,8,9,10. At present, the control of rice pests is highly dependent on frequent applications of broad-spectrum insecticides, and there is widespread concern about their low and inconsistent effects on pest populations and high adverse effects on insect natural enemies of rice pests5.

To meet the world's food demand, there is a growing need for sustainable pest management practices. This means that it requires fundamental changes of current practices. Moreover, it is necessary to adopt rice pest management practices with a markedly reduced reliance on pesticides. A possible solution is "banker plant systems" as part of "ecological engineering"7,11,12. Banker plant systems (as well as ecological engineering systems) involve promotion of plant diversity to enhance pest self-regulatory ecosystem functions, such as, predation and competition, to reduce susceptibility of agricultural crops to native and invasive pests11,13,14. Banker plants may provide resources, such as, shelter, pollen and nectar, or alternative preys15,16,17 to improve the establishment and persistence of beneficial insect populations used to control a specific insect pest. Thus, banker plants are established and managed adjacent to crops, and they may attract insects, which can serve as alternative hosts of arthropod natural enemies when the crop pest populations are low or not available. The first successful banker plant system was developed in 1977 and involved tomato as the banker plant, a parasitoid [*Encarsia formosa* Gahan (Hymenoptera: Aphelinidae)], and a whitefly pest [*Trialeurodes vaporariorum* Westwood (Homoptera: Aleyrodidae)] on tomatoes in greenhouses18. The bank plant systems developed for a series of important combinations of crops and pests are present as a Table 1.

Table 1. Research that have implemented banker plant systems in greenhouse and outdoor vegetable and food crops.

Crop	Banker plant	Alternative host	Natural enemies	Target pests	References
Cucumber	Wheat, Barley	*Rhopalosiphum padi*	*Aphidius colemani*, *Aphidoletes aphidimyza*	*Aphis gossypii*	Bennison and Corless, 1993
Melon	Barley	*Schizaphis graminum*	*A. colemani*, *A. aphidimyza*	*A. gossypii*	Kim and Kim, 2004
Pepper	Broad beans	*Megoura viciae*	*A. aphidimyza*	*Myzus persicae*	Hansen, 1983
Cauliflower	Cabbage, turnip	*Brevicoryne brassicae*, *M. persicae*, *Macrosiphum euphorbiae*	*Diaeretiella rapae*	*Brevicoryne brassicae*	Freuler *et al.*, 2003
Tomato	Papaya	*Trialeurodes variabilis*	*Encarsia Sophia*	*Bemisia tabaci*	Xiao *et al.*, 2011
Vegetable	Ornamental pepper		*Amblyseius swirskii*	*B. tabaci*, *Frankliniella occidentalis*, *Scirtothrips dorsalis*	Xiao *et al.*, 2012
Cangreen bean, Dusky bean	Corn	*Oligonychus pratensis*	*Feltiella acarisuga*	*Tetranychus urticae*	Xiao *et al.*, 2012, 2013
Apple	Elder shrubs	*Aphis sambuc*	syrphids (Diptera: Syrphidae)	*Dysaphis plantaginea*	Brisbosia, 2003
Rice	*Zizania caduciflora*	*Saccharosydne procerus*	*Anagrus nilaparvatae*	*Nilaparvata lugens*	Yu *et al.*, 1999; Yu, 2001
Rice	barnyard grass, *Echinochloa* [illegible]	*Sogatella vibix*	mirid bug, *Cyrtorhinus lividipennis*	*Sogatella furcifera*, *Nilaparvata lugens*	Matsumura and Urano, 2001

In rice systems, banker plant systems have involved planting of sesame (*Sesamum indicum*) as nectar source to promote establishment and persistence of a predatory bug [*Cyrtorhinus lividipennis* (Heteroptera: Miridae)] and of parasitoids [*Anagrus spp.* (Hymenoptera: Mymaridae), *Apanteles ruficrus* (Hymenoptera:Braconidae), *Cotesia chilonis* (Hymenoptera:Braconidae), and *Trichogramma chilonis* (Hymenoptera: Trichogrammatidae)]19,20,21. However, no studies on describing attraction of alternative hosts to parasitoids of rice insect pests were published. *Nilaparvata muiri* is a common planthopper species, which is widely distributed in the rice regions of southern China, Japan, Korea and Vietnam22,23. The grass species, *Leersia sayanuka*, a rice field weed, is one of the main and a preferred host plant for *N. muiri*24.

In this study, we examined critically important elements of the first banker plant system on rice in Asia, in which is a grass species, *L. sayanuka*, is planted adjacent to rice fields to attract a planthopper, *N. muiri*. The hypothesis tested in this study is that *N. muiri* does not attack rice plants, and that is an alternative host for an egg parasitoid, *Anagrus nilaparvatae*, which is the main natural enemy of BPH. As a consequence, we predicted that planting *L. sayanuka* would lead to improved establishment and persistence of *A. nilaparvatae* and therefore enhanced biological control of BPH. Based on laboratory studies, we examined the abilities of: 1) *N. muiri* to complete its life cycle on rice, and 2) BPH to complete its life cycle on *L. sayanuka*. In addition, we studied the functional response by *A. nilaparvatae* to the two planthopper host species (BPH and *N. muiri*). Finally in replicated field study, we quantified rice yields in systems with/without establishment of the banker plant system. The potential of the proposed banker plant system and its complementarity with other ecological engineering approaches were discussed.

RESULTS

Parasitism of *A. nilaparvatae* to eggs of both BPH and *N. muiri* derived from different host plants

There were no significant differences in the parasitism of *A. nilaparvatae* to BPH eggs from two host plants (t-test: $t = 0.484$, $df = 58$, $p = 0.876$). The oviposition of *A. nilaparvatae* female reared from BPH on BPH and on *N. muiri* eggs were 20.1 and 19.1, respectively. The oviposition of each *A. nilaparvatae* female reared from *N. muiri* on BPH and on *N. muiri* eggs was 17.1 and 16.8, respectively. No significant difference was observed in the ability of *A. nilaparvatae* reared from *N. muiri* to parasitize the two species of planthopper (test: $t = 0.404$, $df = 58$, $p = 0.913$) (Figure 1).

Functional responses of *A. nilaparvatae* on eggs from two hosts (BPH or *N. muiri*) reared by different host plants

The number of BPH or *N. muiri* eggs parasitized by *A. nilaparvatae* increased with the density of planthopper eggs in the different treatments to an asymptote. Regression analyses of densities of both BPH and *N. muiri* eggs and the parasitized number showed a significant difference among all treatments. The results were consistent with the equation for Holling's type II functional responses (Table 2). The instantaneous attack rate of *A. nilaparvatae* reared from BPH eggs against BPH was higher than that of *A. nilaparvatae* from *N. muiri* eggs, while the instantaneous attack rates of *A. nilaparvatae* reared from *N. muiri* eggs to BPH and *N. muiri* had no significant difference. The theoretic maximum parasitized eggs numbers of *A. nilaparvatae* reared from the BPH on the two brown planthoppers were not significantly different, while the parasitized egg number of *A. nilaparvatae* from BPH eggs was significantly higher than that of *A. nilaparvatae* from *N. muiri* eggs.

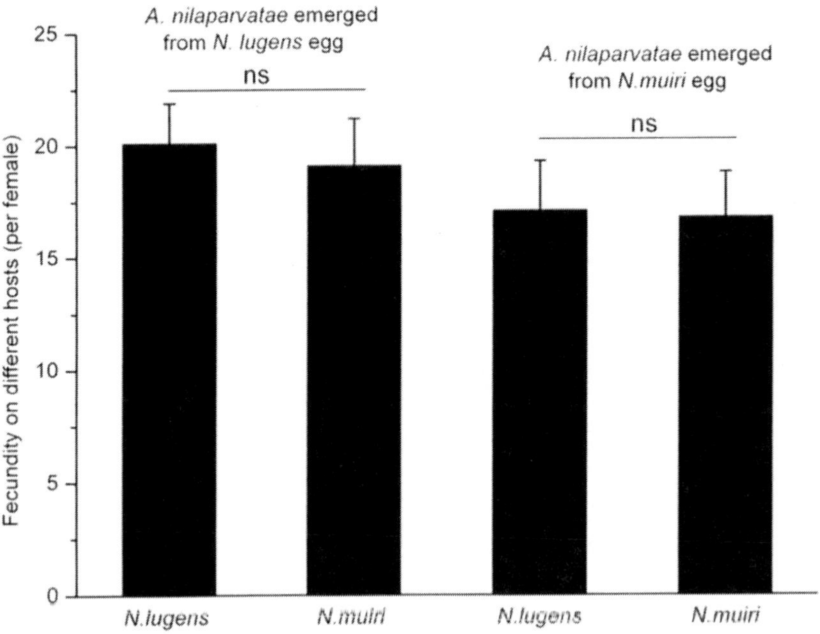

Figure 1: Fecundity of *A. nilaparvatae* from different host plants on different hosts. Ns indicated that there was no significant difference between the *N. lugens* eggs and *N. muiri*eggs treatment by χ^2 testing ($P < 0.05$).

Table 2: Parameter estimates of the functional response of *A. nilaparvatae* reared from different host plants on the eggs of *N. lugens*and *N. muiri.*

Parasitoids	Hosts (eggs)	Instantaneous attack rate (a)	Processing time (Th)	R^2	P	Maximum parasitized eggs (per day) (1/Th)
A. nilaparvatae emerged from *N. lugens* egg	*N. lugens*	0.782±0.348	0.034±0.005	0.3004	<0.001	29.41
	N. muiri	0.339±0.150	0.037±0.008	0.2638	<0.001	27.03
A. nilaparvatae emerged from *N. muiri* egg	*N. lugens*	0.365±0.174	0.046±0.008	0.1910	<0.001	21.74

Oviposition preference of *A. nilaparvatae* reared from different hosts on the eggs of BPH and *N. muiri*

The selectivity index showed that *A. nilaparvatae* derived from hosts had different selectivity toward BPH versus *N. muiri* (Table 3). Between the two planthoppers, *A. nilaparvatae* derived from BPH eggs preferred BPH eggs ($\chi^2 = 6.12 > \chi^2_{0.05}$). With the proportion of BPH eggs to *N. muiri* eggs increased, *A. nilaparvatae*'s preferred *N. lugens* eggs. However, *A. nilaparvatae* derived from *N. muiri* eggs had no obvious preference between BPH eggs and *N. muiri* eggs ($\chi^2 = 1.38 < \chi^2_{0.05}$).

Table 3. Preference test of *A. nilaparvatae* from different host plants on the eggs of *N. lugens*and *N. muiri* and Chi-square test for the hypothesis of no preference.

Parameter	*A. nilaparvatae* emerged from *N. lugens* egg	*A. nilaparvatae* emerged from *N. muiri* egg
Number of replications in each preference test	37	49
α	0.7635±0.0493	0.5381±0.0495
χ^2	$6.12 > \chi^2_{0.05}$	$1.38 < \chi^2_{0.05}$

Field evaluation of the effects of growing *Leersia* planting beside rice plants as a banker plant

The results showed that planting *L. sayanuka* stripes of 10-cm or 50-cm width at the edge of paddy fields kept rice planthopper populations below the threshold density of 15 per hill. The BPH population in the experimental field was significantly lower than that in the control field where the densities of rice planthoppers exceeded the threshold density. There was no significant difference between 10-cm and 50-cm stripes (Table 4). This result suggests that the construction of a 10-cm-wide *L. sayanuka* banker plant system at the edge of a paddy field can achieve effective control of rice planthoppers (mainly BPH).

Table 4. Field evaluation of LNA banker plant system for rice planthoppers management (per hill).

Treatments	Jul 9	Jul 23	Aug 6	Aug 20	Sept 3
Control	9.1±0.9bc	14.3±1.4a	21.9±2.0a	23.2±2.1a	17.7±1.9a
10cm stripe	15.2±1.7a	13.1±1.1a	9.5±1.0b	6.6±0.8b	3.5±0.6b
50cm stripe	12.6±1.4ab	12.3±1.1a	7.1±0.8b	5.7±1.0b	2.4±0.6b
F	4.988	0.728	34.862	49.862	48.401
P	0.011	0.489	<0.001	<0.001	<0.001

Population trend index of BPH and *N. muiri* on different host plants

BPH adults only survived about 2 days on *L. sayanuka* and could not lay eggs. *N. muiri* adults could survive about 5 days on rice plants and could lay only a few eggs (3.6 ± 2.1 eggs per female) and hatching rate was low (23.3%) (Table 5). For the nymphs of *N. muiri*, 32.5% could emerge on rice, which was significantly lower than the emergence rate on *L. sayanuka*. There was no significant difference in the female ratio of *N. muiri* on the two plant hosts. BPH could not complete a generation on *L. sayanuka*, and thus the population trend index was zero. *N. muiri* completed only one generation on rice plants and thus the population trend index was only 0.12, which indicated that *N. muiri* was unable to establish a sustained population. *N. muiri* however had high reproduction ability on *L. sayanuka*, and the population trend index was 44.40% of BPH on rice (Table 4).

Table 5. Parameter and population trend index of *N. lugens* and *N. muiri* on different host plants.

Host plants	Fecundity (per female)		Hatching rate (%)		Emergence rate (%)		Sex ratio (%)		Population trend index	
	N. lugens	*N. muiri*	*N. lugens*	*N. muiri*	*N. lugens*	*N. muiri*	*N. lugens*	*N. muiri*	*N. lugens*	*N. muiri*
Rice	476.3 ± 28.4a (20)*	3.6 ± 1.0b (20)	95.3 ± 3.5 (20)	23.3 ± 6.6b (7)	88.9 ± 1.2 (15)	32.5 ± 3.0b (5)	51.6 ± 3.9 (15)	44.0 ± 4.2 (3)	208.2	0.12
L. sayanuka	0b (20)	316.8 ± 24.9a (20)	—	91.2 ± 1.5a (20)	—	59.7 ± 3.6a (15)	—	53.6 ± 1.3 (5)	0	92.45
t	12.435	−9.488		−14.629		−4.333		−2.538		
P	<0.001	<0.001		<0.001		0.002		0.044		

DISCUSSION

As an innovation for classical biological control, the banker plant system is used to sustain a reproducing population of natural enemies in order to provide long-term pest suppression25,26,27. Even in conditions of low pest density natural enemies could maintain sufficient populations on alternative hosts from the banker plants28. In addition, this method can be more economical than other classical biological control methods that require more manpower and material resources25,29. The use of banker plants which are common and easy to cultivate as in the case of *Leersia* can thus be introduced as an ecological engineering method.

Although a few research suggests that rice crops may be the host of *N. muiri*30, most researchers have shown that it could not sustain on rice[24,31,32]. Our results were consistent with this observation as although *N. muiri* could complete a generation on rice crops, their population trend index was only 0.12. Meanwhile, BPH could not survive on *L. sayanuka*, and thus could not serve an alternate breeding source of the pest.

A. nilaparvatae populations derived from the eggs of BPH and *N. muiri* had the similar parasitzation capacities for the two brown planthoppers. *A. nilaparvatae* from *N. muiri* eggs had no selection preference between the eggs of BPH and *N. muiri*, suggesting that they do not need an adaptive phase to BPH but can directly play an efficient control role. Meanwhile, *A. nilaparvatae* from paddy fields preferred to parasitize the BPH eggs, indicating that *A. nilaparvatae* rarely moves back to *L. sayanuka*, if there are enough BPH pests in rice fields. *A. nilaparvatae* derived from *N. muiri* had similar instantaneous attack rates and handling times for the two planthoppers' eggs. The instantaneous attack rates and handling times for BPH were higher suggesting higher efficiency for the control of BPH.

In the paddy fields of China, *N. muiri* populations are generally high. Luo *et al.*[23] reported that *N. muiri* accounted for over 40% of the total number of *Nilaparvata* insects captured in 2008 and 2009 by using light traps in four provinces and five locations in China. Moreover, the number of *N. muiri* sampled from Jiangxi, Hu'nan, and Zhejiang provinces before September were higher than BPH. Our unpublished data of field experiment also showed that the larger *A. nilaparvatae* populations were often found on *L. sayanuka* (unpublished data). *L. sayanuka* was introducd into the paddy field ecosystem as a banker plant, which can serve as the reservoir of the parasitic wasps for *A. nilaparvatae* for the management of rice planthoppers seems feasible. As *L. sayanuka* is the perennial graminaceuous plant and it can also serve as an overwintering host plant for parasitoids as well as other predators. Thus sufficiently high populations of *A. nilaparvatae* may be available after winter before rice planthopper immigration occurs. The field experiment showed that a 10-cm-wide *L. sayanuka* strips planted at the edges of paddy fields might be sufficient in suppressing rice planthopper population to below the threshold.

Previous studies have indicated that the spatial pattern of non-crop habitat can affect the composition, structure, diversity and dynamic of natural enemies[33,34]. Different species of natural enemies showed different responses to non-crop habitats[35,36], and the number and variety of natural enemies might not be able to improve the biological control ability[37,38]. Caballero-López *et al.*[39] found that the contiguous of crop habitat and habitat diversity have no obvious effect on the number of aphids, parasitoids and beetles, however, the crop area size reduction can enhance carabids. Our study indicated that there was no significant difference between the 50-cm-wide strip and 10-cm-wide strip of *L. sayanuka*.

Environmental factors are also important in biological control using a banker plant. For instance, in the "Oat - *Myzus persicae* - *Aphidius colemani*" banker plant system, efficient biological control was achieved in a short period of the winter crops (*Eruca sativa*). However, it was not effective in the longer period of the summer crops (*Capsicum annuum* L.)40. In addition, the LNA banker plant system might also be challenged with different rice cultivations, rice varieties and rice field locations. Although the parasitoid is extremely susceptible to early season insecticide applications, it is vitally important when banker plants are used and early season insecticide applications should be avoided.

One risk in using the LNA banker plant system might arise from the creeping stems of *Leersia* that might affect rice as a weed. However, *L. sayanuka* grows more slowly than *Leersia hexandra* Swartz which is an important weed in rice ecosystem. In addition, some rice diseases might affect *L. sayanuka* and might serve as a disease reservoir for instance bacterial blight of rice (*Xanthomonas oryzae* pv. Oryzae)41. *Leersia* is also a host for the rice leaf folder, *Cnaphalocrocis medinalis*42,43 and *Hydrellia philippina* Ferino44. Hence, it is necessary to further evaluation the potentials risk of this LNA banker plant system. Spiders and other predators always preyed on *N. muiri*. Our suggested system might notably influence the paddy field ecosystem and nutritional relationships among arthropods. Although *L. sayanuka* may be the host of *C. medinalis* and *Susumia exigua*43, the eggs and larvae of these two rice leaf folders in *L. sayanuka* may be the host of natural enemies for the rice lepidopteran pests, and improve the abundance of egg and larva parasitoids, including *Trichogramma japonicun, Trichogramma chilonis*, and *Apanteles cypris*. Thus, *L. sayanuka* strips planted at the edge of paddy fields also benefits to serve as a natural enemy bank of rice lepidopteran pests, such as rice leaf folder.

METHODS

Plants

A BPH susceptible variety of rice (TN1) was originally obtained from the International Rice Research Institute (IRRI), Philippines. After indoor germination, seedlings were grown in a standard potting mix in a climate controlled room (25 ± 1 °C; 12 h: 12 h = L: D; RH = 80%). Fifteen-day-old rice seedlings were individually transplanted into plastic pots (diameter 10 cm, height 12 cm) and used in laboratory studies.

L. sayanuka plants were originally collected from a wetland field on the outskirts of Hangzhou (120°15′E, 30°11′N), and then those plants were transferred to a greenhouse. At the tiller stage, the primarily tillers were cut and transplanted into plastic pots (diameter 10 cm, height 12 cm) and used in laboratory studies.

Both TN1 and *L. Sayanuka* plants kept insects during cultivation free. After 50-days of plant transplanting, each potted rice plant was trimmed to main stem and two primarily tillers and each potted *L. sayanuka* was trimmed to main stem and four primarily tillers in order to maintain equivalent biomass in the experimental pots.

Insects

Colonies of BPH and *N. muiri* were established from specimens collected from paddy fields at the Jinhua experimental station in Zhejiang Province, China (119°65′E, 29°08′N). The BPH was maintained on 50-day-old TN1 rice plants, while the *N. muiri* was maintained on *L. sayanuka* seedlings in separate cages (90 cm × 80 cm × 80 cm) for four generations in an insectary (27 ± 0.5 °C, 70–90% RH, 12D:12L). The parasitoid colony of *A. nilaparvatae* was established from seedlings of rice and *L. sayanuka*, which contained planthopper eggs parasitized by *A. nilaparvatae. A. nilaparvatae* populations were reared with BPH eggs in TN1 rice plants or reared *N. muiri* eggs in *L. sayanuka* plants for four generations in an insectary (27 ± 0.5 °C, 70–90% RH and 12D:12L).

Population trend index of BPH and *N. muiri* on different host plants

We used an index (I) developed by ref. 32 to evaluate the population growth trends of *N. muiri* and BPH on *L. sayanuka* and rice plants, respectively. Each pot of TN1 and *L. sayanuka* was covered by netted cage and inoculated with a pair of BPH or *N. muiri* adults emerged in 4 hours, with 20 replicates for a planthopper specie in each host plant. The number of survival adults was observed daily. After egg hatching, the nymphs were counted and then removed every two days. When no more newly hatched nymphs were recorded from tested plants for continuous 4 days, the plants were cut and dissected under a dissecting microscope to record the unhatched eggs. The longevity of female and male adults, total fecundity (the sum of hatched nymphs and unhatched eggs), and egg hatching rate were calculated.

Newly hatched nymphs were individually reared on their parents' host plats for further observation. Each potted plant covered by netted cage was inoculated with 10 new hatched nymphs. The survival and emergence of nymphs were recorded daily. Treatments were replicated 15 times and host plants were replaced or supplied with new ones in every 3 days. From the first day of adult emergence, the number and gender of emerged insects were recorded daily until all emerged. Afterwards, the emergence rate and sex rate were calculated.

Parasitism by *A. nilaparvatae* to eggs of both BPH and *N. muiri* derived from different host plants

At the tiller stage, healthy potted TN1 rice plants or *L. sayanuka* plants were selected and stripped of old leaves, and then covered with an inverted transparent disposable plastic cup (diameter 7 cm, height 9 cm), which had a hole (diameter 2 cm) on the cup botton for plants growing. Five gravid female adults of BPH or *N. muiri*, respectively, were introduced into each pot and were removed after 48 hours. A pair of the newly (<4 h) adults of *A. Nilaparvatae*, emerged either from TN1 or *L. sayanuk*a was introduced into each cup. During the experiment, a 10% V/V honey water solution soaked cotton wool swab was placed in every pot as the food source. Parasitoids were removed after 24 h and the number of planthopper eggs, both parasitized and non-parasitized, were counted using a dissecting microscope after 4 days. Each treatment was replicated 30 times and arranged in a complete randomized design in an artificial climate chamber (27.0 ± 0.5 °C, 70–90% RH and 12L:12D).

Functional responses of *A. nilaparvatae* on eggs from two hosts (BPH or *N. muiri*) reared by different host plants

Seedlings used for the functional response experiments were described above. TN1 and *L. sayanuka* potted plants were covered with mylar cages and inoculated with female adults of BPH and *N. muiri*, respectively at different densities (1, 2, 4, 6, 8 and 10 adults per pot) for 48 hours to provide varying eggs densities. A pair of freshly emerged of *A. nilaparvatae* parasiotids was introduced into each cage for 24 hours. The plants in each cage were dissected after 4 days under a dissecting microscope to record the number of healthy and parasitized eggs. During the study, a 10% V/V honey water solution soaked cotton wool swab was placed in every pot as the food source. Treatments were replicated 20 times and conducted in a climate chamber set at 27.0 ± 0.5 °C, 70–90% RH, and 12D:12L.

Oviposition preference of *A. nilaparvatae*

TN1 and *L. sayanuka* seedlings were planted into the same pot and trimmed one tiller in rice and three tillers in *L. sayanuka* to maintain similar biomasses of the two plants in each experiment. One rice plant infested by BPH eggs and another plant infested by *N. muiri* eggs, respectively, were prepared to provide a range of prey egg densities. A pair of freshly emerged of *A. nilaparvatae* (1 d mated) was introduced into each pot for 24 h. During the study, a 10% V/V honey water solution soaked cotton wool swab was placed in every pot as the food source. The TN1 and *L. sayanuka* plants were dissected under a binocular microscope after 4 days to record the number of healthy and parasitized eggs. Treatments were replicated 20 times and conducted in the climate chamber set at 27.0 ± 0.5 °C, 70–90% RH, and 12D:12L.

Field evaluation on the potential of *L. sayanuka* as a banker plant

A field experiment was conducted at the Jinhua experimental station (119°65′E, 29°08′N) of Zhejiang in China, in which we examined rice yields from replicated paddy field plots (20 m in length and spaced at 20 m intervals) under the following three treatments:

A. *L. sayanuka* was planted in a 10 cm wide belt on the edge of paddy fields

B. *L. sayanuka* was planted in a 50 cm wide belt on the edge of paddy fields

C. Control – without *L. sayanuka* and other weeds on the edges of paddy fields

No pesticides were sprayed on paddy field plots, and the three treatments were replicated three times in a complete random split plot. *L. sayanuka* plants were introduced into field based on the experiment design prior to rice (Variety: Yongyou 12) transplanted. After 15 days, 100 female *N. miuri* per square meter were introduced to *L. sayanuka* seedlings, and the BPH population in rice field was assessed using the white enamel plate monthly internals. Five samples were collected respectively at 5, 10 and 15 m away from the bunds and the average population densities were compared among three treatments at the different distances, respectively.

Statistics and analysis

The functional response data for *A. nilaparvatae* were analyzed using the Michael Marquardt method and the parasitoid model of Rogers45and Royama46:

$$Np = Nt\left[1 - \exp\left(-\frac{aTPt}{1 + aThNt}\right)\right] \qquad (1)$$

Where Np = the number of plants parasitized by each female wasp per day, Nt = the host density, a = the search effect (instantaneous attack rate), T = the total search time (one day), P = the parasitoid density, and Th = the treatment time.

The oviposition selectivity index for *A. nilaparvatae* on different hosts was calculated using the equation proposed by Manly *et al.*47:

$$bi = \ln(Ri/Ai)/\sum_{j=1}^{k} \ln(Rj/Aj) \qquad (2)$$

Where b_i is the probability of parasitism on host type i when all hosts provided could be used, R_i is the number of host type i without parasitized at the end of the experiment, Ai is the total number of host type i in the experiment, k is the number of host types (k = 2 in experiment 1. 2. 3).

The population trend index (I) for the two species of rice planthoppers were calculated according to the method of Qiao *et al.*32:

$$I = F \times SE \times SN \times P♀ \qquad (3)$$

Where F is the fecundity of each female adult, SE is the hatching rate, SN is the emergence rate (nymph survival rate), and $P_♀$ is sex ratio of the females.

All statistical tests were performed using SPSS v17. The significance of difference in the related parameters of population trend index of BPH and *N. muiri* was analyzed using the t-test. The percentage was first normalized by the arcsine square root transformation before analysis. The amounts of oviposition of both BPH and *N. muiri* on rice and *L. sayanuka* plants, and the oviposition preference of *A. nilaparvatae* to the eggs on different host plant were subjected to χ^2 testing. One-way ANOVA was used to test for treatment effects of *L. sayanuka* on the population size of rice planthoppers and the Tukey post hoc test applied.

REFERENCES

1. Yuan, L. P. Development of hybrid rice to ensure food security. Rice Sci. 21, 1–2 (2014).
2. Godfray, H. C. & Garnett, T. Food security and sustainable intensification. Philos. Trans. R. Soc. B 369, 101–120 (2014).
3. Lu, Z. & Heong, K. L. Planthoppers: new threats to the sustainability of intensive rice production systems in Asia (eds Heong, K. L. & Hardy, B.) 247–256 (International Rice Research Institute, 2009).
4. Rashid, M. M., Jahan, M. & Islam, K. S. Impact of nitrogen, phosphorus and potassium on brown planthopper and tolerance of its host rice plants. Rice Sci. 23, 119–131 (2016).
5. Heong, K. L., Wong, L. & Reyes, J. H. D. Rice Planthoppers (eds Heong, K. L., Cheng, J. & Escalada, M. M.) 65–76 (Springer Netherlands, 2013).
6. Sogawa, K. Planthopper outbreaks in different paddy ecosystems in Asia: man-made hopper plagues that threatened the green revolution in rice. (ed. Sogawa, K.) 33–63 (Springer Netherlands, 2015).
7. Gurr, G. M., Wratten, S. D., Snyder, W. E. & Read, D. M. Y. Biodiversity and Insect Pests: key issues for sustainable management (eds Gurr, G. M., Wratten, S. D., & Snyder, W. E.) 214–229 (Wiley-blackwell, 2012).
8. Horgan, F. G. et al. Virulence of brown planthopper (Nilaparvata lugens) populations from South and South East Asia against resistant rice varieties. Crop Prot. 78, 222–231 (2015).
9. Cheng, J. Planthoppers: New threats to the sustainability of intensive rice production systems in Asia (eds Heong, K. L. & Hardy, B.) 157–178 (International Rice Research Institute, 2009).
10. Lou, Y. G. & Cheng, J. Basic research on the outbreak mechanism and sustainable management of rice planthoppers. Chin. J. Appl. Entomol. 48, 231–238 (2011).
11. Paoletti, M. G. Ecological engineering for pest management—advances in habitat manipulation for arthropods. Austral Ecol. 30, 613–614 (2005).
12. Settele, J., Biesmeijer, J. & Bommarco, R. Switch to ecological engineering would aid independence. Nature 456, 570 (2008).
13. Cullen, R., Warner, K. D., Jonsson, M. & Wratten, S. D. Economics and adoption of conservation biological control. Biol. Control 45, 272–280 (2008).
14. Lu, Z. et al. Rice planthoppers (eds Heong, K. L., Cheng, J. & Escalada, M. M.) 161–178 (Springer Netherlands, 2015).
15. Hossain, Z., Gurr, G. M., Wratten, S. D. & Raman, A. Habitat manipulation in lucerne Medicago sativa: arthropod population dynamics in harvested and 'refuge' crop strips. J. Appl. Ecol. 39, 445–454 (2002).
16. Wackers, F., Romeis, J. & Van Rijn, P. Nectar and pollen feeding by insect herbivores and implications for multitrophic interactions. Entomol. 52, 301–323 (2007).
17. Gurr, G. M., Zhu, Z. R. & You, M. Rice planthoppers (eds Heong, K. L., Cheng, J. & Escalada, M. M.) 143–160 (Springer Netherlands, 2015).
18. Stacey, D. L. 'Banker' plant production of Encarsia formosa Gahan and its use in the control of glasshouse whitefly on tomatoes. Plant Pathol. 26, 63–66 (1977).
19. Zhu, P. et al. Laboratory screening supports the selection of sesame (Sesamum indicum) to enhance Anagrus spp. parasitoids (Hymenoptera: Mymaridae) of rice planthoppers. Biol. Control 64, 83–89 (2013).
20. Zhu, P. et al. Selection of nectar plants for use in ecological engineering to promote biological control of rice pests by the predatory bug, Cyrtorhinus lividipennis, (Heteroptera: Miridae). Plos One 9, e108669 (2013).
21. Zhu, P. et al. Selective enhancement of parasitoids of rice Lepidoptera pests by sesame (Sesamum indicum) flowers. BioControl 60, 157–167 (2015).

22. Ding, J. A review of studies on classification and species of the genus Nilaparvata distant (Homoptera: delphacidae). J. Nanjing Agric. Univ. 78–85 (1981).
23. Luo, J. et al. Population dynamics of Nilaparvata lugens and its two sibling species under black light trap. Chin. J. Rice Sci. 24, 315–319 (2010).
24. Cui, Y. L., HE, J. C., Luo, J., Lai, F. X. & Fu, Q. Host plants of Nilaparvata muiri China and N. bakeri (Muir), two sibling species of N. lugens (Stål). Chin. J. Rice Sci. 27, 105–110 (2013).
25. Stevend, F. Biological control of arthropod pests using banker plant systems: past progress and future directions. Biol. Control 52, 8–16 (2010).
26. Enkegaard, A. The banker plant method in biological control. Crit. Rev. Plant Sci. 30, 259–278 (2011).
27. Xiao, Y., Mao, R., Shen, G. & Sosborne, L. Banker plant system: a new approach for biological control of arthropod pests. Chin. J. Biol. Control 25, 1157–1170 (2012).
28. Xiao, Y., Avery, P., Chen, J., Mckenzie, C. & Osborne, L. Ornamental pepper as banker plants for establishment of Amblyseius swirskii (Acari: Phytoseiidae) for biological control of multiple pests in greenhouse vegetable production. Biol. Control 63, 279–286 (2012).
29. Xiao, Y. et al. Evaluation of corn plant as potential banker plant for supporting predatory gall midge, Feltiella acarisuga (Diptera: Cecidomyiidae) against Tetranychus urticae (Acari: Tetranychidae) in greenhouse vegetable production. Crop Prot. 30, 1635–1642 (2011).
30. Dupo, A. & Barrion, A. Planthoppers: new threats to the sustainability of intensive rice production systems in Asia (eds Heong, K. L. & Hardy, B.) 3–155 (International Rice Research Institute, 2009).
31. Claridge, M. F., Hollander, J. D. & Morgan, J. C. The status of weed-associated populations of the brown planthopper, Nilaparvata lugens (Stål) - host race or biological species? Zool. J. Linn. Soc. 84, 77–90 (1985).
32. Qiao, H. et al. Fitness of the small brown planthopper (Laodelphax striatellus) on different plants. Chin. J. Rice Sci. 23, 71–78 (2009).
33. Costamagna, A. C., Landis, D. A. & Difonzo, C. D. Suppression of soybean aphid by generailist predators results in a trophic cascade in soybeans. Ecol. Appl. 17, 441–451 (2007).
34. Gardiner, M. M. et al. Landscape diversity enhances biological control of an introduced crop pest in the north-central USA. Ecol. Appl. 19, 143–154 (2009).
35. Gardiner, M. M. et al. Landscape composition influences the activity density of Carabidae and Arachnida in soybean fields. Biol. Control 55, 11–19 (2010).
36. Woodcock, B. A. et al. Impact of habitat type and landscape structure on biomass, species richness and functional diversity of ground beetles. Agric. Ecosyst. Environ. 139, 181–186 (2010).
37. Östman, Ö., Ekbom, B. & Bengtsson, J. Landscape heterogeneity and farming practice influence biological control. Basic Appl. Ecol. 2, 365–371 (2001).
38. Snyder, W. E., Snyder, G. B., Finke, D. L. & Straub, C. S. Predator biodiversity strengthens herbivore suppression. Ecol. Lett. 9, 789–796 (2006).
39. Caballero-López, B. et al. Aphids and their natural enemies are differently affected by habitat features at local and landscape scales. Biol. Control 63, 222–229 (2012).
40. Andorno, A. V. & López, S. N. Biological control of Myzus persicae (Hemiptera: Aphididae) through banker plant system in protected crops. Biol. Control 78, 9–14 (2014).
41. Noda, T. & Yamamoto, T. Reaction of Leersia grasses to Xanthomonas oryzae pv. oryzae collected from Japan and Asian countries. J. Gen. Plant Pathol. 74, 395–401 (2008).
42. Pan, X. & Wang, Y. Studies on the overwinter of rice leafroller (Cnaphalocrocis medinalis Guenée) in Sichuan basin. J. Southwest Agricul. Univ. 1, 73–76 (1986).
43. Khan, Z. R., Abenes, M. L. P. & Fernandez, N. J. Suitability of graminaceous weed species as host plants for rice leaffolders, Cnaphalocrocis medinalis and Marasmia patnalis. Crop Prot. 15, 121–127 (1996).

44. Wei, Z. Preliminary study on the rice pest Hydrellia philippina Ferino. Entomol. Knowl. 17, 49–53 (1980).
45. Rogers, D. Random search and insect population models. J. Anim. Ecol. 41, 369–383 (1972).
46. Royama, T. A comparative study of models for predation and parasitism. Pop. Ecol. 15, 121–121 (1973).
47. Manly, B. F. J. & Cook, L. M. Analysis of a selective predation experiment. Am. Nat. 106, 719–736 (1972).

CHAPTER 8

Insect Pathogenic Bacteria in Integrated Pest Management

Luca Ruiu

Dipartimento di Agraria, Sezione di Patologia Vegetale ed Entomologia, University of Sassari, Via E. de Nicola, 07100 Sassari, Italy

ABSTRACT

The scientific community working in the field of insect pathology is experiencing an increasing academic and industrial interest in the discovery and development of new bioinsecticides as environmentally friendly pest control tools to be integrated, in combination or rotation, with chemicals in pest management programs. In this scientific context, market data report a significant growth of the biopesticide segment. Acquisition of new technologies by multinational Ag-tech companies is the center of the present industrial environment. This trend is in line with the requirements of new regulations on Integrated Pest Management. After a few decades of research on microbial pest management dominated by *Bacillus thuringiensis* (*Bt*), novel bacterial species with innovative modes of action are being discovered and developed into new products. Significant cases include the entomopathogenic nematode symbionts *Photorhabdus* spp. and *Xenorhabdus* spp., *Serratia* species, *Yersinia entomophaga*, *Pseudomonas entomophila*, and the recently discovered Betaproteobacteria species *Burkholderia* spp. and *Chromobacterium* spp. Lastly, Actinobacteria species like *Streptomyces* spp.

and *Saccharopolyspora* spp. have gained high commercial interest for the production of a variety of metabolites acting as potent insecticides. With the aim to give a timely picture of the cutting-edge advancements in this renewed research field, different representative cases are reported and discussed.

Keywords: IPM; biopesticide; biological control; entomopathogen; *Bacillus thuringiensis*; *Photorhabdus*; *Xenorhabdus*; *Serratia*; *Streptomyces*; Spinosad

1. INTRODUCTION

Bacteria are widespread in the environment and they have evolved a variety of interactions with insects including essential symbiosis [1]. While many bacterial species inhabit bodies of insects establishing different levels of mutualistic relationships, only a limited number of them behave as insect pathogens. The latter have evolved a multiplicity of strategies to invade the host, to overcome its immune responses, to infect and to kill it. The mechanisms leading to these kinds of interactions are presumed to have ancient origin and to have developed throughout a long co-evolution process [2]. In line with this concept, a variety of insecticidal toxins produced by certain spore forming entomopathogenic bacteria, have a similar structure and mode of action. This is the case for protein toxins produced by *Bacillus thuringiensis* Berliner (*Bt*) and localized in parasporal bodies [3]. These toxins are normally very specific to a limited range of targets, while in other cases bacteria produce metabolites that show a broader insecticidal spectrum.

Important information to understand the molecular mechanisms involved in diverse pathogen-host interactions are being produced as a result of modern "omic" studies. However, many aspects are still unrevealed and after few decades of microbial pest management dominated by *B. thuringiensis*, novel bacterial species with innovative modes of action have been discovered and formulated as new biopesticidal products [4].

The entomopathogenic bacteria domain has traditionally been well represented by members of the *Bacillaceae* family, such as *B. thuringiensis*, *Lysinibacillus sphaericus* (Meyer & Neide) Ahmed *et al.*, *Paenibacillus* spp. and *Brevibacillus laterosporus* Laubach. The entomopathogens belonging to the Gammaproteobacteria class, which includes the entomopathogenic nematode symbionts *Photorhabdus* spp.

and *Xenorhabdus* spp., *Serratia* species, *Yersinia entomophaga* Hurst and *Pseudomonas entomophila* (Mulet *et al.*), are also important.

More recent is the discovery of Betaproteobacteria species that show broad-spectrum insecticidal properties. This group includes specific strains of *Burkholderia* spp. and *Chromobacterium* spp. Lastly, certain Actinobacteria species have gained high scientific and commercial interest in relation to the production of a variety of metabolites acting as potent insecticides. This is the case for *Streptomyces* and *Saccharopolyspora* species.

As a result of continuous industrial and academic screening activities, the discovery of new bacterial species and insecticidal metabolites is expected in the near future [5,6]. This trend is also the result of modern legislative frameworks fostering the use of bioinsecticides in Integrated Pest Management (IPM) programs.

The purpose of the present review is to give an overall and concise picture of the knowledge advancements of insect pathogenic bacteria and of their use as bio-insecticidal products for IPM.

2. INSECT PATHOGENIC BACTERIA

2.1. Bacillaceae

2.1.1. *Bacillus thuringiensis*

Bacillus thuringiensis (*Bt*) is the most studied entomopathogenic species and some of its crystal producing strains have certainly represented the main active substances used for the microbial pest management during the last decades [7]. The pathogenic action of this bacterium normally occurs after ingestion of spores and crystalline inclusions containing insecticidal δ-endotoxins that specifically interact with receptors in the insect midgut epithelial cells [8]. It is largely demonstrated that these toxins, mostly represented by Cry proteins, after being solubilized and activated in the insect midgut, act through a pore-forming mechanism determining the disruption of natural cell membrane permeability, with consequent cell lysis followed by gut paralysis and death [9].

Most commercially available formulations are based on spore-crystal mixtures with effectiveness against different pest species. *B. thuringiensis* subsp. *kurstaki* (*Btk*) is generally used against young Lepidopteran larvae and includes different strains with significant commercial interest like HD-1, SA-11, SA-12, PB 54, ABTS-351 and EG2348. Strains of *B. thuringiensis* subsp. *aizawai* (*Bta*) (*i.e.*, ABTS-1857) are also used against armyworms and diamondback moth larvae. Besides, strains belonging to the subsp. *israelensis* (*Bti*) and *tenebrionis* (*Btt*) have been employed for the management of mosquitoes and simulids, and against coleoptera, respectively [10]. Continuous research activities have led to the isolation of many strains and to the discovery and characterization of various *Bt* insecticidal toxins produced in different bacterial stages (*i.e.*, Cyt, VIP) [3]. The insecticidal potential of bacterial strains closely related to *B. thuringiensis* has also been demonstrated [11].

Additionally, the integration of *cry* genes into genetically modified plants has been successfully implemented to confer resistance to specific crop pests.

2.1.2. *Lysinibacillus sphaericus*

Entomopathogenic strains belonging to the *L. sphaericus* (formerly *Bacillus sphaericus*) species group are featured by the production of spherical endospores closely associated with parasporal crystals containing an equimolar ratio of binary protein toxins (BinA and BinB) [12]. The insecticidal mode of action includes damages to the microvillar epithelial cells in the midgut comparable to the ones known for *B. thuringiensis* [13]. In addition, vegetative cells of certain strains produce mosquitocidal toxins (Mtx proteins).

The main targets of commercial formulations based on *L. sphaericus* strains are mosquitoes, blackflies and non-biting midges.

2.1.3. *Paenibacillus* spp.

The genus *Paenibacillus* includes different species showing pathogenicity against insects like the causative agent of the honeybee disease American Foulbrood (AFB), *P. larvae* subsp. *larvae* [14].

On the other hand, the spore-formers *P. popilliae* (Dutky) Pettersson *et al.* and *P. lentimorbus* (Dutky) Pettersson *et al.* are the causal agents of milky disease in phytophagous coleopteran larvae. The production of parasporal inclusions within the sporangial cells has been observed in *P. popilliae*, even if they are not directly responsible for the insecticidal action. However, homology between a 80 kDa parasporal protein this species produces and *Bt* Cry toxins has been demonstrated [15]. After spores are ingested by the host, they germinate in the midgut. The following pathogenicity seems to be in relation to the septicemia caused by vegetative cells.

2.1.4. *Brevibacillus laterosporus*

Brevibacillus (former *Bacillus*) *laterosporus* is a pathogen of invertebrates and a broad spectrum antimicrobial species [16]. During sporulation it produces a typical canoe-shaped parasporal body (CSPB) firmly associated with the spore coat, which gives this species a unique morphological feature. The insecticidal action of different *B. laterosporus* strains has been reported against insects in different orders, including Coleoptera, Lepidoptera and Diptera, and against mollusks, nematodes, phytopathogenic bacteria and fungi. In relation to its antifungal and antibacterial properties, due to the production of antibiotics, it has also found use in medicine.

The whole genome of *B. laterosporus* has recently been published [17,18], which reveals the potential to produce different toxins.

Certain strains showing toxicity against the corn rootworms (*Diabrotrica* spp.) and other coleopteran larvae, produce insecticidal secreted proteins (ISPs) that act as binary toxins in the insect midgut and have high homology with *B. thuringiensis* vegetative insecticidal proteins (VIPs) [19].

Specific strains toxic to mosquitoes produce parasporal inclusion bodies reminiscent of those produced by *B. thuringiensis*. These bodies contain proteins and their implication in the mosquitocidal action has been reported [20]. Spores of a strain lacking parasporal crystals are highly toxic to the house fly *Musca domestica* L., and the mode of action implies histopathological changes in the midgut with disruption of the microvillar epithelium [21,22].

2.2. Clostridiaceae

Clostridium bifermentans

A strain of *C. bifermentans* (Weinberg and Séguin) Bergey *et al.* serovar *malaysia* (C.b.m.), isolated in Malaysia, shows high toxicity against mosquitoes and black flies. During sporulation, this bacterium produces three major proteins involved in the insecticidal action [23]. These include the mosquitocidal protein Cbm71 showing homology to *B. thuringiensis* delta endotoxins [24].

2.3. Gammaproteobacteria

2.3.1. *Photorhabdus* spp. and *Xenorhabdus* spp.

The entomopathogenic members of the genera *Photorhabdus* and *Xenorhabdus* are represented by endosymbionts of insecticidal nematodes. The first are typically associated with entomopathogenic nematodes in the genus *Heterorhabditis*, while the second to *Steinernema* species. The pathogenic action usually involves the release of symbiotic bacteria in the insect hemocoel once the nematodes have actively entered the insect body. Here the bacteria proliferate producing various antimicrobial compounds to contrast the growth of other microorganisms. They also release different enzymes that contribute to the degradation processes in the hemocoel, thus creating an ideal environment for the development of the nematode population.

A variety of bacterial virulence factors are involved in the interaction with the susceptible host.

Different *Photorhabdus* and *Xenorhabdus* species producing an insecticidal toxins complex (Tc) have high potential for pest management [25]. Generally, the Tcs are high-molecular weight and multi-subunit proteins that include three components, A, B and C, orally active against different insects [26]. While the mode of action is not completely understood, all these components are normally needed to achieve full toxicity [27].

Another example of insecticidal proteins produced by these bacterial species is represented by the *Photorhabdus* insect related (Pir) proteins, produced by *P. luminescens* (Thomas and Poinar), that show similarity to *B. thuringiensis* delta-endotoxins and have been proposed to be mimics of the juvenile hormone esterases (JHEs) interfering with insect development regulation [28].

In addition to insecticidal toxins and various metabolites, these bacterial endosymbionts have evolved different mechanisms to face the insect immune response. For instance, it has been shown that certain *Photorhabdus* species exploit lipopolysaccharide modifications to resist the action of insect antimicrobial peptides (AMPs) [29], while *X. nematophila* (Poinar and Thomas) interferes with the expression mechanisms of host AMPs [30].

In addition to *P. luminescens* and *X. nematophila*, most studied species include *P. asymbiotica* Fischer-Le Saux *et al.*, *P.temperata* Fischer-Le Saux *et al.*, *X. beddingii* (Akhurst) Akhurst and Boemare, *X. bovienii* (Akhurst) Akhurst and Boemare, *X. japonica* Nishimura, and *X.*

poinarii (Akhurst) Akhurst and Boemare. Besides, new species are continuously being identified and characterized [31].

So far, the commercial use of these bacterial endosymbionts is related to the employment of the nematode species with which they are associated.

2.3.2. ***Serratia*** **spp.**

The association of *Serratia* spp. with insects or with entomopathogenic nematodes is well documented [32,33,34]. Different species in this genus produce a variety of virulence factors [35]. Common is the production of toxin complexes analogous to those produced by *Xenorhabdus* spp. and *Photorhabdus* spp. *S. entomophila* Grimont *et al.*, a pathogen of the grass grub, *Costelytra zealandica* (White) (Coleoptera: Scarabaeidae) [36], produces Sep proteins (SepA, SepB, SepC), a group of insecticidal toxins showing similarities to the insecticidal toxins of *P. luminescens* [37].

On the other hand, the recent genome sequencing of *S. nematodiphila* Zhang *et al.* also highlighted other pathogenic factors of *Serratia* species. Among these was a variety of secreted extracellular enzymes such as proteases, lipases, and chitinases [38].

It has recently been demonstrated that the pathogenicity of *S. marcescens* Bizio is increased by the action of a serralysin metalloprotease it secretes and that this bacterium is able to suppress cellular immunity by decreasing the adhesive properties of immunosurveillance cells of the insect host [39].

2.3.3. *Yersinia entomophaga*

Isolated from the New Zealand grass grub, *C. zealandica*, *Y. entomophaga* is a non-spore-forming entomopathogenic bacterium characterized by the production of an insecticidal toxin complex (Yen-Tc) showing similarity to those produced by *Photorabdus* spp. [40]. These complexes include three Yen protein families, A, B and C, and two chitinases (Chi1 and Chi2) [41].

The broad insecticidal range of these toxins and their post-ingestion histopathological action in the insect midgut epithelium have been reported [42].

Promising are the studies conducted in field conditions with insecticidal formulations containing *Y. entomophaga* against the pasture pest porina (*Wiseana* spp. larvae) [43].

2.3.4. *Pseudomonas entomophila*

P. entomophila is a ubiquitous bacterium that orally infects larvae of insects in different orders determining extensive gut cell damages. Host-pathogen interactions have been studied in experiments with *Drosophila melanogaster* Meigen (Diptera: Drosophilidae), which highlighted a specific post-ingestion immune response [44]. Recent complete genome sequencing of *P. entomophila* revealed a specific secretion system and the associated toxins probably responsible for its entompathogenic properties [45].

2.4. Betaproteobacteria

2.4.1. *Burkholderia* spp.

Different insect species harbor symbiotic bacteria of the genus *Burkholderia*, mostly in association withspecific gut regions [46,47]. In addition to these mutualistic relationships with insects, *Burkholderia* sp. has recently been reported to affect ovipositon and fecundity of the bean bug *Riptortus pedestris* (Fabricius) (Hemiptera: Alydidae) [48]. The pontential of *Burkholderia* species as biocontrol agents against different plant pathogens has also been reported [49,50]. More recently, the insecticidal properties of a new strain isolated in soil from Japan and identified as *B. rinojensis* sp. nov., were discovered [51]. Whole cell broth cultures of this bacterial strain, named A396, show oral toxicity and contact effects against the beet armyworm *Spodoptera exigua* Hübner (Lepidoptera: Noctuidae) and the two-spotted spider mite *Tetranychus urticae* Koch (Acari: Tetranychidae). Insecticidal and miticidal properties are maintained after heat-treatment, hence commercial formulations against a variety of chewing and sucking insects and mites are based on heat-killed cells and spent fermentation media. Different bacterial metabolites might be involved in the insecticidal action. Interestingly, a recent study highlighted the biochemical properties of natural compounds produced by fermentation of a *Burkholderia* sp. strain showing high homology to *B. rinojensis* [52].

2.4.2. *Chromobacterium* spp.

A strain of *Chromobacterium subtsugae* Martin *et al.*, isolated from a soil sample in Maryland (USA) and named PRAA4-1T, was discovered to show high insecticidal activity against insect species in different orders, including the Diamondback moth *Plutella xylostella* L. (Lepidoptera: Plutellidae), the Sweet potato whitefly *Bemisia tabaci* Gennadium

(Rhynchota: Aleyrodidae), the Southern green stink bug *Nezara viridula* L. (Rhynchota: Pentatomidae), the Southern corn rootworm *Diabrotica undecimpunctata* Mannerheim (Coleoptera:Chrysomelidae spectrum), the Western corn rootworm *Diabrotica virgifera* Le Conte (Coleoptera: Chrysomelidae), the Colorado potato beetle *Leptinotarsa decemlineata* Say (Coleoptera: Chrysomelidae), and the Small hive beetle *Aethina tumida* Murray (Coleoptera: Nitidulidae) [53,54]. The broad spectrum activity of this strain is related to multiple modes of action probably involving different chemical compounds produced by the bacterium. Among the bacterial metabolites, *C. subtsugae* synthesizes the tryptophan derivative violacein, which confers a typical violet color to its colonies. In addition to this, various molecules produced by this species have been characterized and associated with the insecticidal action [55]. Bioactive compounds were reported to be associated to the stationary growth phase [56], and the heat-stability of insecticidal toxins was also demonstrated [57]. The active ingredient of available commercial formulations is represented by *C. subtsugae* strain PRAA4-1T and spent fermentation media.

More recently, a new strain identified as *Chromobacterium* (Csp_P), isolated from the midgut of *Aedes aegypti* L. (Diptera: Culicidae) was shown to be able to colonize the insect midgut and to display entomopathogenic and anti-pathogen properties [58].

2.5. Actinobacteria

2.5.1. *Streptomyces* spp.

Different *Streptomyces* spp. are associated with herbivorous insects that take advantage of their cellulolytic properties [59]. Other species and strains in this genus produce a variety of metabolites acting as potent toxins against either phytopathogenic microbials or insect pests [60]. Among the first discovered insecticidal substances produced by *Streptomyces* species are flavensomycin [61], antimycin A [62], piericidins [63], macrotetralides [64] and prasinons [65]. Later on, the insecticidal and anthelmintic activities of avermectins produced by the soil Actinomycete *Streptomyces avermitilis* MS & Dwas, were discovered [66]. These macrocyclic lactone derivatives target the gamma-aminobutyric acid (GABA) receptor in the insect peripheral nervous system. The enhancement of GABA binding generates a cascade of events resulting in the inhibition of neurotransmission and paralysis of the neuromuscular systems [67].

Insecticides based on avermectins include a mixture of avermectin B1a and avermectin B1b, known as abamectin, that act by contact and ingestion and have a limited plant translaminar activity. Analogous substances produced by *Streptomyces* species include Emamectin, especially toxic to Lepidoptera, and Milbemectin, isolated from *S. hygroscopicus* Jensen. A variety of other secondary metabolites produced by diverse *Streptomyces* species, have been isolated and characterized so far. Commercialized products have been very successful against ectoparasites and endoparasites with medical and veterinary importance.

2.5.2. *Saccharopolyspora spinosa*

Saccharopolyspora spinosa Mertz and Yao was discovered during a screening program where the insecticidal activity of the isolate A83543 emerged [68]. Subsequent assays highlighted the broad toxicity of specific compounds isolated from the fermentation broth that were given the generic name of spynosins. The major component is spinosyn A, whose structure comprises a tetracyclic polyketide aglycone to which a neutral saccharide substituent is attached [69]. A variety of spinosyn analogs have been isolated and many studies have been conducted to investigate the pathway of spinosyn biosynthesis, which led to the characterization of specific gene clusters [70].

Since the first experimentations, spinosyns exhibited broad-spectrum activity against insect species in different orders, especially Lepidoptera and Diptera [71]. The natural *S. spinosa* fermentation-derived mixtures were named "spinosad" and contain spinosyn A and spinosyn D, as major and minor component, respectively. The biological activity of numerous semisynthetic derivatives has been studied [72].

The insecticidal mode of action of spinosyns is not completely understood, but is considered unique in comparison with other insecticides. It has been demonstrated their interaction with g-aminobutyric acid receptors and nicotinic acetylcholine receptors, eventually leading to the disruption of neuronal activity and consequent insect paralysis and death [73]. Despite their broad spectrum of activity against insects, spinosyns are associated with a low risk toward non-target species, including mammals and various aquatic organisms, in comparison with other insecticides [74].

The continuous research and industrial interest in this field has recently led to the discovery of a variety of spinosyn-related compounds produced by another *Saccharopolyspora* species, *S. pogona* [75].

3. USE OF ENTOMOPATHOGENIC BACTERIA IN INTEGRATED PEST MANAGEMENT

Insect pathogenic bacteria and their derived products represent the active substances of various "biopesticides". There is a range of definitions for this term, but it essentially includes mass reared living organisms (natural insect predators and parasitoids), nematodes and micro-organisms (bacteria, fungi, microsporidia, virus), natural compounds from plant extracts, and semiochemicals (e.g., insect pheromones). Besides, in countries where their use is permitted, biopesticides comprise genetically modified plants that express genes conferring resistance against pests or diseases (plant incorporated products).

Biological control agents, like bacterial entomopathogens, are generally recognized as lower risk substances than conventional chemical pesticides, and various benefits are associated with their use. For instance, their mode of action is normally more complex than conventional chemicals, targeting a diversity of action sites, which makes the development of resistant pests less likely. Although entomopathogenic bacteria can be used as stand-alone products for pest management in organic farming, their use in rotation or combination with chemicals is strongly encouraged to achieve full efficacy and eco-sustainability. Many studies have highlighted compatibility and synergistic effects of entomopathogenic bacteria and chemical substances [76,77,78]. Among the other advantages of including biopesticides in pest management programs are their safety for workers, the reduction of residues on crop and the flexibility on harvests, due to minimal or no pre-harvest interval.

The effectiveness of entomopathogenic bacteria is often associated with a proper application in the field. For instance, in the case of products that act by ingestion (*i.e.*, *Bt* based products), timing is critical to ensure that the bacterial toxins remain stable in the environment until they are ingested by their target insect stage [7]. Another aspect is to ensure a proper coverage of substrates (e.g., foliage) frequented or eaten by insects. This has led to the development of special processing and formulation of bacteria-based bioinsecticides, with the aim of maximizing shelf-life, improving dispersion and adhesion, reducing spray drift and above all enhancing efficacy. A variety of adjuvants and additives for microbial formulations have been developed by the industry. These include dispersants, surfactants, wetters, spreaders, drift control agents, pH buffers, antifoam agents, carriers, phagostimulants and attractants [79]. Depending on the application target and on the adverse environment conditions, a choice of solid and liquid formulations is available. The first include dusts, granules, briquettes and wettable powders (WPs), while liquid suspensions may consist of suspension concentrates (SCs), emulsions and encapsulations. Advanced technologies aiming at increasing residual effects comprise microencapsulations and microgranules.

The variety of entomopathogenic bacteria based active substances authorized for commercialization and of their relative commercial products is significantly increasing worldwide. Consequently, their market is expanding into new segments, thus broadening their use for pest management, traditionally relegated to niche contexts (forest, public health, protected crops) [6].

Besides optimizing efficacy, modern pest management strategies tend to minimize the impact on the environment and on non-target organisms. This is also in line with the need to meet current regulations on the maximum residue levels (MRLs) for synthetic pesticides. The management of pest counter-adaptation (resistance) to pesticidal products is another concern in the agro-ecosystem. The integration of bio-based pesticides in pest management programs in many cases represents an important resource to face these challenges. During the last years, the number of microbial products available has grown significantly and many efforts are dedicated to increasing the awareness and to fostering the adoption of biopesticides in integrated pest management programs. This includes the implementation of worldwide industrial understandings like the Biopesticide Industry Alliance (BPIA) and the International Biocontrol Manufacturers' Association (IBMA). According to this trend, a database of available biopesticides is maintained within the context of the biopesticide program of the Interregional Research Project Number 4 (IR-4) at Rutgers University (U.S.).

Many biopesticides have recently gained interest within the legislative framework of most important regions like USA and Europe, fostering the use of low risk active substances in agriculture. In the United States, for the pre-market approval (registration) of any pesticide, the Federal Insecticide, Fungicide, and Rodenticide Act (FIFRA) requires a specific evaluation by the Environment Protection Agency (EPA). Since 1994, to facilitate the registration of biopesticides a Biopesticides and Pollution Prevention Division was established in the Office of Pesticide Programs. In Europe, to facilitate procedures for registration of new products, previously based on EU Directive 91/414, new criteria have recently been implemented by Regulation (EC) No 1107/2009. Besides, specific data requirements and the principles for evaluation and authorization of new plant protection products (PPPs) have been established by recent Regulation (EU) No 283 and 284/2013, and Regulation (EU) No 546/2011, respectively. On this basis, any new active substance is first authorized at the EU level, while the formulated product is subjected to approval at Member State level.

4. MARKET OVERVIEW

Based on a recent report published by BCC Research LLC (CHM029E, June 2014), the global market for pesticides, including both the synthetic pesticides and the biopesticides segments, is estimated to grow at a compound annual growth rate (CAGR) of 6.3% and to reach more than $80 billion by 2019.

The fastest growing segment of this market is represented by biopesticides (around 4%) that are expected to increase at more than twice the rate of synthetic pesticides. This segment comprises both biochemical (*i.e.*, semiochemicals, plant extracts) and microbial biopesticides. Among the latter are included products based on bacteria, fungi, virus and nematodes, targeting different kinds of plant pests and parasites. A different category is represented by biocides, including products for public health like microbials targeting mosquitoes and flies, whose commercialization is differently regulated. This market is also expected to experience substantial growth in future (Research and Markets, ID: 2692355, October 2013).

The increased industrial interest in biopesticides is shaping a changing global scenario featured by strategic partnerships, mergers and acquisitions of companies. Significant examples of this trend are the recent acquisitions of the US-based biological company AgraQuest and of the German company Prophyta GmbH by Bayer CropScience. Similarly, BASF acquired Becker Underwood Inc., while Syngenta agreed to acquire the US-based biotechnology company Pasteuria Bioscience Inc. and the Swiss Devgen. On the other side, Monsanto and Novozymes have established an historical BioAg Alliance committed to the discovery and development of new microbial solutions for agriculture.

5. CONCLUSIONS

The need to feed a growing human population requires continuous advancements in pest management systems limiting production losses in major crops [80]. On the other side, the land available for cultivation and farming on Earth is limited, which requires the development of new technologies supporting improvements in productivity. Safeguarding the environment and human health, and the need to manage the development of insect resistance to pesticides, are additional concerns. For all these reasons, the integration of bio-based insecticides in combination or rotation with synthetic formulations is strongly recommended. This is in line with the expectations of fruit and vegetable consumers and with the requirements of a modern legislative framework on the use of IPM in agriculture.

Following this trend, an increasing academic and industrial interest in the discovery and development of new bioinsecticides is being experienced by the scientific community working in the field of insect pathology. The relative market segments and the industrial interest in this field are also growing at a significant rate. As a result, researches on entomopathogenic bacteria are gaining momentum and new discoveries are expected in the near future. Due to the regulatory issues related to the pre-market authorization of new active substances, the availability of bio-based products, including entomopathogenic bacteria, is still limited to certain crop pests. However, both research funding bodies and industry are progressing in the direction of increasing investments in this field, which will result in a continuous expansion of the repertoire of insect pathogenic bacteria-based formulations available for integrated pest management.

ACKNOWLEDGMENTS

The author is sincerely thankful to Sardegna Ricerche (Project IndBase, Aiuti per Progetti di Ricerca e Sviluppo 2012, Bioecopest Srl, POR FESR SARDEGNA 2007–2013) and to Fondazione Banco di Sardegna (Project 2014: Studi per il controllo biologico di parassiti nocivi di primaria importanza per l'agricoltura in Sardegna) for financial support.

CONFLICTS OF INTEREST

The author declares no conflict of interest.

REFERENCES

1. Feldhaar, H. Bacterial symbionts as mediators of ecologically important traits of insect hosts. Ecol. Entomol. 2011, 36, 533–543.
2. Vilcinskas, A. Coevolution between pathogen-derived proteinases and proteinase inhibitors of host insects. Virulence2010, 1, 206–214.
3. De Maagd, R.A.; Bravo, A.; Berry, C.; Crickmore, N.; Schnepf, H.E. Structure, diversity, and evolution of protein toxins from spore forming entomopathogenic bacteria. Annu. Rev. Genet. 2003, 37, 409–433.
4. Ruiu, L.; Satta, A.; Floris, I. Emerging entomopathogenic bacteria for insect pest management. Bull. Insectol. 2013, 66, 181–186.
5. Lacey, L.A.; Frutos, R.; Kaya, H.K.; Vail, P. Insect pathogens as biological control agents: Do they have a future? Biol. Control 2001, 21, 230–248.
6. Glare, T.; Caradus, J.; Gelernter, W.; Jackson, T.; Keyhani, N.; Kohl, J.; Marrone, P.; Morin, L.; Stewart, A. Have biopesticides come of age? Trends Biotechnol. 2012, 30, 250–258.
7. Vega, F.E.; Kaya, H.K. Insect Pathology, 2nd ed.; Elsevier: London, UK, 2012; p. 504.
8. Pigott, C.R.; Ellar, D.J. Role of receptors in Bacillus thuringiensis crystal toxin activity. Microbiol. Mol. Biol. Rev. 2007, 71, 255–281.
9. Bravo, A.; Gill, S.S.; Soberon, M. Mode of action of Bacillus thuringiensis Cry and Cyt toxins and their potential for insect control. Toxicon 2007, 49, 423–435.
10. Glare, T.R.; O'Callaghan, M. Bacillus thuringiensis: Biology, Ecology and Safety; Wiley: Chichester, UK, 2000.
11. Ruiu, L.; Falchi, G.; Floris, I.; Marche, M.G.; Mura, M.E.; Satta, A. Pathogenicity and characterization of a novel Bacillus cereus sensu lato isolate toxic to the Mediterranean fruit fly Ceratitis capitata Wied. J. Invertebr. Pathol. 2015.
12. Baumann, P.; Clark, M.A.; Baumann, L.; Broadwell, A.H. Bacillus sphaericus as a mosquito pathogen: Properties of the organism and its toxins. Microbiol. Rev. 1991, 55, 425–436.
13. Charles, J.F.; Silva-Filha, M.H.; Nielsen-LeRoux, C. Mode of action of Bacillus sphaericus on mosquito larvae: Incidence on resistance. In Entomopathogenic Bacteria: From Laboratory to Field Application; Charles, J.F., Delecluse, A., Nielsen-LeRoux, C., Eds.; Kluwer Academic Publishers: London, UK, 2000; pp. 237–252.
14. Davidson, E.A. Ultrastructures of American foulbrood disease pathogenesis in larvae of the worker honey bee Apis mellifera. J. Invertebr. Pathol. 1973, 21, 53–61.
15. Zhang, J.; Hodgman, T.C.; Krieger, L.; Schnetter, W.; Schairer, H.U. Cloning and analysis of the cry gene from Bacillus popilliae. J. Bacteriol. 1997, 179, 4336–4341.
16. Ruiu, L. Brevibacillus laterosporus, a pathogen of invertebrates and a broad-spectrum antimicrobial species. Insects2013, 4, 476–492.
17. Djukic, M.; Poehlein, A.; Thürmer, A.; Daniel, R. Genome sequence of Brevibacillus laterosporus LMG 15441, a pathogen of invertebrates. J. Bacteriol. 2011, 193, 5535–5536.
18. Sharma, V.; Singh, P.K.; Midha, S.; Ranjan, M.; Korpole, S.; Patil, P.B. Genome sequence of Brevibacillus laterosporus strain GI-9. J. Bacteriol. 2012.
19. Warren, G.W. Vegetative insecticidal proteins: Novel proteins for control of corn pests. In Advances in Insect Control: The Role of Transgenic Plants; Carozzi, N.B., Koziel, M.G., Eds.; Taylor & Francis: London, UK, 1997; pp. 109–121.
20. Zubasheva, M.V.; Ganushkina, L.A.; Smirnova, T.A.; Azizbekyan, R.R. Larvicidal activity of crystal-forming strains of Brevibacillus laterosporus. Appl. Biochem. Microbiol. 2010, 46, 755–762.
21. Ruiu, L.; Floris, I.; Satta, A.; Ellar, D.J. Toxicity of a Brevibacillus laterosporus strain lacking parasporal crystals against Musca domestica and Aedes aegypti. Biol. Control 2007, 43, 136–143.

22. Ruiu, L.; Satta, A.; Floris, I. Observations on house fly larvae midgut ultrastructure after Brevibacillus laterosporus ingestion. J. Invertebr. Pathol. 2012, 111, 211–216.
23. Nicolas, L.; Hamon, S.; Frachon, E.; Sebald, M.; de Barjac, H. Partial inactivation of the mosquitocidal activity of Clostridium bifermentans serovar malaysia by extracellular proteinases. Appl. Microbiol. Biotechnol. 1990, 34, 36–41.
24. Barloy, F.; Delècluse, A.; Nicolas, L.; Lecadet, M.M. Cloning and Expression of the first anaerobic toxin gene from Clostridium bifermentans subsp. malaysia encoding a new mosquitocidal protein with homologies to Bacillus thuringiensis delta-endotoxins. J. Bacteriol. 1996, 178, 3099–3105.
25. Waterfield, N.R.; Bowen, D.J.; Fetherston, J.D.; Perry, R.D.; Ffrench-Constant, R.H. The TC genes of Photorhabdus: A growing family. Trends Microbiol. 2001, 9, 185–191.
26. Ffrench-Constant, R.; Waterfield, N. An ABC guide to the bacterial toxin complexes. Adv. Appl. Microbiol. 2006, 58, 169–183.
27. Ffrench-Constant, R.H.; Dowling, A.; Waterfield, N.R. Insecticidal toxins from Photorhabdus bacteria and their potential use in agriculture. Toxicon 2007, 49, 436–451.
28. Waterfield, N.; Kamita, S.G.; Hammock, B.D.; Ffrench-Constant, R. The Photorhabdus Pir toxins are similar to a developmentally regulated insect protein but show no juvenile hormone esterase activity. FEMS Microbiol. Lett. 2005, 245, 47–52.
29. Eleftherianos, I.; Millichap, P.J.; Ffrench-Constant, R.H.; Reynolds, S.E. RNAi suppression of recognition protein mediated immune responses in the tobacco hornworm Manduca sexta causes increased susceptibility to the insect pathogen Photorhabdus. Dev. Comparat. Immunol. 2006, 30, 1099–1107.
30. Ji, D.; Kim, Y. An entomopathogenic bacterium, Xenorhabdus nematophila, inhibits the expression of an antimicrobial peptide, cecropin, of the beet armyworm, Spodoptera exigua. J. Insect Physiol. 2004, 50, 489–496.
31. Tailliez, P.; Pagès, S.; Ginibre, N.; Boemare, N. New insight into diversity in the genus Xenorhabdus, including the description of ten novel species. Int. J. Syst. Evolut. Microbiol. 2006, 56, 2805–2818.
32. Zhang, C.-X.; Yang, S.-Y.; Xu, M.-X.; Sun, J.; Liu, H.; Liu, J.R.; Liu, H.; Kan, F.; Sun, J.; Lai, R.; et al. Serratia nematodiphila sp. nov., associated symbiotically with the entomopathogenic nematode Heterorhabditidoides chongmingensis (Rhabditida: Rhabditidae). Int. J. Syst. Evolut. Microbiol. 2000, 59, 1603–1608.
33. Abebe, E.; Abebe-Akele, F.; Morrison, J. An insect pathogenic symbiosis between a Caenorhabditis and Serratia. Virulence 2011, 2, 158–161.
34. Torres-Barragan, A.; Suazo, A.; Buhler, W.; Cardoza, Y. Studies on the entomopathogenicity and bacterial associates of the nematode Oscheius carolinensis. Biol. Control 2011, 59, 123–129.
35. Castagnola, A.; Stock, S.P. Common virulence factors and tissue targets of entomopathogenic bacterial for biological control of Lepidopteran pests. Insects 2014, 5, 139–166.
36. Jackson, T.A.; Pearson, J.F.; O'Callaghan, M.; Mahanty, H.K.; Willocks, M. Pathogen to product development of Serratia entomophila Enterobacteriaceae as a commercial biological control agent for the New Zealand grass grub Costelytra zealandica. In Use of Pathogens in Scarab Pest Management; Jackson, T.A., Glare, T.R., Eds.; Intercept Ltd.: Andover, UK, 1992; pp. 191–198.
37. Hurst, M.R.; Glare, T.R.; Jackson, T.A.; Ronson, C.W. Plasmid-located pathogenicity determinants of Serratia entomophila, the causal agent of amber disease of grass grub, show similarity to the insecticidal toxins of Photorhabdus luminescens. J. Bacteriol. 2000, 182, 5127–5138.
38. Kwak, Y.; Khan, A.R.; Shin, J.H. Genome sequence of Serratia nematodiphila DSM 21420(T), a symbiotic bacterium from entomopathogenic nematode. J. Biotechnol. 2015, 193, 1–2.
39. Ishii, K.; Adachi, T.; Hamamoto, H.; Sekimizu, K. Serratia marcescens suppresses host cellular immunity via the production of an adhesion-inhibitory factor against immunosurveillance cells. J. Biol. Chem. 2014, 289, 5876–5888.

40. Hurst, M.R.H.; Jones, S.A.; Binglin, T.; Harper, L.A.; Jackson, T.A.; Glare, T.R. The main virulence determinant of Yersinia entomophaga MH96 is a broad-host-range toxin complex active against insects. J. Bacteriol. 2011, 193, 1966–1980.
41. Landsberg, M.J.; Jones, S.A.; Rothnagel, R.; Busby, J.N.; Marshall, S.D.G.; Simpson, R.M.; Lott, J.S.; Hankamer, M.; Hurst, M.R.H. 3D structure of the Yersinia entomophaga toxin complex and implications for insecticidal activity. Proc. Natl. Acad. Sci. USA 2011, 108, 20544–20549.
42. Marshall, S.D.G.; Hares, M.C.; Jones, S.A.; Harper, L.A.; Vernon, J.R.; Harland, D.P.; Jackson, T.A.; Hurst, M.R.H. Histopathological effects of the Yen-Tc toxin complex from Yersinia entomophaga MH96 (Enterobacteriaceae) on the Costelytra zealandica (Coleoptera: Scarabaeidae) larval midgut. Appl. Environ. Microbiol. 2012, 78, 4835–4847.
43. Ferguson, C.M.; Barton, D.M.; Harpe, L.A.; Swaminathan, J.; van Koten, C.; Hurst, M.R.H. Survival of Yersinia entomophaga MH96 in a pasture ecosystem and effects on pest and non-target invertebrate populations. New Zealand Plant Prot. 2012, 65, 166–173.
44. Vodovar, N.; Vinals, M.; Liehl, P.; Basset, A.; Degrouard, J.; Spellman, P.; Boccard, F.; Lemaitre, B. Drosophila host defense after oral infection by and entomopathogenic Pseudomonas species. Proc. Natl. Acad. Sci. USA 2005, 102, 11414–11419.
45. Vodovar, N.; Vallenet, D.; Cruveiller, S.; Rouy, Z.; Barbe, V.; Acosta, C.; Cattolico, L.; Jubin, C.; Lajus, A.; Segurens, B.; et al. Complete genome sequence of the entomopathogenic and metabolically versatile soil bacterium Pseudomonas entomophila. Nat. Biotechnol. 2006, 24, 673–679.
46. Kim, J.K.; Kim, N.H.; Jang, H.A.; Kikuchi, Y.; Kim, C.H.; Fukatsu, T.; Lee, B.L. Specific midgut region controlling the symbiont population in an insect-microbe gut symbiotic association. Appl. Environ. Microbiol. 2013, 79, 7229–7233.
47. Martinson, V.G.; Danforth, B.N.; Minckley, R.L.; Rueppell, O.; Tingek, S.; Moran, N.A. A simple and distinctive microbiota associated with honey bees and bumble bees. Mol. Ecol. 2011, 20, 619–628.
48. Kil, Y.J.; Seo, M.J.; Kang, D.K.; Oh, S.N.; Cho, H.S.; Youn, Y.N.; Yasunaga-Aoki, C.; Yu, Y.M. Effects of Enterobacteria (Burkholderia sp.) on development of Riptortus pedestris. J. Fac. Agric. Kyushu Univ. 2014, 59, 77–84.
49. Burkhead, K.D.; Schisler, D.A.; Slininger, P.J. Pyrrolnitrin production by biological-control agent Pseudomonas cepacia B37W in culture and in colonized wounds of potatoes. Appl. Environ. Microbiol. 1994, 60, 2031–2039.
50. Janisiewicz, W.J.; Roitman, J. Biological control of blue mold and gray mold on apple and pear with Pseudomonas cepacia. Phytopathology 1988, 78, 1697–1700.
51. Cordova-Kreylos, A.L.; Fernandez, L.E.; Koivunen, M.; Yang, A.; Flor-Weiler, L.; Marrone, P.G. Isolation and characterization of Burkholderia rinojensis sp. nov., a non-Burkholderia cepacia complex soil bacterium with insecticidal and miticidal activities. Appl. Environ. Microbiol. 2013, 79, 7669–7678.
52. He, H.; Ratnayake, A.S.; Janso, J.E.; He, M.; Yang, H.Y.; Loganzo, F.; Shor, B.; O'Donnell, C.J.; Koehn, F.E. Cytotoxic spliceostatins from Burkholderia sp. and their semisynthetic analogues. J. Nat. Prod. 2014, 77, 1864–1870.
53. Martin, P.A.W.; Hirose, E.; Aldrich, J.R. Toxicity of Chromobacterium subtsugae to Southern stink bug (Heteroptera: Pentatomidae) and corn rootworm (Coleoptera: Chrysomelidae). J. Econ. Entomol. 2007, 100, 680–684.
54. Martin, P.A.W.; Gundersen-Rindal, D.; Blackburn, M.; Buyer, J. Chromobacterium subtsugae sp. nov., a betaproteobacterium toxic to Colorado potato beetle and other insect pests. Int. J. Syst. Evolut. Microbiol. 2007, 57, 993–999.
55. Asolkar, R.; Huang, H.; Koivunen, M.; Marrone, P. Chromobacterium Bioactive Compositions and Metabolites. US Patent Application Publication, US 14/293,728, 9 October 2014.
56. Koivunen, M.; Chanbusarakum, L.; Fernandez, L.; Asolkar, R.; Tan, E.; Wallner, D.; Marrone, P. Development of a new microbial insecticide based on Chromobacterium subtsugae. IOBC/wprs Bull. 2009, 45, 183–186.

57. Martin, P.A.W.; Shropshire, A.D.S.; Gundersen-Rindal, D.E.; Blackburn, M.B. Chromobacterium subtsugae sp. nov. and Use for Control of Insect Pests. U.S. Patent Application Pubblication, 2007/0172463 A1, 26 July 2007.
58. Ramirez, J.L.; Short, S.M.; Bahia, A.C.; Saraiva, R.G.; Dong, Y.; Kang, S.; Tripathi, A.; Mlambo, G.; Dimopoulos, G. Chromobacterium Csp_P reduces malaria and dengue infection in vector mosquitoes and has entomopathogenic and in vitro anti-pathogen activities. PLOS Pathog. 2014, 10, e1004398.
59. Book, A.J.; Lewin, G.R.; McDonald, B.R.; Takasuka, T.E.; Doering, D.T.; Adams, A.S.; Blodgett, J.A.V.; Clardy, J.; Raffa, K.F.; Fox, B.G.; et al. Cellulolytic Streptomyces strains associated with herbivorous insects share a phylogenetically linked capacity to degrade lignocelluloses. Appl. Environ. Microbiol. 2014, 80, 4692–4701.
60. Copping, G.L.; Menn, J.J. Biopesticides: A review of their action, applications and efficacy. Pest Manag. Sci. 2000, 56, 651–676.
61. Craveri, R.; Giolitti, G. An antibiotic with fungicidal and insecticidal activity produced by Streptomyces. Nature 1957, 179, 1307.
62. Kido, G.S.; Spyhalski, E. Antimycin A, an antibiotic with insecticidal and miticidal properties. Science 1950, 112, 172–173.
63. Takahaski, N.; Suzuki, A.; Kimura, Y.; Miyamoto, S.; Tamura, S.; Mitsui, T.; Fukami, J. Isolation, structure and physiological activities of piericidin B, natural insecticide produced by a Streptomyces. Agric. Biol. Chem. 1968, 32, 1115–1122.
64. Oishi, H.; Sugawa, T.; Okutomi, T.; Suzuki, K.; Hayashi, T.; Sawada, M.; Ando, K. Insecticidal activity of macrotetrolide antibiotics. J. Antibiot. 1970, 23, 105–106.
65. Box, S.J.; Cole, M.; Yeoman, G.H. Prasinons A and B: Potent insecticides from Streptomyces prasinus. Appl. Microbiol.1973, 29, 699–704.
66. Turner, M.J.; Schaeffer, J.M. Mode of action of ivermectin. In Ivermectin and Abamectin; Cambell, W.C., Ed.; Springer-Verlag: New York, NY, USA, 1989; pp. 73–88.
67. Bloomquist, J.R. Ion Channels as Targets for Insecticides. Annu. Rev. Entomol. 1996, 41, 163–190.
68. Mertz, F.P.; Yao, R.C. Saccharopolyspora spinosa sp. nov. isolated from soil collected in a sugar mill rum still. Int. J. Syst. Bacteriol. 1990, 40, 34–39.
69. Kirst, H.A.; Michel, K.H.; Martin, J.W.; Creemer, L.C.; Chio, E.H.; Yao, R.C.; Nakatsukasa, W.M.; Boeck, L.D.; Occolowitz, J.L.; Paschal, J.W.; et al. A83543A-D, unique fermentation-derived tetracyclic macrolides. Tetrahedron Lett.1991, 32, 4839–4842.
70. Waldron, C.; Madduri, K.; Crawford, K.; Merlo, D.J.; Treadway, P.; Broughton, M.C.; Baltz, R.H. A cluster of genes for the biosynthesis of spinosyns, novel macrolide insect control agents produced by Saccharopolyspora spinosa. Antonie Leeuwenhoek 2000, 78, 385–390.
71. Kirst, H.A. The spinosyn family of insecticides: Realizing the potential of natural products research. J. Antibiot. 2010, 63, 101–111.
72. Salgado, V.L.; Sparks, T.C. The spinosyns: Chemistry, biochemistry, mode of action, and resistance. In Comprehensive Molecular Insect Science; Gilbert, L.J., Iatrou, K., Gill, S.S., Eds.; Elsevier: Oxford, UK, 2005; Volume 6, pp. 137–173.
73. Perry, T.; McKenzie, J.A.; Batterham, P.A. Da6 knockout strain of Drosophila melanogaster confers a high level of resistance to spinosad. Insect Biochem. Mol. Biol. 2007, 37, 184–188.
74. Sparks, T.C.; Crouse, G.D.; Durst, G. Natural products as insecticides: The biology, biochemistry and quantitative structure-activity relationships of spinosyns and spinosoids. Pest Manag. Sci. 2001, 57, 896–905.
75. Lewer, P.; Hahn, D.R.; Karr, L.L.; Duebelbeis, D.O.; Gilbert, J.R.; Crouse, G.D.; Worden, T.; Sparks, T.C.; McKamey, P.; Edwards, R.; et al. Discovery of the butenyl-spinosyn insecticides: Novel macrolides from the new bacterial strain Saccharopolyspora pogona. Bioorg. Med. Chem. 2009, 17, 4185–4196.

76. Morris, O.N. Susceptibility of some forest insects to mixtures of commercial Bacillus thuringiensis and chemical insecticides, and sensitivity of the pathogen to the insecticides. Can. Entomol. 1972, 104, 1419–1425.
77. Seleena, P.; Lee, H.L.; Chiang, Y.F. Compatibility of Bacillus thuringiensis serovar israelensis and chemical insecticides for the control of Aedes mosquitoes. J. Vector Ecol. 1999, 24, 216–223.
78. Musser, F.R.; Nyrop, J.P.; Shelton, A.M. Integrating biological and chemical controls in decision making: European corn borer (Lepidoptera: Crambidae) control in sweet corn as an example. J. Econ. Entomol. 2006, 99, 1538–1549.
79. Satinder, K.B.; Verma, M.; Tyagi, R.D.; Valéro, J.R. Recent advances in downstream processing and formulations of Bacillus thuringiensis based biopesticides. Process Biochem. 2006, 41, 323–342.
80. Oerke, E.C.; Dehne, H.W. Safeguarding production losses in major crops and the role of crop protection. Crop Prot.2004, 23, 275–285.

CHAPTER 9

Insect Pests of Sweetpotato in Uganda: Farmers' Perceptions of Their Importance and Control Practices

Joshua Sikhu Okonya[1]*, Robert OM Mwanga[2] , Katja Syndikus[3] and Jürgen Kroschel[4]

[1] *Global Program of Integrated Crop and Systems Research, International Potato Center (CIP), Plot 47, Ntinda II Road, Naguru, Box 22274, Kampala, Uganda.*

[2] *International Potato Center (CIP), Plot 47, Ntinda II Road, Naguru, Box 22274, Kampala, Uganda.*

[3] *IMOswiss AG, Weststrasse 51, CH-8570 Weinfelden, Switzerland.*

[4] *Global Program of Integrated Crop and Systems Research, International Potato Center (CIP), Avenida La Molina 1895, Lima 12, Peru*

ABSTRACT

Insect pests are among the most important constraints limiting sweetpotato (*Ipomoea batatas*) production in Africa. However, there is inadequate information about farmers' knowledge, perceptions and practices in the management of key insect pests. This has hindered development of effective pest management approaches for smallholder farmers. A standard questionnaire was used to interview individual sweetpotato farmers (n = 192) about their perception and management practices regarding insect pests in six major sweetpotato producing districts of Uganda. The majority (93%) of farmers perceived insect pests to be a very serious problem. With the exception of Masindi and Wakiso districts where the sweetpotato butterfly (*Acraea acerata*) was the number one constraint, sweetpotato weevils (*Cylas puncticollis* and *C. brunneus*) were ranked as the most important insect pests. Insecticide use in sweetpotato fields was very low being highest (28–38% of households) in districts where *A. acerata* infestation is the biggest problem. On average, 65% and 87% of the farmers took no action to control *A. acerata* and *Cylas* spp., respectively. Farmers were more conversant with the presence of and damage by *A. acerata* than of *Cylas* spp. as they thought that *Cylas* spp. root damage was brought about by a prolonged dry season. Different levels of field resistance (ability of a variety to tolerate damage) of sweetpotato landraces to *A. acerata* (eight landraces) and *Cylas* spp. (six landraces) were reported by farmers in all the six districts. This perceived level of resistance to insect damage by landraces needs to be investigated. To improve farmers' capabilities for sweetpotato insect pest management, it is crucial to train them in the basic knowledge of insect pest biology and control.

Keywords : Sweetpotato butterfly *Acraea acerata* African sweetpotato weevils *Cylas puncticollis Cylas brunneus* Production constraints *Ipomoea batatas* Integrated pest management

INTRODUCTION

Sweetpotato (*Ipomoea batatus* L. Lam.) is the world's sixth most important food crop consumed after rice (*Oryza sativa L.*), wheat (*Triticum aestivum* L.), potato (*Solanum tuberosum* L.), maize (*Zea mays* L.), and cassava (*Manihot esculenta* Crantz) (CIP 2010). It is also the third most important root crop grown in eastern Africa after cassava and potato (FAO 2011). Sweetpotato is both a staple and a food security crop in eastern and southern Africa, and is mainly grown by smallholder women farmers (Mutuura et al. 1992; Bashaasha et al. 1995; Andrade et al. 2009). Sweetpotato is also grown for its vines as planting material; leaves are often eaten as a vegetable while shoots and roots are used as animal feed in many countries. In Uganda and western Kenya, the sale of fresh sweetpotato roots, vines and processed foods in both local and urban markets is becoming increasingly popular thus contributing to household cash income (Abidin 2004; Kaguongo et al. 2012). Orange-fleshed sweetpotato is also a rich source of beta-carotene, a precursor of bio-available vitamin A, and has potential of combating Vitamin A deficiency among rural resource-constrained farmers in many developing countries (Jalal et al. 1998; Jaarsveld et al. 2005; Low et al. 2007; Mwanga et al. 2003a; Burri 2011).

Sweetpotato production in Uganda declined from 2.84 million tons in 2010 to 2.55 million tons in 2011 (FAO 2011). Uganda also dropped from being the second to being the fourth largest producer of sweetpotato in the world after China, Tanzania and Nigeria (FAO 2011). This reduction in sweetpotato production could be due to many biotic stresses including insect pests and diseases (viral and fungal) that have led to the decline in vine and root quality and root yields. Research conducted in most southern and eastern Africa has shown that insects are among the most economically important pests of sweetpotato (Smit 1997). The most serious and commonly reported insect pest species in Uganda (Smit 1997; Hakiza et al. 2000; Abidin 2004; Ebregt et al. 2005), Kenya (Smit 1997; Nderitu et al. 2009), Ethiopia (Azerefegne et al. 2001), Rwanda (Smit et al. 1997) and Nigeria (Girma 1994; Tewe et al. 2003) are caterpillars of the sweetpotato butterfly, *Acraea acerata* Hew. (Lepidoptera: Nymphalidae), the African sweetpotato weevils, *Cylas brunneus* F. and *C. puncticollis* Boheman (Coleoptera: Brentidae), the clearwing moth *Synanthedon* spp. (Lepidoptera: Sesiidae), the sweetpotato hornworm, *Agrius convolvuli* L. (Lepidoptera: Sphingidae) and vectors of the sweetpotato virus diseases, such as the whitefly *Bemisia tabaci* Gennadius (Hemiptera: Aleyrodidae) (Fuglie 2007). *Acraea acerata* and *Cylas* spp. occur in 15 and 23 African countries, respectively (CAB International 2005).

The two African *Cylas* species often occur together in fields and cause huge yield losses of up to 100% (Girma 1994; Smit 1997; Chalfant et al. 1990). Root and stem damage by *Cylas* species is of great economic importance as it leads to a reduction in root yield, and root and vine quality. Due to egg laying and extensive larval tunneling by *Cylas* species, the plant produces bitter defense terpenes and phenolic compounds that make roots unsuitable for both human and animal consumption (Chalfant et al. 1990; Ames et al. 1996).

The *A. acerata* larvae feed on leaves of the sweetpotato plant and heavy infestations can lead to complete plant defoliation. Defoliation of young plants by *A. acerata* larvae can sometimes result in failure of the crop to re-establish and hence reduced yields. As a result, farmers continue to record huge losses due to these pests. Since sweetpotato is an important food security crop, its low production has a bearing impact on national food security. Understanding sweetpotato production constraints, particularly insect pests and farmers' methods of managing *A. acerata* and *Cylas* species, could be useful in designing an effective integrated pest management (IPM) strategy. If sweetpotato IPM research is to be useful and sustainable, it is important to understand farmers' perceptions and their management methods.

Assessing farmers' perceptions of crop production constraints has been used as a tool for documenting pest status and designing pest management options suitable for a particular community (Smit and Matengo 1995; Obopile et al. 2008). Soleri et al. (2000) emphasized the need for integrating methods from both biological and social sciences to understand farmers' selection criteria of crop varieties, since objectives of farmers while selecting a particular maize variety, differed from what research scientists had normally assumed. In a similar way, it would be equally important for crop entomologists to integrate farmers' perspectives of insect pest management in the development of any intervention measure for local use. Bonhof et al. (2001) used farmer participatory rural appraisals for maize farmers at the Kenyan coast to understand the pest status and control strategies for the maize stemborer (*Chilo partellus* Swinhoe). In central Zambia, Mukanga et al. (2011) solicited farmers' views on various management practices of maize ear rots caused by several fungal pathogens before resistant varieties to ear rots could be developed. Lebesa et al. (2012) and Midega et al. (2012) undertook household surveys in western Kenya using questionnaires to determine the pest status of herbivorous blister beetles (*Hycleus* spp.) and cotton (*Gossypium hirsutum* L.) pests. Tounou et al. (2013), similarly used questionnaires to interview farmers to determine geographic distribution and importance of stemborers on maize in southern Togo.

The amount of edible sweetpotato roots and vines damaged by insect pest infestations is significant and calls for an effective management strategy. To develop an integrated pest management strategy that is appropriate for resource poor farmers, information is needed on the current distribution, importance and control methods of sweetpotato pests in various agro-ecologies of Uganda. In Uganda, detailed studies on sweetpotato insect pests are limited. Since the farm surveys conducted by Bashaasha et al. (1995) between 1989 and 1992 in nine different agro-ecological zones, no recent quantitative surveys have been undertaken in the country. The pest status and distribution of key insect pests are expected to vary considerably with the current climate variability. There is, therefore, need to find feasible solutions to agricultural production constraints by incorporating farmer views into research for development programs.

The International Potato Center (CIP) through its global program of integrated crop and systems research seeks to develop an effective IPM strategy for key insect pests of sweetpotato according to climatic factors and pest severity. In this regard, a questionnaire was developed to capture the information and to document the present insect pest distribution, farmers' knowledge and coping strategies to control insect pests of sweetpotato. This study was specifically carried out to 1) determine farmers' perceptions of importance of sweetpotato insect pests and their distribution, 2) analyze the major sweetpotato production constraints and farmers' coping strategies for the control of sweetpotato insect pests in various agro-ecologies of Uganda, 3) identify local sweetpotato landraces that have some field resistance to damage by *A. acerata* and *Cylas* spp.

MATERIALS AND METHODS

Study area

The study was conducted between August and October 2011 in six districts of Uganda. Major sweetpotato growing districts were purposively selected from six different agro-ecological zones to obtain a wide range of household perceptions. These were Gulu (2°46′48″N, 32°18′0″E), Kabale (1° 15′ 0″ S, 29° 59′ 24″ E), Kasese (0° 11′ 0″ N, 30° 5′ 0″ E), Masindi (1° 41′ 1″ N, 32° 43′ 20″ E), Wakiso (0°23′53″N, 32°28′41″E) and Soroti (1° 42′ 54″ N, 33° 36′ 40″ E) (Figure 1). In each district, two sub-counties were purposively selected based on intensity of sweetpotato farming as perceived by the district agriculture office for this observational study.

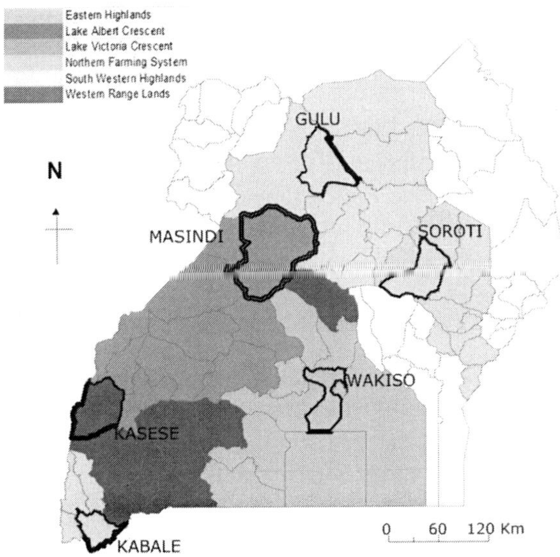

Figure 1. Map of uganda showing the six study districts with their corresponding agro-ecological zones; gulu in the northern farming system, masindi in the lake albert crescent, soroti in the eastern highlands, wakiso in the lake victoria crescent, kasese in the western range highlands and kabale in the south western highlands.

The questionnaire was pre-tested with five sweetpotato farmers in Luweero district, not included in the sample, three weeks before the study. After the pilot test, changes were made in the expression of some questions to be asked. The survey team consisted of two entomologists, one research assistant and twelve agricultural extension workers.

Farmers who had grown sweetpotato in 2011 and preferably still had it in their fields were randomly selected at regular intervals of 5–10 km distance between sites or until a sweetpotato growing household was found on motorable roads within the subcounties. A total of 32, both female and male farmers, were interviewed face-to-face in their homesteads in each district using a standard questionnaire – partly structured and partly open. All interviews were conducted in the local language of that district with the help of an enumerator. Data was collected on: sweetpotato cropping systems, constraints in sweetpotato production, sweetpotato insect pests and, their control measures and field resistance of sweetpotato varieties to the three major insect pests (*A. acerata, C. brunneus* and *C. puncticollis*).

Constraints to sweetpotato production were identified by asking farmers to mention all the constraints they had experienced, and ranking them in order of importance of each constraint (for instance most destructive, occurs frequently) in the last cropping season. Farmers were shown colored photographs and/or vials of insects (larval and adult stages) in alcohol to ensure they made the correct identification of the pest. Farmers' compared the perceived field resistance (level of damage) of their presently cultivated local varieties to damage by the most important insect pests.

Quantitative data were subjected to analysis of variance (ANOVA) using General Linear Model (GLM) in SAS software (SAS Institute Inc 2008). Means were compared using Fisher's LSD Test. Descriptive statistics (percentages and mean values) were the main statistical tools employed to analyze qualitative data.

RESULTS AND DISCUSSION

Farm household characteristics and some aspects of sweetpotato production

Sweetpotato production in the six districts of Uganda is a female dominated activity with males representing only 39% of 192 respondents. In Soroti district, however, male respondents out-numbered female respondents by 25%; this is probably because sweetpotato is more of a commercial than food security crop in this district. Men have been reported in many African countries to dominate cash crop production (World Bank et al. 2009). This domain change of sweetpotato, from being a female controlled crop to a male crop in Soroti due to commercialization, has been reported for other crops like coffee (*Coffea* spp.) in Uganda (Kasente et al. 2002) and cocoa (*Theobroma cacao* L.) in Ghana (Doss 2001). In Uganda, 51% of the population is female (UNHS 2010) and women are mostly responsible for household food crop production; in particular the sweetpotato crop which is referred to as a "female" crop (Bashaasha et al. 1995). Consequently, efforts geared towards promoting and disseminating new sweetpotato technologies should mainly target women.

Most of the sweetpotato farmers were in their active age (mean of 40 to 46 years) which is a good sign for food security in Uganda (Table 1). There is a huge potential to increase production if more investments are made into developing pest and disease resistant high yielding sweetpotato varieties as part of developing integrated pest management strategies. Mean household size across districts was 7.5 members; this is higher than the national average by 2.5 members (UNHS 2010). On average, Gulu district had the biggest households with 10 members, while Kabale district had the smallest household size of six members (Table 1).

Table 1. Household Characteristics Of Interviewed Sweetpotato Farmers In The Six Districts Of Uganda And Their Sweetpotato Production Practices/ Techniques, August-October 2011

Household characteristics	Gulu	Kabale	Kasese	Masindi	Soroti	Wakiso	p-value (N/A = not applicable)
Female respondents (%)	53.1	62.5	59.4	75.0	37.5	78.1	N/A
Female headed households (%)	28.1	28.1	21.9	28.1	15.6	43.8	N/A
No formal education (%)	28.1	25.0	18.8	9.4	6.3	20.0	N/A
≤7 years of formal education (%)	40.6	65.6	56.3	46.9	71.9	60.0	N/A
>7 years of formal education (%)	31.3	9.4	25.0	43.8	21.9	20.0	N/A
Mean elevation (m a.s.l.)	1091^{cd}	2090^{a}	1006^{e}	1124^{c}	1079^{d}	1191^{b}	<0.0001
Mean age of respondent (years)	43.7 ± 2.6^{a}	40.0 ± 2.0^{a}	41.2 ± 2.4^{a}	44.9 ± 2.4^{a}	44.1 ± 2.0^{a}	45.8 ± 2.4^{a}	0.7024
Mean household size (persons)	10.2 ± 1.3^{a}	5.9 ± 0.5^{c}	6.6 ± 0.5^{bc}	6.5 ± 0.6^{c}	8.6 ± 0.7^{ab}	7.5 ± 0.8^{bc}	0.0002
Mean rotation duration (months)	20.4 ± 2.2^{ab}	11.0 ± 1.7^{c}	6.6 ± 0.8^{c}	17.2 ± 1.4^{b}	23.3 ± 2.1^{a}	9.1 ± 1.1^{c}	<0.0001
Mean years of growing sweetpotato	25.1 ± 2.7^{ab}	19.6 ± 1.9^{abc}	16.8 ± 2.0^{c}	26.3 ± 2.3^{a}	23.8 ± 1.9^{ab}	19.7 ± 2.4^{bc}	0.0179
Mean sweetpotato acreage in 2011 (ha)	0.3 ± 0.1^{ab}	0.2 ± 0.0^{b}	0.2 ± 0.0^{b}	0.2 ± 0.0^{b}	0.3 ± 0.0^{ab}	0.4 ± 0.1^{a}	0.1393
Arable land devoted to sweetpotato (%)	12.3 ± 3.1^{b}	30.5 ± 5.0^{a}	17.9 ± 3.8^{b}	14.1 ± 2.0^{b}	20.0 ± 3.5^{b}	31.1 ± 5.1^{a}	<0.0001
Mean total land holding (ha)	100.2 ± 82.3^{a}	2.9 ± 1.7^{a}	2.4 ± 0.5^{a}	6.6 ± 2.8^{a}	2.4 ± 0.3^{a}	1.3 ± 0.2^{a}	0.2625
Total land cropped (%)	59.5 ± 6.3^{c}	85.8 ± 5.6^{ab}	85.4 ± 3.9^{ab}	63.6 ± 5.9^{c}	77.8 ± 5.6^{b}	93.9 ± 3.0^{a}	<0.0001

Means followed by the same letter in the same row are not significantly different ($p \geq 0.05$, Fisher's least significant difference). Values are means ± SE.

Mean elevation of the surveyed homesteads was highest in Kabale district (2090 m a.s.l.) and lowest in Kasese district (1005 m a.s.l.). Sweetpotato had been grown for more than 16 years in all the six districts. The average period of growing sweetpotato was longest (26 years) for Masindi followed by Gulu district (25 years). The average sweetpotato acreage in 2011 was highest in Soroti district (0.4 ha) and lowest in Kabale, Kasese and Masindi districts (0.2 ha). Sweetpotato is a commercial crop in Soroti district and this could explain the allocation of relatively larger portions of arable land to sweetpotato production.

Constraints to sweetpotato production

Identification of factors limiting production and provision of environmentally-friendly options for integrated crop management is inevitable if sweetpotato production among smallholder farmers in Uganda is to be increased. In the districts visited, setbacks to sweetpotato production were many, however, insect pests attacking roots and leaves were the most important (Table 2). Poor yields of local varieties and degraded soils came second in importance while rodents or rats were ranked the third most important constraint. Other constraints that were unique to a district included delayed rainfall and weeds in Wakiso, shortage of land and small size of roots in Kabale, high cost of labor, floods, millipedes (Diplopoda) and poor roads in Kasese, extreme rainfall, high transport cost, lack of capital to plant large fields and high labor demands for the sweetpotato crop in Gulu districts.

Table 2. Top Five Most Important Constraints To Sweetpotato Production As Ranked By Farmers In The Study Districts Of Uganda, August-October 2011

Constraints to sweetpotato production	Rank for each constraint (% households)*					
	1st	2nd	3rd	4th	5th	Mean
No constraint mentioned/experienced	0.5	8.3	26.6	55.7	81.8	34.6
Insect pests	33.9	42.2	33.3	18.8	5.7	26.8
Poor yields of local varieties/soils	12.0	10.4	6.3	1.6	1.6	6.4
Rats and rodents	8.3	8.9	6.3	4.2	1.0	5.7
Drought/prolonged dry seasons	10.9	3.1	4.2	2.6	0.5	4.3
High cost of labor/labor intensive/shortage of labor/weeds	6.3	2.1	2.6	1.6	1.6	2.8
Lack of market	3.6	5.2	2.1	2.1	0.5	2.7
Viral diseases	2.6	3.1	3.1	2.6	1.0	2.5
Shortage of planting material	5.2	1.6	3.1	1.0	0.0	2.2
Wild game (elephants, hippos, pigs)	4.2	2.6	1.6	1.0	1.0	2.1
Others	1.5	2.6	2.1	3.1	1.0	2.1
Millipedes	1.6	1.6	3.6	0.5	2.6	2.0
Land shortage	2.1	2.1	2.6	0.5	0.5	1.6
Floods/excess rainfall/storms	1.0	1.6	1.0	1.0	0.5	1.0
Lack of money to hire labor/plant large fields/build drying places	2.1	1.6	0.0	0.5	0.0	0.8
Domestic animals (cattle, goats)	1.0	1.0	0.0	1.0	0.5	0.7
Changed onset and cessation of rainfall in the seasons	1.6	0.5	0.5	0.5	0.0	0.6
Alternaria blight disease	1.0	0.5	0.0	1.0	0.0	0.5
Root rots	0.0	1.0	1.0	0.5	0.0	0.5

*Some columns do not add up to 100% due to rounding off.

Sweetpotato insect pests

Among the sweetpotato insect pests reported by farmers, *Cylas* spp. were ranked number one by 87% of the households followed by caterpillars of *A. acerata* (60.7%) (Table 3). Caterpillars of *A. acerata* in Masindi and Wakiso districts were perceived to be more damaging to sweetpotato than the *Cylas* spp. Insect pests that appeared to be local included *Agrius convolvuli* in Gulu and Soroti districts, and the sweetpotato armyworm (*Spodoptera* spp.) in Soroti district. *A. convolvuli* occasionally appeared in both districts as in 2009 and 2010 where it caused serious sweetpotato damage leading to food insecurity (IPC 2010). *Spodoptera* spp. had occurred in Soroti in the past three consecutive years (2007 to 2010). Small black ants ("*Munyeera*" in Luganda) were mentioned to construct nests inside the stem base of sweetpotato plants leading to vine damage. Ants are not pests *per se* but rather predators of *Cylas* spp. larvae feeding inside the stem base.

Table 3. Major Insect Pests Experienced By Farmers In Sweetpotato (% Households) And Rank

Insect pest reported	% households*						
	Gulu	Kabale	Kasese	Masindi	Soroti	Wakiso	Overall mean
Sweetpotato weevils (*Cylas* spp.)	97 (1)	56 (1)	84 (1)	100 (2)	94 (1)	91 (1)	87.0
Sweetpotato butterfly (*Acraea acerata*)	28 (2)	50 (2)	88 (2)	88 (1)	13 (2)	97 (2)	60.7
Sweetpotato hornworm (*Agrius convolvuli*)	33	0	0	0	25	0	9.7
Armyworm (*Spodoptera* spp.)	0	0	0	0	34	0	5.7
Others (ants, whiteflies)	0	0	9	3	3	0	2.5

*percentage values add to more than 100 due to multiple responses. Rank (in parentheses) for the top two most important insect pests.

Our results are in accordance with those reported by Girma and Belehu (1994) in Ethiopia where insect pests and specifically *A. acerata* and *Cylas* spp. are most important. Farmers reported that *A. acerata* was important only in some years with the last distinct outbreak in Soroti being in February 2010. Farmers in Masindi, however, said that *A. acerata* started to be a problem in 1993 until now.

It is important to note that farmers tend to have a high perception of damage caused by insect pests, hence, rank them highly (Smit 1997). This must have been the case in the current study since results from a subsequent field survey of farmers' sweetpotato fields in Kabale, Buliisa and Masindi districts reported lower insect pest infestation rates and densities (Okonya and Kroschel 2013a, b). *B. tabaci* is an important insect vector of the sweetpotato virus disease component (*Sweetpotato chlorotic stunt virus*) in Uganda. However, most farmers did not consider *B. tabaci* an important insect pest because they do not have the knowledge to relate the presence of the insect (vector) to virus symptoms or crop damage and yield loss. Such knowledge gaps call for better training on IPM through the national extension services.

There was a general belief by farmers that once sweetpotato is attacked by *A. acerata*, root yield will be poor. The African locust (*Locusta migratoria migratorioides* (R. & F.) appeared in Soroti district in 2008/2009 cropping season and completely defoliated sweetpotato plants.

Farmers management practices of sweetpotato pests

The main methods used by farmers in managing sweetpotato insect pests on their farms included chemical insecticide application, ash application, hand-picking or a combination of two or more physical and cultural strategies (Table 4). Control strategies for caterpillars of *A. acerata* included use of chemical insecticides of mainly permethrin, dimethoate and cypermethrin by 24% of the households; application of wood ash, hand-picking of caterpillars and a combination of two or three of the above mentioned methods. Early harvesting was the most common method used to limit the damage caused by *Cylas* spp., however, mulching, re-hilling to cover soil cracks, crop rotation and insecticide application were being used but on a very low scale. On average, 65% and 87% of the sweetpotato farmers did not control *A. acerata* and *Cylas* spp., respectively.

Table 4. Control methods for the two (***a*** **.** ***Acerata*** and ***cylas*** spp.) Most important insect pests of sweetpotato (% households)

Control strategy for *A. acerata* and *Cylas* spp.	% households						Overall mean
	Gulu	Kabale	Kasese	Masindi	Soroti	Wakiso	
a) for *A. acerata*							
Chemical insecticides	2	20	14	50	0	56	24
Ash application	0	0	7	8	0	0	3
Hand picking	0	0	0	8	0	6	2
Chemical insecticides and hand picking	0	0	0	8	0	0	1
Chemical insecticides and ash application	0	0	0	0	0	13	2
Hand picking and ash application	0	0	0	8	0	0	1
Chemical insecticides, hand picking and ash application	0	0	0	0	0	6	1
None	98	80	79	17	100	19	65
b) for *Cylas* spp.							
Chemical insecticides	0	0	0	6	0	8	2
Chemical insecticides and re hilling	7	0	7	6	0	0	3
Crop rotation	7	8	0	0	0	0	3
Early harvesting	0	0	7	0	7	8	4
Mulching and re hilling	0	0	0	0	0	8	1
None	87	92	87	88	93	77	87

Use of chemical insecticides was relatively low; being highest in Wakiso district followed by Masindi district but absent in Soroti district (Table. 4). Insecticide application was highest in districts which ranked *A. acerata* as the main insect pest damaging sweetpotato. It was evident that some farmers implemented more than one control strategy to reduce field infestation by insect pests. Aiming at improving the effectiveness of available control methods is therefore desired. The low use of insecticides in sweetpotato could partly be due to the high cost of insecticides which subsistence farmers cannot afford but also lack of knowledge about pest biology of especially *Cylas* spp. Many farmers did not know how to control *Cylas* spp. However, farmers who applied insecticides to control *A. acerata* observed reduced damage by *Cylas* spp. as well. This therefore encouraged farmers to apply insecticides two months after planting even in the absence of *A. acerata*.

Due to the severity of pest infestation during outbreaks, farmers usually received insecticides from agriculture extension workers or local authorities to spray against *A. acerata* in Kabale district and *A. convolvuli* in Gulu and Soroti districts. Use of chemical insecticides against *A. acerata* could be because farmers took the pest seriously, pest and pest damage were more visible or pesticides are effective. It should be noted, however, that use of chemical insecticides is not a permanent solution as it can be disastrous to human health due to poor handling, elimination of natural enemies for the pest and is out of reach for most of the resource constrained poor farmers (Croft and Brown 1975).

Major sweetpotato varieties grown and farmers perceptions of their field resistance to insect pests

Sweetpotato variety mixtures were a common practice by at least 89% of interviewed farm households. With the exception of households in Kabale and Soroti districts, farmers in the remaining districts sometimes planted vines affected by insect pests due to lack of planting materials, especially at the end of a prolonged dry season. The most popular grown varieties were Larila in Gulu, Rwabafuluki in Kabale, Rwatoro in Kasese, Dimbuka in Masindi, Kampala and Araka in Soroti and NASPOT 1 in Wakiso district (Table 5). Farmers identified eight landraces (Mukono, Kigabali/Magabali, Red mamba, Kiryenamwami, Dimbuka, Boy, Setyabule and Mbale), which have shown some form of field resistance to vine damage by *A. acerata* and six landraces (Ochol/Ocuc, Rwatoro, Muhamoud, Dimbuka, Kyebandula and Opaku) with some form of resistance to field root damage by *Cylas* spp. Information on resistance levels of landraces needs to be taken with caution since no variety has been reported to be resistant in laboratory no choice experiments to *Cylas* spp. for instance (Mwanga et al. 2003a, b). This

Table 5. Perceived Field Resistance Of Major Local Sweetpotato Varieties Or Landraces To ***A . Acerata*** And ***Cylas*** Spp. In 2011

District	Local name of Sweetpotato variety/landrace	% Households growing the variety	Resistence level to *Acraea acerata* damage (% responses)			Resistence level to *Cylas* spp. damage (% responses)		
			Low	Moderate	High	Low	Moderate	High
	Lalira	22	0	15	9	0	13	11
	Alero	12	0	6	3	0	7	4
	Ochol/Ocuc	12	0	6	3	1	7	1
GULU	Adoch	11	0	6	15	0	2	9
	Mukiga	3	0	3	0	0	2	2
	Others	40	0	18	18	0	18	21
	Rwabafuluki, Kandazi/Mulungi	22	0	4	25	0	7	17
	Mukazi	17	0	0	11	0	17	14
KABALE	Mukono	8	4	0	11	0	7	3
	Kidodo	6	0	4	11	0	3	0
	Kigabali/Magabali	4	4	0	4	0	7	0
	Others	43	0	0	25	0	17	7
	Rwatoro	13	0	2	9	2	2	6
	Red mamba	10	2	4	4	0	4	0
	Rosemary	10	0	4	7	0	6	6
	Kiryenamwami	8	2	2	7	0	4	4
KASESE	Muhamoud	6	0	4	7	2	2	6
	Bitambi	6	0	2	2	0	2	2
	Kyebandula	6	0	2	2	0	4	2
	Others	41	11	0	26	4	21	17
	Dimbuka	22	1	7	18	1	16	10
	Nakato/Nyakato	14	0	6	12	0	14	5
	Kahogo/New Kawogo	9	0	4	9	0	5	4
MASINDI	Kyebandula	5	0	1	4	1	1	0
	Suwedi	3	0	3	1	0	3	3
	Kabakumba	3	0	0	3	0	0	3
	Others	44	0	4	25	0	18	16
	Kampala	18	0	0	11	0	6	12
	Araka	18	0	5	16	0	8	12
	Ateseke	7	0	5	11	0	6	4
SOROTI	Opaku	5	0	0	0	1	4	1
	Letesi/Latesi	4	0	0	0	0	3	1
	Mwambi	4	0	5	11	0	3	1
	Boy	4	5	0	0	0	0	4
	Others	40	0	16	16	1	16	17
	Naspot 1	31	0	16	23	0	18	35
	Dimbuka	23	2	5	16	3	3	15
	Setyabule	13	2	5	7	0	3	3
WAKISO	Mbale	10	2	0	7	0	3	5
	New Kawogo	6	0	2	2	0	5	5
	Others	17	0	7	5	0	5	0

notwithstanding, these reports of resistance of landraces to insects need to be investigated further as they may provide potential sources of resistance to these two most economically important pests of sweetpotato. Various authors have found differences in *Cylas* spp. damage among cultivars (Mwanga et al. 2003b; Stathers et al. 2003). Factors such as quantity of root latex, depth of rooting and amount of foliage have been reported to contribute to reduced *Cylas* spp. field damage (Mwanga et al. 2001; Stathers et al. 2003). Anyanga et al. (2013) found that chemical compounds in the root latex were responsible for the host plant resistance to *Cylas* spp. damage of "New Kawogo" sweetpotato variety. This variety (New Kawogo) was also mentioned by farmers in Masindi and Wakiso districts in this study to be resistant to *Cylas* spp. damage.

The study shows that sweetpotato varieties susceptible to insect pests are cultivated by farmers across the six districts. This is due to the fact that high resistance to either *A. acerata* or *Cylas* spp. has not been found in the world sweetpotato germplasm collection (Mwanga et al. 2009), which could have been used to develop and release resistant varieties in Uganda. Evaluations for resistance against *C. puncticollis* of transgenic sweetpotato expressing Cry3Ca1, Cry7Aa1 and ET33-34 proteins from *Bacillus thuringiensis* (*Bt*) at the National Crop Resources Research Institute, Namulonge, Uganda are underway (Rukarwa et al. 2013).

CONCLUSION AND OUTLOOK

This study has provided insight into sweetpotato production in the main production districts of Uganda. Insect pests are a major production constraint in all the districts surveyed. The majority of farmers perceive insect pests to be a very serious problem. With the exception of Masindi and Wakiso districts where *A. acerata* is the number one constraint, sweetpotato weevils (*C. puncticollis* and *C. brunneus*) are ranked as the most important insect pests. However, many sweetpotato farmers take no measures to control *Cylas* spp. but invest in the use of chemical insecticides to control *A. acerata*, which has a high priority in Masindi and Wakiso districts. Promoting the use of cultural control methods such as mulching, re-hilling to cover soil cracks, use of clean planting material, crop rotation, taking time period and crops planted before into account, have potential to reduce damage by *Cylas* spp. Further, it would be important to invest in research to develop additional control measures. Biological control using the entomopathogen *Beauveria bassiana* has shown to reduce the *Cylas formicarius* Fabricius damage in Cuba alone or in combination with sexual pheromones (Lagnaoui et al. 2000) Farmers lack knowledge on *Cylas* spp. biology, damage and control. Hence, training about the biology and ecology of this important pest can help farmers to make informed crop management decisions. Baseline information gained from this study will assist the international agricultural research system (NARS), and non-governmental organizations (NGO's) in designing IPM strategies that are based on the needs of smallholder farmers and their sweetpotato production systems. This information will also be useful in setting research priorities.

ACKNOWLEDGEMENT

The authors are thankful to the farmers who participated in the survey, the enumerators for translating the questions into local languages, the district agriculture officers and extension workers in the surveyed districts for assisting in site selection. The assistance of Ssempa Zaamu (CIP-Uganda) during data collection is highly appreciated. This work was supported by a grant from the German Federal Ministry for Economic Cooperation and Development (BMZ), which is highly acknowledged.

COMPETING INTERESTS

The authors declare that they have no competing interests.

AUTHORS' CONTRIBUTIONS

JSO designed the questionnaire, collected and analyzed the data, and drafted the manuscript. KS participated in designing the questionnaire, data collection and edited the manuscript; ROMM edited the manuscript and provided overall administrative support for the study. JK conceived and designed the study, revised and polished the manuscript, and supervised the whole study; all authors read and approved the final manuscript.

REFERENCES

1. Abidin PE: *Sweetpotato breeding for northeastern Uganda: farmer varieties, farmer-participatory selection, and stability of performance.* 2004. Dissertation, Wageningen University
2. Ames T, Smit NEJM, Braun AR, O'Sullivan JN, Skoglund LG: *Sweetpotato: major pests, diseases and nutritional disorders.* International Potato Center (CIP), Lima, Peru; 1996.
3. Andrade M, Barker I, Cole D, Dapaah H, Elliott H, Fuentes S, Grüneberg W, Kapinga R, Kroschel J, Labarta R, Lemaga B, Loechl C, Low J, Lynam J, Mwanga R, Ortiz O, Oswald A, Thiele G: *Unleashing the potential of sweetpotato in sub-Saharan Africa: current challenges and way forward.* International Potato Center (CIP), Lima, Peru; 2009.

4. Anyanga MO, Muyinza H, Talwana H, Hall DR, Farman DI, Ssemakula GN, Mwanga ROM, Stevenson PC: Resistance to the weevils *Cylas puncticollis* and *Cylas brunneus* conferred by sweetpotato root surface compounds. *J Agric Food Chem* 2013, 61(34):8141-8147.
5. Azerefegne F, Solbreck C, Ives AR: Environmental forcing and high amplitude fluctuations in the population dynamics of the tropical butterfly *Acraea acerata* (Lepidoptera: Nymphalidae). *J Anim Ecol* 2001, 70: 1032-1045.
6. Bashaasha B, Mwanga ROM, Ocitti p'Obwoya C, Ewell PT: *Sweetpotato in farming and food systems of Uganda. A farm survey report.* 1995. Sub-Saharan African region, Nairobi, Kenya/National Agricultural Research Organization, Kampala, Uganda
7. Bonhof MJ, van Huis A, Kiros FG, Dibogo N: Farmers' perceptions of importance, control methods and natural enemies of maize stemborers at the Kenya coast. *Insect Sci Appl* 2001, 21: 33-43.
8. Burri BJ: Evaluating sweetpotato as an intervention food to prevent vitamin A deficiency. *Compr Rev Food Sci Food Saf* 2011, 10: 118-130.
9. CAB International: *Distribution maps of plant pests. 1st revision A: map no. 279. Cylas puncticollis (Boheman).* 2005, 1-2.
10. Chalfant RB, Jansson RK, Dakshina RS, Schalk JM: Ecology and management of sweet potato insects. *Annu Rev Entomol* 1990, 35: 157-180.
11. CIP (International Potato Center): *Facts and figures about sweetpotato 2010.* 2010. . Accessed 10 July 2013 http://cipotato.org/resources/publications/fact-sheets-flyer-leaflet/facts-and-figures-about-sweetpotato/
12. Croft BA, Brown AWA: Responses of arthropod natural enemies to insecticides. *Annu Rev Entomol* 1975, 20: 285-335.
13. Doss C: Designing agricultural technology for African women farmers: lessons from 25 years of experience. *World Dev* 2001, 29: 2075-2092.
14. Ebregt E, Struik PC, Odongo B, Abidin PE: Pest damage in sweetpotato, groundnut and maize in north-eastern Uganda with special reference to damage by millipedes (Diplopoda). *NJAS-Wagen J Life Sci* 2005, 53: 49-69.
15. FAO (Food and Agriculture Organization): *Production quantity of primary crops, country statistics for eastern Africa.* 2011. . Accessed 29 April 2013 http://faostat3.fao.org/faostat-gateway/go/to/download/Q/QC/E
16. Fuglie KO: Priorities for sweetpotato research in developing countries: results of a survey. *Hortscience* 2007, 42: 1200-1206.
17. Girma A: Review of research on sweetpotato pest management in Ethiopia. In *Proceedings of a regional workshop on root and tuber pest management in east and southern Africa. Mombasa, Kenya, 10–14 August 1992.* Edited by: Allard GB, Skoglund LG, Neuenschwander P, Murphy RJ. Nairobi: IIBC; 1994:53-56.
18. Girma A, Belehu T: Sweetpotato production constraints and research in Ethiopia. In *Proceedings of a regional workshop on root and tuber pest management in east and southern Africa. Mombasa, Kenya, 10–14 August 1992.* Edited by: Allard GB, Skoglund LG, Neuenschwander P, Murphy RJ. Nairobi: IIBC; 1994:49-51.
19. Hakiza JJ, Turyamureeba G, Kakuhenzire RM, Odongo B, Mwanga ROM: Potato and sweetpotato improvement in Uganda: a historical perspective. *Proceedings of the 5th African potato association (APA) conference, 29 May - 2 June 2000, Makerere University, Kampala* 2000, 47-58.
20. IPC (Integrated food security Phase Classification): *Uganda food security brief. Food security and agricultural livelihoods cluster.* Food and Agriculture Organization, Kampala, Uganda; 2010.
21. Jaarsveld PJ, Faber M, Tanumihardjo SA, Nestel P, Lombard CJ, Benadé AJS: β-carotene–rich orange-fleshed sweet potato improves the vitamin A status of primary school children assessed with the modified-relative-dose–response test1,2,3. *Am J Clin Nutr* 2005, 81: 1080-1087.
22. Jalal F, Nesheim MC, Agus Z, Sanjur D, Habicht JP: Serum retinol concentrations in children are affected by food sources of β-carotene, fat intake, and anthelmintic drug treatment. *Am J Clin Nutr* 1998, 68: 623-629.

23. Kaguongo W, Ortmann G, Wale E, Darroch M, Low J: Factors influencing adoption and intensity of adoption of orange flesh sweetpotato varieties: evidence from an extension intervention in Nyanza and western provinces, Kenya. *Afr J Agric Res* 2012, 7: 493-503.
24. Kasente D, Lockwood M, Vivian J, Whithead A: Gender and the expansion of non-traditional agricultural exports in Uganda. In *Shifting burdens: gender and agrarian change under neoliberalism*. Edited by: Razavi S. Bloomfield, CT, USA: Kumarian Press Inc; 2002:35-65.
25. Lagnaoui FC, Alcázar J, Morales F: A sustainable pest management strategy for the sweetpotato weevil in Cuba: a success story. *Food and Fertilizer Technology Center, Extension Bulletin* 2000, 493: 1-7.
26. Lebesa LN, Khan ZR, Krueger K, Bruce JA, Hassanali A, Pickett JA: Farmers' knowledge and perceptions of blister beetles, *Hycleus* spp. (Coleoptera: Meloidae), as pest herbivores of *Desmodium* legumes in western Kenya. *Int J Pest Manag* 2012, 58: 165-174.
27. Low JW, Arimond M, Osman N, Cunguara B, Zano F, Tschirley D: A food-based approach introducing orange-fleshed sweetpotatoes increased vitamin A intake and serum retinol concentrations in young children in rural Mozambique. *J Nutr* 2007, 137: 1320-1327.
28. Midega CAO, Nyang'au IM, Pittchar J, Birkett MA, Pickett JA, Borges M, Khan ZR: Farmers' perceptions of cotton pests and their management in Western Kenya. *Crop Prot* 2012, 42: 193-201.
29. Mukanga M, Derera J, Tongoona P, Laing MD: Farmers' perceptions and management of maize ear rots and their implications for breeding for resistance. *Afr J Agric Res* 2011, 6: 4544-4554.
30. Mutuura J, Ewell PT, Abubaker A, Munga T, Ajanga S, Irungu J, Omari F, Maobe S: Sweetpotato in the food systems of Kenya. Results of a socio-economic survey. *Proceedings of KARI/CIP technical workshop on collaborative research, Nairobi* 1992.
31. Mwanga ROM, Odongo B, Ocitti p'Obwoya C: Release of five sweetpotato cultivars in Uganda. *HortSci* 2001, 36: 385-386.
32. Mwanga ROM, Niringiye C, Odongo B, Opio AF, Lemaga B, Kyaddondo B, Sserunjogi L, Owori C, Kapinga R: Partnership: an approach to promote orange-fleshed sweetpotato, experiences from Uganda. In *Proceedings of the 13th international society for tropical root crops (ISTRC) symposium. Tropical root and tuber crops: opportunities for poverty alleviation and sustainable livelihoods in the developing world. November 10–14, 2003, Arusha* Edited by: Kapinga R, Kingamkono R, Msabaha M, Ndunguru J, Lemaga B, Tusiime G. 2003, 770-779.
33. Mwanga ROM, Odongo B, Turyamureeba G, Alajo A, Yencho GC, Gibson RW, Smit NEJM, Carey EE: Release of six sweetpotato cultivars. *HortSci* 2003, 3: 475-476.
34. Mwanga ROM, Odongo B, Niringiye C, Alajo A, Kigozi B, Makumbi R, Lugwana E, Namakula J, Mpembe I, Kapinga R, Lemaga B, Nsumba J, Tumwegamire T, Yencho CG: 'NASPOT 7', 'NASPOT 8', 'NASPOT 9 O', 'NASPOT 10 O', and 'Dimbuka-Bukulula' Sweetpotato. *Hortscience* 2009, 44: 828-832.
35. Nderitu J, Silai M, Nyamasyo G, Kasina M: Insect species associated with sweetpotatoes (*Ipomoea batatas* (L.) Lam.) in eastern Kenya. *Int J Sustain Crop Prod* 2009, 4: 14-18.
36. Obopile MD, Munthali C, Matilo B: Farmers' knowledge, perceptions and management of vegetable pests and diseases in Botswana. *Crop Prot* 2008, 27: 1220-1224.
37. Okonya JS, Kroschel J: Incidence, abundance and damage by the sweet potato butterfly (*Acraea acerata* Hew.) and the African sweet potato weevils (*Cylas* spp.) across an altitude gradient in Kabale district, Uganda. *Int J AgriScience* 2013, 3: 814-824.

38. Okonya JS, Kroschel J: Pest status of *Acraea acerata* Hew. and *Cylas* spp. in sweetpotato (*Ipomoea batatas* (L.) Lam.) and incidence of natural enemies in the Lake Albert crescent agro-ecological zone of Uganda. *Int J Insect Sci* 2013, 5: 41-46.
39. Rukarwa RJ, Prentice K, Ormachea M, Kreuze JF, Tovar J, Mukasa SB, Ssemakula G, Mwanga ROM, Ghislain M: Evaluation of bioassays for testing *Bt* sweetpotato events against sweetpotato weevils. *Afr Crop Sci J* 2013, 21: 235-244.
40. SAS Institute Inc: *The SAS system for windows v9.2*. SAS Institute Inc, Cary, North Carolina; 2008.
41. Smit NEJM: *Integrated pest management for sweetpotato in Eastern Africa*. 1997. Dissertation, Wageningen University
42. Smit NEJM, Matengo LO: Farmers' cultural practices and their effects on pest control in sweetpotato in South Nyanza Kenya. *Int J Pest Manage* 1995, 41: 2-7.
43. Smit NEJM, Lugojja F, Ogenga L, Morris W: The sweetpotato butterfly (*Acraea acerata*): a review. *Int J Pest Manage* 1997, 43: 275-278.
44. Soleri D, Smith SE, Cleveland DA: Evaluating the potential for farmer and plant breeder collaboration: a case study of farmer maize selection in Oaxaca, Mexico. *Euphytica* 2000, 116: 41-57.
45. Stathers TE, Rees D, Kabi S, Mbilinyi L, Smit N, Kiozya H, Jeremiah S, Nyango A, Jeffries D: Sweetpotato infestation by *Cylas* spp. in east Africa. I cultivar differences in field infestation and the role of plant factors. *Int J Pest Manage* 2003, 49: 131-140.
46. Tewe OO, Ojeniyi FE, Abu OA: *Sweetpotato production, utilization, and marketing in Nigeria*. International Potato Center (CIP), Lima, Peru; 2003.
47. Tounou AK, Agboka K, Agbodzavu KM, Wegbe K: Maize stemborers distribution, their natural enemies and farmers' perception on climate change and stemborers in southern Togo. *J Appl Biosci* 2013, 64: 4773-4786.
48. UNHS (Uganda National Household Survey): *Abridged report for the socioeconomic module*. Uganda Bureau of Statistics (UBOS), Kampala, Uganda; 2010.
49. World Bank: *Gender in agriculture: sourcebook*. World Bank Publications, Washington, DC; 2009.

CHAPTER 10

A Modelling Methodology to Assess the Effect of Insect Pest Control on Agro-Ecosystems

Nian-Feng Wan[1,2], Xiang-Yun Ji[1] , Jie-Xian Jiang[1] & Bo Li[2]

[1]Eco-environment Protection Research Institute, Shanghai Academy of Agricultural Sciences, Shanghai Key Laboratory of Protected Horticultural Technology, Shanghai 201403, China,

[2] Ministry of Education Key Laboratory for Biodiversity Science and Ecological Engineering, Institute of Biodiversity Science, Fudan University, Shanghai 200438, China.

ABSTRACT

The extensive use of chemical pesticides for pest management in agricultural systems can entail risks to the complex ecosystems consisting of economic, ecological and social subsystems. To analyze the negative and positive effects of external or internal disturbances on complex ecosystems, we proposed an ecological two-sidedness approach which has been applied to the design of pest-controlling strategies for pesticide pollution management. However, catastrophe theory has not been initially applied to this approach. Thus, we used an approach of integrating ecological two-sidedness with a multi-criterion evaluation method of catastrophe theory to analyze the complexity of agro-ecosystems disturbed by the insecticides and screen out the best insect pest-controlling strategy in cabbage production. The results showed that the order of the values of evaluation index ($R_{CC/CP}$) for three strategies in cabbage production was "applying frequency vibration lamps and environment-friendly insecticides 8 times" (0.80) < "applying trap devices and environment-friendly insecticides 9 times" (0.83) < "applying common insecticides 14 times" (1.08). The treatment "applying frequency vibration lamps and environment-friendly insecticides 8 times" was considered as the best insect pest-controlling strategy in cabbage production in Shanghai, China.

INTRODUCTION

Chemical insecticides are often used to repel insect pests in agro-ecosystems. However, the extensive use of chemical insecticides has resulted in many environmental problems, such as ground water contamination, loss of non-target species, reduction of beneficial species, insect pest resistance to insecticides, pest rebound and secondary pest outbreaks and residues in food. In order to protect the environment and ensure food safety, researchers have explored the alternatives to chemical control, and found that biological control, physical control and cultural control are the potential ways of reducing chemical pesticides in agro-ecosystems1,2,3,4,5. For example, many countries like Indonesia6, Norway7, Netherlands8, Sweden8, France8, Spain8 and Belgium9 have already achieved a 50% reduction in the use of chemical insecticides.

Cabbage (*Brassica oleracea*) is regarded as the most important member of the Cruciferae or mustard family, and has been one of the major vegetable crops of the world. The world's first-largest producer of cabbage is China where the annual cultivated area and yield is approximately 100.0×10^4 ha and 10.5×10^{10} kg, respectively. In practice, we have found that *Spodoptera exigua* (Hübner), *Spodoptera litura* (Fabricius), *Plutella xylostella* (Linnaeus), *Pieris rapae* (Linnaeus) and aphids are the most serious insect pests. To control these pests, farmers become accustomed to applying chemical insecticides. Official records indicate that chemical insecticide application in cabbage production in Shanghai of China was more than 7.0 $kg \cdot ha^{-1}$ per year (far higher than the national average), which aroused wide attention of the public.

In order to avoid insecticide contamination and ensure food safety, potential insecticide-reducing strategies have been realized in cabbage production at the source in Shanghai suburbs, such as the installation of insect-killing lamps and trap devices (sex attractant and color plastic strips), and the rotation of the use of different classes of environment-friendly insecticides. However, little is known about which of the insecticide-reducing strategies is the optimal or how to identify the best pest-controlling strategy. Therefore, how to evaluate and determine which technique is the best is a challenging task.

To understand the complexity of the agro-ecosystems, Wan et al.10proposed ecological two-sidedness concept to analyze the positive and negative effects of any disturbances on complex ecosystems consisting of economic, ecological and social subsystems[11,12] and developed the $R_{CC/CP}$ index borrowed from other existing methodologies to evaluate the superiority of different technologies for pest control[13,14,15,16,17,18]. It seems that catastrophe theory can be potentially used to evaluate the complex ecosystems disturbed by pest-controlling strategies.

Catastrophe theory is a special topic within the broader domain of nonlinear dynamics systems that was developed by Rene Thom[19] to study phenomena with non-mechanical discontinuity. Insect pest management can be considered as a process by which sudden and discontinuous changes of state in each of different types of phenomenon in disturbed agro-ecosystems are generated, such as death of insects, resource consumption, pesticide residues and cost input. Thus, catastrophe theory might be a feasible way to shed light on the effects of different pest-controlling strategies on the complex agro-ecosystems. Therefore, this paper used ecological two-sidedness concept based on catastrophe theory to assess the ecological engineering of insect pest-controlling strategies.

RESULTS

Values of evaluation indicators

The aim of this study was to screen out the best pest-controlling strategy for cabbage production in Shanghai suburbs. "Negative effect" caused by the candidate strategies was measured by comprehensive negative effect indicators and "positive effect" was measured by comprehensive positive effect indicators (Table 1). According to the reference materials, reference standard values (x_0) of each factor are listed in Tables 2 and 3; and in accordance with the experimental results, the actual values of the indicators (x_i) of comprehensive cost and comprehensive profit for three strategies are also presented in Tables 2 and 3.

Table 1. Comprehensive cost and comprehensive profit indicators for insect pest-controlling strategies in cabbage production

Sub-object layer (B layer)	Indicator layer (C layer)	Sub-object layer (B layer)	Indicator layer (C layer)
Economic cost (u)	Labor cost (u_1) Insecticide cost (v_1) Cost of non-chemical control (w_1) Instrument depreciation cost (t_1)	Economic profit (u')	Income from cabbage production (u'_1) Time-saving(v'_1) Increased value per unit investment of cabbage commodity (w'_1) Production value created by every labor force (t'_1)
Ecological cost (v)	Killing natural enemy (u_2) Insecticide residue in cabbages (v_2) Active insecticide component dosage (w_2) —	Ecological profit (v')	Beneficial: harmful arthropod ratio (u'_2) Neutral: harmful arthropod ratio (v'_2) Arthropod diversity index (w'_2) Insecticide safety (t'_2)
Social cost (w)	Insecticide stress on society (u_3) Irrigation water consumption (v_3)	Social profit (w')	Percentage of farmer acceptance (u'_3) Life cycle of technology (v'_3)

Note: "—" means no indicator.

Table 2. Indicator data of comprehensive negative effect for insect pest-controlling strategies in cabbage production

			Reference standard value	Treatment one		Treatment two		Treatment three	
Comprehensive cost	Unit	Indicator description	x_0	x_i	x_i/x_0	x_i	x_i/x_0	x_i	x_i/x_0
Economic cost	RMB·ha⁻¹	Labor cost	7800.00	5400.00	0.69	5700.00	0.73	6600.00	0.85
	RMB·ha⁻¹	Insecticide cost	1200.00	661.80	0.55	667.80	0.56	720.00	0.60
	RMB·ha⁻¹	Cost of non-chemical control	1000.00	48.00	0.05	364.95	0.36	0.00	0.00
	RMB·ha⁻¹·a⁻¹	Instrument depreciation cost	190.00	96.00	0.51	108.00	0.57	168.00	0.88
Ecological cost	%	Killing natural enemy	50.00	10.30 ± 1.43b	0.21	8.48 ± 0.97b	0.17	45.35 ± 1.51a	0.91
	%	Insecticide residue in cabbages	10.00	4.27 ± 0.12b	0.43	4.57 ± 0.28b	0.46	9.97 ± 0.35a	1.00
	kg·hm⁻²	Active insecticide component dosage	2.00	0.70	0.35	0.71	0.36	1.47	0.74
Social cost	kg·ha⁻¹	Insecticide stress on society	10.00	5.25	0.53	5.40	0.54	7.20	0.72
	m³·ha⁻¹	Irrigation water consumption	9000.00	7500.00	0.83	7500.00	0.83	7500.00	0.83

Table 3. Indicator data of comprehensive positive effect for insect pest-controlling strategies in cabbage production

			Reference standard value	Treatment One		Treatment Two		Treatment Three	
Comprehensive cost	Unit	Indicator description	x_0	x_i	x_i/x_0	x_i	x_i/x_0	x_i	x_i/x_0
Economic profit	RMB·ha⁻¹	Cabbage income	80000.00	75274.50 ± 71.30a	0.94	75175.50 ± 73.80a	0.94	74003.50 ± 258.90a	0.93
	workday·ha⁻¹	Time-saving	150.00	120.00	0.80	105.00	0.70	60.00	0.40
	—	Increased value per unit investment of cabbage commodity	1.00	0.82	0.82	0.76	0.76	0.61	0.61
	RMB·ha⁻¹	Production value created by every labor force	300.00	278.80	0.93	263.80	0.88	224.30	0.75
Ecological profit	—	Beneficial, harmful arthropod ratio	16.00	15.67 ± 0.66a	0.98	15.93 ± 0.49a	1.00	8.83 ± 0.20b	0.55
	—	Neutral, harmful arthropod ratio	16.00	13.43 ± 0.58a	0.84	14.33 ± 0.35a	0.90	8.57 ± 0.29b	0.54
	—	Arthropod diversity index	5.00	4.46 ± 0.04a	0.89	4.56 ± 0.03a	0.91	3.74 ± 0.09b	0.75
	mg·kg⁻¹	Insecticide safety	2500.00	2446.20	0.98	2446.20	0.98	541.50	0.22
Social profit	%	Percentage of farmer acceptance	100.00	88.46	0.88	84.62	0.85	80.77	0.81
	a	Life cycle of technology	10.00	6.00	0.60	5.00	0.50	1.00	0.10

Note: "–" means symbol without unit; increased value per unit investment of cabbage commodity = (cabbage income of technology optimization area in unit area – cabbage income of conventional pest-controlling area in unit area) ÷ investment value of technology for s cabbage production in unit area; production value created by every labor force, refers to the economic income of cabbages created by every workday.

Computational results

In light of ecological two-sidedness concept and catastrophe theory, comprehensive cost (cc) (composed of economic, ecological and social costs) and comprehensive profit (cp) (composed of economic, ecological and social profits) were considered as state variables separately; and different indicators as different control variables. The objective index u_{cc} of comprehensive cost or u_{cp} of comprehensive profit (*A* layer), which is called state variable in catastrophe theory, is the classification of pest-controlling strategies. There are three indicators in the second evaluation level (*B* layer), including economic cost (u) or economic profit (u'), ecological cost (v) or ecological profit (v'), and social cost (w) or social profit (w'). In the root evaluation level (*C* layer), nine indicators for comprehensive cost and ten indicators for comprehensive profit were involved.

The initial data (x_i) from the *C* layer were transformed into dimensionless data with value from 0 to 1. To get dimensionless data (Tables 2 and 3), each of initial data was divided by each reference standard data (x_0). There were two control variables for social cost and social profit which were both adopted by cusp catastrophe model, three control variables for ecological cost which were adopted by swallowtail catastrophe, and four control variables for economic cost, economic profit and ecological profit which were all adopted by butterfly catastrophe model. The evaluation results of the different strategies in *C* layer are presented in Table 4 with the analysis of catastrophe models. According to the complementary principle, the evaluation results of the different strategies in *B* layer and *A* layer were obtained and are given in Tables 5 and 6. Hence, $W_{cc/cp}$, matrix of ratio of comprehensive cost to comprehensive profit ($R_{cc/cp}$) for three treatments was obtained by $W_{cc/cp} = (R_{cc/cp,1}, R_{cc/cp,2}, R_{cc/cp,3})^T = (0.75/0.94, 0.77/0.93, 0.85/0.79)^T = (0.80, 0.83, 1.08)^T$ (Table 6). According to ecological two-sidedness concept, the lower the value of $R_{cc/cp}$ was, the more reasonable the candidate strategy was10,15. Therefore, the order of superiority of pest-controlling strategies was treatment one > treatment two > treatment three, i.e., the treatment "applying frequency vibration lamps and environment-friendly insecticides 8 times" should be identified as the best insect pest-controlling strategy in cabbage production in Shanghai, China.

Table 4. Catastrophe progression values of indicator layer for insect pest-controlling strategies in cabbage production

Sub-object layer (*B* layer)	Indicator layer (*C* layer)	Treatment one	Treatment two	Treatment three
Economic cost (u)	Labor cost (u_1)	0.83	0.85	0.92
	Insecticide cost (v_1)	0.82	0.82	0.84
	Cost of non-chemical control (w_1)	0.47	0.77	0.00
	Instrument depreciation cost (t_1)	0.87	0.89	0.97
Ecological cost (v)	Killing natural enemy (u_2)	0.46	0.41	0.95
	Insecticide residue in cabbages (v_2)	0.75	0.77	1.00
	Active insecticide component dosage (w_2)	0.77	0.77	0.93
Social cost (w)	Insecticide stress on society (u_3)	0.73	0.73	0.85
	Irrigation water consumption (v_3)	0.94	0.94	0.94
Economic profit (u')	Cabbage income (u'_1)	0.97	0.97	0.96
	Saving time (v'_1)	0.93	0.89	0.74
	Increased value per unit investment of cabbage commodity (w'_1)	0.95	0.93	0.88
	Production value created by every labor force (t'_1)	0.99	0.97	0.94
Ecological profit (v')	Beneficial: harmful arthropod ratio (u'_2)	0.99	1.00	0.74
	Neutral: harmful arthropod ratio (v'_2)	0.94	0.97	0.81
	Arthropod diversity index (w'_2)	0.97	0.98	0.94
	Insecticide safety (t'_2)	1.00	1.00	0.74
Social profit (w')	Percentage of farmer acceptance (u'_3)	0.94	0.92	0.90
	Life cycle of technology (v'_3)	0.84	0.79	0.46

Table 5. Catastrophe progression values of sub-object layer for insect pest-controlling strategies in cabbage production

Object layer (*A* layer)	Sub-object layer (*B* layer)	Treatment one	Treatment two	Treatment three
Comprehensive cost (u_{cc})	Economic cost (u)	0.75	0.83	0.68
	Ecological cost (v)	0.66	0.65	0.96
	Social cost (w)	0.84	0.84	0.90
Comprehensive profit (u_{cp})	Economic profit (u')	0.96	0.94	0.88
	Ecological profit (v')	0.98	0.99	0.81
	Social profit (w')	0.89	0.86	0.68

Table 6. Catastrophe progression values of object layer for insect pest-controlling strategies in cabbage production

Object layer (*A* layer)	Treatment one	Treatment two	Treatment three
Comprehensive cost (u_{cc})	0.75	0.77	0.85
Comprehensive profit (u_{cp})	0.94	0.93	0.79

Experimental results

The increasing order of labor cost was treatment one (applying frequency vibration lamps and environment-friendly insecticides 8 times) < treatment two (applying trap devices and environment-friendly insecticides 9 times) < treatment three (control areas, no lamps, no trap devices, applying common insecticides 14 times). The orders of insecticide cost, instrument depreciation cost, active insecticide component dosage, and insecticide stress on society were all the same. The insecticide residues in cabbages were significantly greater in the control than other treatments but not significantly different among treatments, and the same order was presented in the indicator of the efficacy of killing natural enemies. Cost of non-chemical control was just inputted in treatment one and treatment two; three treatments were given the same irrigation water consumption (Table 2).

The order of income from cabbage production was treatment one > treatment two > treatment three; the order of insecticide safety was treatment one = treatment two > treatment three, and the same orders were for time-saving, increased value per unit investment of cabbage commodity, production value created by every labor force and percentage of farmer acceptance and life cycle of technology. The ratio of the abundance of beneficial arthropods to the one of harmful arthropods was significantly higher both in treatment one and treatment two than the control (Table 3), and the same orders were presented in the ratio of the abundance of neutral arthropods to the one of harmful arthropods and the arthropod diversity.

DISCUSSION

Chemical insecticides might be the most effective, convenient and reliable method for pest suppression once the pest situation is out-of-hand. However, chemical insecticides often cause serious environmental and food safety problems, so it might be best to minimize the use of chemical insecticides, or use them carefully or selectively[3,4,5]. In order to control insect pests and to avoid the soil contamination by chemical insecticides, researchers have explored the strategies such as biological control[20], physical control[21] and cultural control[22,23,24]. Here we used the strategies of optimizing insecticide variety structure (the use of eight slightly toxic insecticides was rotated, and the four moderately toxic insecticides, 48% Chlorpyrifos EC, 20% fenvalerate EC, 18% Bisultap AS and 2.5% Deltamethrin, were excluded), installing trap devices and installing lamps to control insect pests in cabbage fields.

Different strategies use varying doses of insecticides, and might affect the richness and abundance of arthropods. Our study indicated that arthropod communities were more diverse and stable, and species richness and abundance of beneficial arthropods were significantly higher in the two treatments with lamps or trap devices than in control. Compared to those in the control, in the strategy of "lamps + environment-friendly insecticides 8 times" and the strategy of "trap devices + environment-friendly insecticides 9 times", the arthropod diversity increased respectively by 19.3% and 21.9%, the beneficial: harmful arthropod ratio increased by 77.5% and 80.4%, the beneficial: harmful arthropod ratio increased by 56.7% and 67.2%, while the active ingredients of insecticides decreased respectively by 52.0% and 51.6% in the two strategies.

One of the most important sections of catastrophe theory is that catastrophe progression method has no need to assign the weight to each indicator and has a simple calculation procedure, which has been widely applied to the comprehensive evaluation[25,26,27]. Here we used this method to assess the ecological engineering of insect pest-controlling strategies in agro ecosystems, which can help to solve the problem of pesticide pollution. The information provided here is useful for the application of catastrophe theory to the field of pest management assessment. However, further attention should be paid to the establishment of the fuzzy membership functions of underlying indicators, the dissociation of the indicator system and the priority of indicators from each level.

Ecological two-sidedness approach proposed by Wan et al.[10] has been applied to the ecological assessment of vegetable ecosystems[15], paddy ecosystems[13,17] and protected horticultural fields[14,16], as well as to the assessment of the combined reduction of chemical pesticides and chemical fertilizers for low-carbon agriculture[18]. Here we adopted this theory to analyze the effects of different pest-controlling strategies on cabbage-producing ecosystems, and established a comprehensive evaluation system based on catastrophe theory, but it might also have the potential to apply other theories to treat such problems.

The methodology developed here reduces the subjectivity of decision-makers, farmers and experts. The maximum value of the $R_{cc/cp}$ was 1.08 in the control, implying that the control was the worst strategy for pest control in cabbage production in Shanghai. Our conclusion here was similar to that of the expert group from local practices. However, it is still necessary to further determine if a more scientific and reasonable evaluation system has been identified, and to validate that system in practice. Also, we need to further improve both the strategies and the assessment methods to optimize the identification of pest-controlling strategies for the cabbage production.

METHODS

Experimented materials

Cabbage variety (Meimao II), produced by Xiehe company of Japan; frequency vibration insecticidal lamps (PS-15 II type), manufactured by Jiaduo company limited of Henan province of China; manual pesticide sprayers (3WBS-16 type), manufactured by Qiangye sprayer factory of Luozhuang district of Linyi city of Shandong province of China; trap devices (tobacco cutworm sex pheromone, beet armyworm sex pheromone, diamond back moth sex pheromone, plastic tubes for suspending sex pheromone, yellow and blue plastic strips), manufactured by Zhangzhou Enjoy Technology Company Limited of China. Insecticides included 10% Imidacloprid WP, 10% Chlorfenapyr SC, 20% Tebufenozide SC, 15% Indoxacarb SC, 40% Phoxim EC, 25% Buprofezin WP, 48% Chlorpyrifos EC, 2.5% Deltamethrin EC, 4.5% Beta cypermethrin EC, 18% Bisultap AS, 20% Fenvalerate EC and ACNPV SC.

Experiment design and method

Experiments were conducted by adopting a series of individual 1 ha plots in cabbage fields in Modern Agriculture Park of Songjiang District, Shanghai of China (121.13°E, 30.94°N) from April to October in 2006. Three treatments were included in the experiment design. In treatment one, 1.0-m high frequency vibration lamps were installed 30 days after the transplanting of cabbage. In treatment two, lamps were not adopted but trap devices including sex pheromone lures and color (yellow and blue) trap strips were used throughout the entire growing period of cabbages after transplanting. However, lamps or trap devices were not installed in the control (treatment three) in which conventional insect pest-controlling strategies were used by farmers with empirical planting. Each treatment and the control were replicated three times.

Slightly toxic insecticides (10% Imidacloprid WP, 10% Chlorfenapyr SC, 20% Tebufenozide SC, 15% Indoxacarb SC, 40% Phoxim EC, 25% Buprofezin WP, 4.5% Beta cypermethrin EC and ACNPV SC) were used in treatments one and two, while the moderately toxic insecticides (48% Chlorpyrifos EC, 2.5% Deltamethrin EC, 20% Fenvalerate EC and 18% Bisultap AS) and slightly toxic ones (10% Imidacloprid WP and 40% Phoxim EC) were done in the control. In treatment and control fields, insecticides were appropriately used in accordance with the population dynamics of insect pests.

The total growing period of cabbages was approximately 100 days, seeds sown in early July, seedlings transplanted at a density of 45000 plants·ha^{-1} in early August, cabbages harvested in middle October. The sex pheromone was placed inside a capillary vessel (0.1-m long and 1.5-mm caliber). One each tobacco cutworm, beet armyworm and diamondback moth sex pheromone trap was installed at 20 m intervals in each direction in each plot of treatment two, was hung 1.0 m above the ground level, and was replaced once per month. Seventy-five yellow plastic strips per hectare and seventy-five blue plastic strips per hectare were installed with equidistance in each plot of treatment two, and were suspended 0.2 m above the plants. Color strips' dimensions were all 0.3-m long, 0.4-m wide, 1.5-mm thick, with sticky jelly well distributed on. Mini pests like aphids and whiteflies tend towards yellow things while thrips tend towards blue things so that they stick to death on the plastic strips, and the aim of pest control can be achieved.

All the treatments were given similar management of water and fertilizers during the whole period of cabbage growth. Weeds were managed by hand pulling. Arthropods were systematically scouted and sampled at about 20–30 d intervals in all fields. "Z" style sampling with 5 dots of 100 plants was employed at random from seedling stage to harvesting stage16. To ensure the continuity and reliability of the statistics, we routinely recorded the service condition of the agricultural resources in each of the plots, the planting and growing management procedures, cost accounting factors, the sales status of cabbages, etc. All data presented were analyzed with Microsoft Excel and SPSS16.0 softwares.

Determination of evaluation indicators

According to the idea of ecological two-sidedness approach, the agro-ecosystems disturbed by the insect pest-controlling strategies either acquire profit or incur a cost (Figure 1). The complex ecosystem is composed of three subsystems (economic, ecological and social subsystems) with mutual restrictions. The relevant effect indicators should be considered when measuring the standard in the complex ecosystem. These indicators need to address the following questions: whether the economic subsystem is beneficial, whether the ecological subsystem is stable, and whether the social subsystem is rational17. According to above goals, a specific indicator system should be confirmed that maximizes the comprehensive profit and minimizes the comprehensive cost. As the complex ecosystem in cabbage fields is measured by indicators of the three subsystems, constructing an indicator system for pest control should abide by the following criteria: (1) indicators should reflect economic, ecological and social aspects, economic cost and profit, resources and time consumption, acceptance degree, as well as the impact of candidate strategies on society and nature17; (2) indicators should reflect the short- and long-term profits, partial and whole cost that candidate strategies can bring15; (3) quantitative indicators should be adopted as far as possible, and subjective deviation should be avoided17; (4) comprehensive indicators should be used and each indicator should keep independent without directly interactive relationships16,28; (5) the indicators should complement each other, and the structure produced by the indicators should reflect fully and even completely its various functional characteristics and be constituted completely29; (6) indicators should reflect the diversity, stability and structure characteristics of arthropods14. According to the above criteria, the evaluation indicators for insect pest-controlling strategies in cabbage fields were summarized and listed in Table 1. Indicator standards refer to safe and health standards of agricultural produce issued by Shanghai Standardization Association.

Different strategies might lead to different economic costs which include the labor cost, insecticide cost, instrument depreciation cost, cost of non-chemical control (lamp and electric cost, trap device cost), etc. Positive economic effect indicators were split into either direct or indirect. Direct effects included cabbage income, cabbage commodity increase value per unit investment and production value created by the labor force, while indirect effects included saving time (the lowest person workdays per hectare was 390.0 in cabbage production).

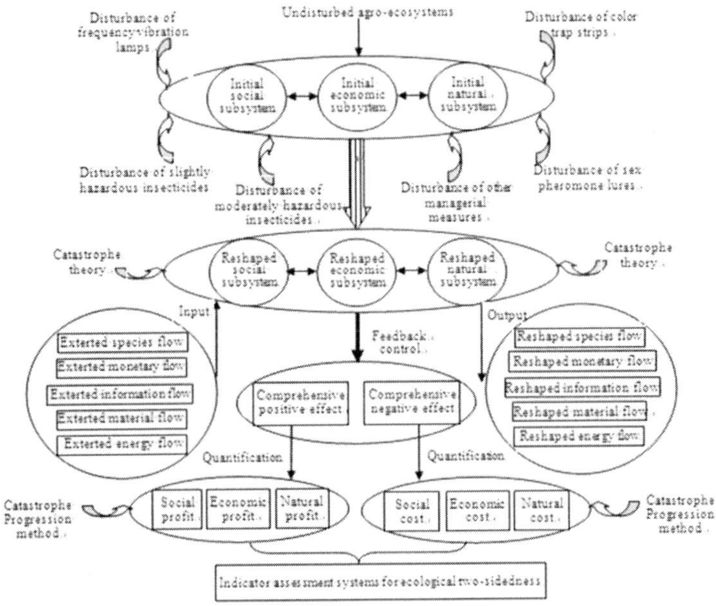

Figure 1: The framework of understanding the agro-ecosystems disturbed by the insect pest-controlling strategies.

Negative ecological effect indicators reflected the adverse impact of strategies on the ecological subsystem. The effect of each strategy on natural enemies was measured by the difference between 3.3 and the actual natural enemy diversity in all treatments and control fields. The greater the percentage of difference between the two diversity values, the stronger the effect of the strategy for damaging natural enemies15. Enzyme inhibition rate method was adopted to detect pesticide residues in cabbage produce. The active insecticide component dosage played an important role in the preservation of natural environment and the maintenance of food safety.

Positive ecological indicators reflected the profitable effect of strategies on the ecological subsystem. Beneficial and harmful arthropods are the base of the food web in the tritrophic interactions in agro-ecosystems, both of which play important roles in the regulation of harmful arthropods; arthropod diversity maintained the diversity and stability of agro-ecosystems; insecticide safety was measured by the geometric averages of the LD_{50} for acute oral toxicity of insecticides on rats.

Social effect indicators reflected the impact of strategies on the social subsystem. A questionnaire survey was conducted to obtain information regarding farmers' acceptance of various strategies. Life cycle of technology had a profound effect on agricultural production which involved some other sections of social production. The indicator of insecticide stress on society was measured according to the total insecticide dosage, as the more insecticides were used, the greater stress of the insecticides on society might be. Irrigation water consumption was measured in reference to actual volume of water consumed.

Evaluation theory and methods

Catastrophe theory, a new mathematical discipline, is a special branch of dynamical systems theory and can be adopted to evaluate the status and analyze the development trends, known as "the mathematical revolution after calculus"30. Catastrophe theory has been widely used, and one common application is the use of the catastrophe progression method derived from catastrophe theory to solve multiple criteria decision making problems27. In the multi-criteria evaluation method based on catastrophe theory, the dependency of state variables on control variables is determined by the catastrophe fuzzy membership functions, rather than weights assigned by the users, and different control variables have different impacts on state variables25.

The general evaluation process of catastrophe progression method consists of the following steps. First, the complex ecosystem is divided into three subsystems (economic, ecological and social subsystems) with different evaluation indicators in accordance with the inner mechanisms of the complex ecosystem. Next, the initial data from the underlying layers are normalized by the integration of catastrophe theory and fuzzy mathematics, and to realize this, multidimensional catastrophe fuzzy membership functions should assign values ranging from 0 to 1 to resolve the incomparability of various initial data generated by differences in the data span and dimensions. Finally, the total catastrophe fuzzy membership functions of the complex ecosystems are obtained by the normalized data25.

Thom has presented seven fundamental types of catastrophe models, among which the models of elliptic umbilic catastrophe, hyperbolic umbilic catastrophe and parabolic umbilic catastrophe are under the behavior variation of two dimensions and the models of fold catastrophe, cusp catastrophe, swallowtail catastrophe and butterfly catastrophe are commonly applied. The four potential functions and normalization formulas are as follows:

$$\text{Fold catastrophe}: f(x) = x^3 + ux \qquad x_u = u^{1/2}$$
$$\text{Cusp catastrophe}: f(x) = x^4 + ux^2 + vx \qquad x_u = u^{1/2}, x_v = v^{1/3}$$
$$\text{Swallowtail catastrophe}: f(x) = x^5 + ux^3 + vx^2 + wx$$
$$x_u = u^{1/2}, x_v = v^{1/3}, x_w = w^{1/4}$$
$$\text{Butterfly catastrophe}: f(x) = x^6 + ux^4 + vx^3 + wx^2 + tx$$
$$x_u = u^{1/2}, x_v = v^{1/3}, x_w = w^{1/4}, x_t = t^{1/5}$$

where $f(x)$ is the potential function of the state variable x, and u, v, w and t are control variables of the state variable.

The catastrophe progression of each control variable can be computed from the initial fuzzy subordinate function, based on normalization formulas. When performing recursive computations, non-complementary principle (the principle of minimum values) or complementary principle (the principle of mean values) is selected after the determination of whether or not the indicators are complementary to each other or interchangeable within one subsystem25. The non-complementary principle means that the control variables of a system, such as u, v, w, and t cannot offset each other. Therefore, when finding the value of the state variable x using the normalization formulas, the smallest of the state variable values corresponding to the control variables (x_u, x_v, x_w and x_t) is selected as the state variable value of the whole system. The complementary principle means that the control variables complement each other so that each of them tends to reach the average value, i.e., $x = (x_u + x_v + x_w + x_t)/4$. Hierarchically doing the calculation in the same way, the value of the overall catastrophe subordinate function can be obtained27.

REFERENCES

1. Aufauvre, J. et al. Parasite-insecticide interactions: a case study of Nosema ceranae and fipronil synergy on honeybee. Sci. Rep. 2, 326 (2013).
2. Brühl, C. A., Schmidt, T., Pieper, S. & Alscher, A. Terrestrial pesticide exposure of amphibians: an underestimated cause of global decline? Sci. Rep. 3, 1135 (2013).
3. Gergs, A., Zenker, A., Grimm, V. & Preuss, T. G. Chemical and natural stressors combined: from cryptic effects to population extinction. Sci. Rep. 3, 2036 (2013).
4. Gordon, J. R., Goodman, M. H., Potter, M. F. & Haynes, K. F. Population variation in and selection for resistance to pyrethroid-neonicotinoid insecticides in the bed bug. Sci. Rep. 4, 3836 (2014).
5. Hossard, L. et al. Effects of halving pesticide use on wheat production. Sci. Rep. 4, 4405 (2014).
6. Oka, I. N. Success and challenges of the Indonesian national integrated pest management program in the rice-based cropping system. Crop Prot. 10, 163–165 (1991).

7. Sathre, M. G., Hege, M. O. & Hofsvang, T. Action programmers for pesticide risk reduction and pesticide use in different crops in Norway. Crop Prot. 18, 207–215 (1999).
8. Jansma, J. E., van Keulen, H. & Zadoks, J. C. Crop protection in the year 2000: a comparison of current policies towards agrochemical usage in four West European countries. Crop Prot. 12, 483–489 (1993).
9. Smet, B. D., Claeys, S. & Vagenende, B. The sum of spread equivalents: a pesticide risk index used in environmental policy in Flanders, Belgium. Crop Prot. 24, 363–374 (2005).
10. Wan, N. F., Jiang, J. X. & Li, B. Modeling ecological two-sidedness for complex ecosystems. Ecol. Model. 287, 36–43 (2014a).
11. Ma, S. J. & Wang, R. S. The social-economic-natural complex ecosystem. Acta Ecol. Sin. 4, 1–9 (1984).
12. Holling, C. S. Understanding the complexity of economic, ecological, and social systems. Ecosystems 4, 390–405 (2001).
13. Wan, N. F., Jiang, J. X., Xu, J. X. & Wu, J. C. Application of the analytic hierarchy process to pest management in the rice fields of Shanghai City. Acta Ecol. Sin. 25, 2997–3002 (2005).
14. Wan, N. F., Jiang, J. X. & Ji, X. Y. Application of polygon synthesis for pesticide reduction evaluation in protected horticultural vegetable fields in Shanghai. Acta Sci. Circum. 28, 582–588 (2008).
15. Jiang, J. X. & Wan, N. F. A model for ecological assessment to pesticide pollution management. Ecol. Model. 220, 1844–1851 (2009).
16. Wan, N. F., Jiang, J. X., Ji, X. Y. & Deng, J. Y. Application of analytic hierarchy process-based model of ratio of comprehensive profit (RCCCP) in pest management. Ecol. Econ. 68, 888–895 (2009).
17. Wan, N. F., Jiang, J. X., Ji, X. Y., Huang, K. H. & Li, B. An ecoengineering assessment index for chemical pesticide pollution management strategies to complex agro-ecosystems. Ecol. Eng. 52, 203–210 (2013a).
18. Wan, N. F., Ji, X. Y., Jiang, J. X., Qiao, H. X. & Huang, K. H. A methodological approach to assess the combined reduction of chemical pesticides and chemical fertilizers for low-carbon agriculture. Ecol. Indic. 24, 344–352 (2013b).
19. Kozak, J. J. & Benhamt, C. J. Denaturation: an example of a catastrophe. Proc. Natl. Acad. Sci. U.S.A. 71, 1977–1981 (1974).
20. Bhagat, D., Samanta, S. K. & Bhattacharya, S. Efficient management of fruit Pests by pheromone nanogels. Sci. Rep. 3, 1294 (2013).
21. Grasswitz, T. R. & Fimbres, O. Efficacy of a physical method for control of direct pests of apples and peaches. J. Appl. Entomol. 137, 790–800 (2013).
22. Wan, N. F. et al. Ecological engineering of ground cover vegetation enhances the diversity and stability of peach orchard canopy arthropod communities. Ecol. Eng. 70, 175–182 (2014b
23. Wan, N. F. et al. Ecological engineering of ground cover vegetation promotes biocontrol services in peach orchards. Ecol. Eng. 64, 62–65 (2014c).
24. Wan, N. F., Ji, X. Y. & Jiang, J. X. Testing the enemies hypothesis in peach orchards in two different geographic areas in eastern China: the role of ground cover vegetation. PLoS One 9, e99850 (2014d).
25. Wang, W., Liu, S., Zhang, S. & Chen, J. Assessment of a model of pollution disaster in near-shore coastal waters based on catastrophe theory. Ecol. Model. 222, 307–322 (2011).
26. Wu, J., Sun, J., Liang, L. & Zha, Y. Determination of weights for ultimate cross efficiency using Shannon entropy. Expert Syst. Appl. 38, 5162–5165 (2011).
27. Zhang, L., Liu, X., Li, D. & Fu, Z. Evaluation of the rural informatization level in four Chinese regions: a methodology based on catastrophe theory. Math. Comput. Model. 58, 868–876 (2013).
28. Young, A. Change and constancy: analysis of publications in agroforestry systems. Agroforestry Syst. 13, 195–202 (1991).

29. Wu, G. Nitrogen, phosphorus and potassium recycling in an agroforestry ecosystem of Huanghuaihai plain: with Paulownia elongata intercropped wheat and maize as an example. J. Environ. Sci. 10, 189–196 (1998).
30. Wu, Q., Xie, K. & Chen, Z. A catastrophe model on the evaluation and classification of mine disaster rescue measures. Syst. Eng. Proce. 4, 484–489 (2012).

CHAPTER 11

The Rise and Demise of Integrated Pest Management in Rice in Indonesia

Craig Thorburn

Centre for Geography and Environmental Science, Monash University, Victoria 3800, Australia

ABSTRACT

Indonesia's 11-year (1989–1999) National Integrated Pest Management Program was a spectacularly successful example of wide-scale adoption of integrated pest management (IPM) principles and practice in a developing country. This program introduced the innovative Farmer Field School model of agro-ecosystem-based experiential learning, subsequently adapted to different crops and agricultural systems in countries throughout the world. Since the termination of the program in 1999, Indonesia has undergone profound changes as the country enters a new era of democratic reform. Government support for the national IPM program has wavered during this period, and pesticide producers and traders have taken advantage of the policy vacuum to mount an aggressive marketing campaign in the countryside. These factors have contributed to a reappearance of the pesticide-induced resurgent pest problems that led to the establishment of the National IPM Program in the first place.

Keywords: integrated pest management (IPM); rice; rice brown planthopper (BPH); *nilaparvata lugens* stål; insecticide-induced resurgence; Indonesia

1. INTRODUCTION

Rice is one of the most important human food crops, occupying one-tenth of the world's arable land and accounting for around 20 per cent of total global grain production. In some Asian countries rice comprises up to one third of total planted area. Since World War II, the total global area planted in rice has increased by 67 percent, yields have increased by 95 percent, and total world production trebled [1]. Compared to other grains, however, a relatively small proportion of the world's rice crop is traded internationally [2]. Therefore, self-sufficiency in rice is perceived as key to ensuring national food security in countries where rice is the main staple [3].

Indonesia, like most other South and Southeast Asian countries, undertook an aggressive rice intensification program from the 1960s onward, employing Green Revolution "packages" of high-yielding varieties (HYV), chemical fertilizers, insecticides, improved irrigation and agricultural extension [4,5]. During the 1970s and '80s, Indonesia became one of the greatest success stories of this approach, increasing average rice yields from two to four and a half tons per hectare (the highest in tropical Asia); transforming from being the world's largest rice importer in the 1960s, to achieving self-sufficiency in rice by 1984 [6,7].

However, dependency on a small number of HYV varieties combined with prophylactic application of chemical pesticides rendered the country's rice farmers vulnerable to plagues of insecticide-induced resurgent pests, particularly the rice brown planthopper (BPH, *Nilaparvata lugens* Stål), which threatened to undermine the impressive production gains of the Green Revolution varieties and techniques.

Beginning in the mid-1980s, Indonesia introduced integrated pest management (IPM) as a new crop management approach, developing the Farmer Field School (FFS) model of empowering farmers to make informed crop management decisions. Both IPM and the FFS model represented radical departures from the agricultural intensification and extension models that preceded them, shifting from the top-down dissemination of knowledge, instructions and inputs, to a farmer-driven crop management strategy based on ecosystem health and the simple principle of "grow a healthy crop". The program achieved prodigious results during its 14-year duration, and helped to fundamentally transform the relationship between Indonesia's agricultural bureaucracy and the country's millions of rice farmers.

However, during the political and social upheaval leading up to and following the resignation of Indonesia's strongman president Suharto in 1998, national and local government support for IPM principals and programs in Indonesia has wavered. Today, a decade and a half into Indonesia's democratic *reformasi* era, rice farmers are using more pesticides than ever, with the consequence that the country is again experiencing devastating BPH infestations in many of its key rice-producing areas.

This article relates the story of the successful promotion of IPM in rice in Indonesia, followed by a discussion of the national agriculture bureaucracy's retreat from IPM approaches and concomitant resurgence of the BPH pest. The agroecology of rice itself plays a central part in this story, as does the role of Indonesia's rice farmers. Before discussing the introduction of IPM in rice in Indonesia, it is useful to briefly review the history of the development and spread of HYV rice varieties, as well as Indonesia's experiences with Green Revolution technologies.

The article draws on several years' observation and conversations with key protagonists of Indonesia's national IPM program. The author lived and worked in Indonesia throughout most of the period described in the article. Although he never worked directly for the program, his work with Indonesian and international NGOs and bilateral aid projects put him in close contact with many key players. Beginning in 2006, several years after the termination of the national IPM program in 1999, he was encouraged by colleagues who formerly worked with the program—many of whom were still engaged in the promotion of sustainable agriculture and farmer empowerment in Indonesia—to write the story of the national IPM program, in particular the evolution of the Farmer Field School approach. In 2006 and again in 2012, the author spent several weeks traversing the island of Java meeting with IPM farmer group members, activists, academics, and pest control workers from the Ministry of Agriculture and regional agricultural agencies, to attempt to assess the resilience and sustainability of the IPM program's message, approaches and institutional infrastructure. This article is the result of these explorations.

A caveat is in order regarding statistical and technical data used in—or absent from—this report. Accurate statistical data is notoriously difficult to come by in Indonesia. Many factors contribute to this situation, including the generally poor quality of government data [8], the perceived political sensitivity of rice production statistics and information on rice pest infestations and crop losses, and a legacy of suspicion and mistrust among Indonesian government bureaucracies toward foreign researchers [9]. Some fairly accurate macro-level data were available from reports produced by or for the National IPM Program during the years of its existence. Descriptions and analysis of the more recent increase in insecticide use and concomitant BPH resurgence draw from direct observations, conversations with Indonesian farmers, academics, students and a few government officials and fieldworkers. Studies of this phenomenon undertaken by Indonesian academics rely entirely on small-scale field surveys in a smattering of affected areas. To date, the results of these studies have only been published on researchers' websites and some Indonesian newspapers and magazines.

2. THE (HUMAN) ECOLOGY OF RICE

Rice has fed more people over a longer period than has any other crop, and is presently the source of one quarter of global per capita caloric intake [10]. Chang [11] (p. 143) postulates that the cultivation of Asian rice began in "a broad belt that extended from the Ganges plains below the foothills of the Himalayas, across Upper Burma, northern Thailand, and Laos, to North Vietnam and south China between 10,000 and 4000 BC".

In the early Neolithic era, rice was cultivated in clearings in the forests, direct seeded and dependent on rainfall for water. The techniques of puddling the soil and transplanting seedlings probably developed independently in several locations in China as early as the 2nd or 3rd millennium BC [10]. These innovations, along with the discovery of early maturing and photoperiod insensitive varieties, led to rice becoming the most highly productive grain crop that came to dominate agricultural production throughout Asia [12]. Migrants from the Asian mainland brought wetland rice cultivation to the Philippines during the second millennium BC, and Deutero-Malays carried the practice to the Indonesian archipelago in about 1500 BC.

Wet-land rice cultivation is much more demanding than other staples such as wheat, barley or millet, both in terms of labor and also in human ingenuity, inventiveness and organization. It also supports high concentrations of human population [13]. The Indonesian island of Java, for instance, comprises less than seven per cent of the country's total land area, but produces over 53 per cent of its rice, and is home to 57 per cent of the country's population [14,15].

The extensive geographical distribution of wet-land rice cultivation, combined with its long ecological history, has resulted in the creation of a vast man-made ecozone. Despite being mainly a monocrop system, the ecological complexity of wet-land rice is unique among the world's agro-ecosystems. The arthropod species richness of tropical rice fields surpasses that of most natural (as opposed to man-made) temperate systems [16]. The rice field agro-ecosystem is particularly stable as a result of an extremely rich web of generalist predators. An early-season abundance of detritus- and plankton-feeding insects provides a well-dispersed food source for a diverse community of generalist predators, allowing predator populations to develop well in advance of rice-pest populations. Pest infestations, therefore, usually remain well below economically damaging levels [17]. While numerous host-specific parasites and parasitoids play crucial roles in these complex webs, it is the early season prevalence of generalist predators—spiders, water striders, dragonfly and damselfly nymphs and adults, and ground beetles—that makes the wet-rice agro-ecosystem so uniquely stable and robust [18,19].

These in-field ecological webs can be enhanced or disrupted by many different habitat factors—*i.e.*, spatial planting patterns, staggered or synchronous planting, fallow periods or inter-cropping, water-use patterns, the organic content of paddy soil and water, whether stubble is burned or left to decay, and the use of insecticides and other agricultural chemicals. Variations in any of these factors affect predator and prey communities living on the plants, on the water surface, and in the water [16].

Countless generations of intense interaction with these systems has produced a wealth of place-specific farmer's knowledge of biological and cultural controls. Much of this knowledge has been discounted by the promoters of agricultural modernization. One of the prominent features of the science-based Green Revolution in rice has been standardization—of varieties, and of crop management knowledge and practice. To date, most scientific research relating to arthropods in tropical rice (and management strategies produced) has been directed at only a small handful of "pest" species, without examining the biotic linkages to the rest of the system [16].

3. THE GREEN REVOLUTION IN RICE

The term "Green Revolution" was first coined by William Gaud, Director of USAID, in a speech to the Society for International Development in Washington, DC. He said:

These and other developments in the field of agriculture contain the makings of a new revolution. It is not a violent Red Revolution like that of the Soviets, nor is it a White Revolution like that of the Shah of Iran. I call it the Green Revolution. [20]

A highly emotive term, the "revolution" in cereal production in developing countries is alternately credited with preventing a billion famine deaths [21], or causing a host of dire social and environmental problems that threaten the livelihoods of billions more [22,23].

Regardless of ideological standpoint, few would dispute Dalrymple's [24] (p. 1067) claim that "high-yielding varieties (HYVs) ... of wheat and rice have spread more widely, more quickly, than any other technological innovation in the history of agriculture". Since the International Rice Research Institute (IRRI) released its first successful semi-dwarf variety IR 8 in 1966, the total area planted in HYV rice in Asia, Latin America, the Near East and Africa has increased yearly [1,10]. In Indonesia, more than 85 per cent of wet-rice *sawah* is planted in High-Yielding Varieties [25]. Combined with improved irrigation, HYVs seeds have been the most important factor in the country's spectacular threefold increase in rice production [5,7].

Spearheaded by the Rockefeller Foundation and USAID-supported Consultative Group on International Agricultural Research (CGIAR), the Green Revolution epitomizes the cold-war-era approach to rural development through technological, rather than social, engineering. The basic goal of CGIAR cereal research and dissemination efforts was to develop crop varieties and improved cultural practices that could increase production at a rate that exceeded population growth in developing countries. To achieve the necessary yield increases, the approach depends on intensive cropping, synchronized maturation, heightened reliance on mechanization, precise water control, and intense and precisely timed chemical inputs [26].

At the core of Green Revolution package stand the short, stout stalks of semi-dwarf HYVs. Early efforts to improve yields ran into problems when increased grain weight caused plants to lodge (bend or topple over) [24]. To address this problem, plant breeders began searching for short varieties of wheat and rice to cross with their high-producing strands. Other desirable characteristics were early maturity and photo-insensitivity, which would support multiple cropping. The first breakthrough in rice came from crossing a semi-dwarf variety from Taiwan, *Dee-geo-woo-den* (DGWG), with a tall, vigorous Dutch strain developed in Java in the 1940s, known as *Peta*. The marriage of these two remarkably different rice varieties launched the Green Revolution in rice in South and South-East Asia [27]. The resulting cross, known officially as IR-8-288-3 and popularly as IR 8, tillered profusely in response to applications of nitrogen, producing numerous heavy panicles of grain. Its thick culms and great straw strength prevented lodging. In field trials at IRRI, IR 8 produced yields of up to 10.13 tons of paddy per hectare [4].

IR 8 was released for distribution in 1966, and rapidly adopted throughout Asia. Its release was quickly followed by several new varieties developed by IRRI, the University of the Philippines, and the Indonesian government's own rice-breeding program. Throughout the 1970s, the area planted in IRRI and IRRI-derived high-yielding varieties steadily increased, reaching 67 per cent of rice planted in Indonesia by the end of the decade [5].

To successfully grow high-yielding varieties, farmers are required to carefully control water, necessitating major national investments in irrigation infrastructure. The timing and volumes of chemical fertilizer and other inputs are crucial, again requiring major government investments in production, storage and distribution facilities for seed, fertilizer and chemical pesticides. As well, a national agricultural extension program was needed to instruct and inform farmers in the new methods and techniques. The programs were often accompanied by provision of credit to enable farmers to purchase the necessary inputs [5,7].

By the mid-1970s, when the "revolution" had been underway for a decade, IRRI scientists were becoming concerned that the loss of genetic diversity was creating vulnerability to pest and disease outbreaks. Virtually all semi-dwarf varieties carried the same dwarfing gene from DGWG, while three quarters of the Indonesian HYVs could be traced to the same maternal parent, *Cina* (one of the component varieties of the Dutch crossbreed *Peta*) [27].

The first problems began to appear at IRRI just a first few years into the breeding program, when test plots began evincing BPH infestations and symptoms of hopperburn (see below). Rice brown planthopper is a common insect in rice fields throughout Asia, but had never caused major problems in tropical rice prior to the development and spread of Green Revolution approaches. With the advent of HYV seeds and technological packages, this began to change. By the mid-1970s, BPH outbreaks were beginning to cause crop losses in Indonesia and other Asian countries that had adopted the new varieties [28]. At IRRI, research priorities shifted to developing and disseminating new BPH-resistant rice varieties—an approach that worked for a while [16]. However, within a few seasons, new BPH biotypes appeared that could feed on the resistant varieties, signaling that new approaches would be needed [4].

As previously mentioned, during the 1970s and '80s Indonesia achieved the greatest success of any country in adopting and disseminating HYV rice varieties. As associated problems began to surface, therefore, Indonesia was also the most vulnerable of any country. The following section describes Indonesia's rice intensification program from its outset in the early 1960s, through attainment of rice self-sufficiency in 1984, until its near-unravelling only two years later due to major BPH outbreaks in the "rice bowls" of northern Java.

4. THE BIMAS MASS GUIDANCE RICE INTENSIFICATION PROGRAM

BIMAS (for *Bimbingan Massal*, or Mass Guidance) was the name given to Indonesia's national rice intensification program, as well as the special coordinating body and governmental network established to direct it. Over the years, it spawned a number of sub-programs, with acronyms such as INMAS (Mass Intensification), INSUS (Special Intensification) and SUPRA-INSUS (which required groups of 50 to 100 farmers with contiguous fields to make joint decisions). As well, the BIMAS acronym and approach has also been adapted to other crops and commodities besides rice. The following section focuses on the original, rice intensification BIMAS program.

Although the BIMAS program was initiated two years before the military coup that brought President Suharto's New Order government to power, its title encapsulates two of the most fundamental precepts of the new regime's character, "mass" and "guidance". *Massal*, the second half of the BIMAS acronym, conjures the notorious "Floating Mass" policy introduced by Indonesia's military leaders in the wake of the 1965 coup and genocide, which prohibited any organized political activity in villages or rural areas. The notion underlying the Floating Mass was that Indonesia's un-sophisticated multitudes must not be distracted from the task of development by politics [29]. According to General Ali Moertopo, one of the main architects of New Order socio-political policy, prior to the 1965 coup, the Indonesian people—"who consist mostly of honest, good-willed and simple peasants"—had been led into "disorder and fanaticism" by "fall[ing] prey to the political and ideological interests of [political] parties". Politics, therefore, was seen as a dangerous and divisive factor, best left to trusted and competent leaders who could guide the political process to "support the realization of development in all fields of human life in order to increase the people's welfare' [30] (pp. 83–84).

If it was politics that nearly tore the republic apart, Indonesia's new leaders were convinced that development would stitch it back together again. Development became the new national project. It was the job of politicians and the state bureaucracy to guide the Indonesian people toward this all-encompassing goal [31]. The country's millions of microfundia rice farmers became pliable atoms in a floating mass, needful of strong guidance if they were to perform their important function of feeding the nation.

To guide and manage the development process, the New Order government assembled a vast national bureaucracy extending to the furthest reaches of the archipelago, down to the village level. The hierarchical structure of central government, provinces, districts, sub-districts and villages, mirrored the Indonesian military's territorial command structure. The BIMAS program, like all other national development programs, was structured along these lines. The program delivered credit, agricultural inputs, and guidance conceived by the center and filtered down through layers of territorial structures.

"*Bimbingan*"—the other half of the BIMAS acronym—means "leadership" or "guidance" in Indonesian. Under the BIMAS program, Indonesia's rice farmers were led and guided to deliver the huge increases in rice production needed to feed the growing nation. The initial target was to reach self-sufficiency in rice by the end of the first (1968–1973) Five Year Development Plan [32].

BIMAS was, at its core, a farm credit program, but the name came to encompass the entire range of Green Revolution packages of improved seeds, chemical fertilizer, improved irrigation, logistical support, and agricultural extension that fuelled the rice self-sufficiency endeavor. BIMAS became the emblematic program of the New Order government—and surely one of its most successful [7].

BIMAS exemplified the regime's technocratic nature. The introductory paragraph of first article on the new program to appear internationally effuses:

The BIMAS programmer of agricultural extension was conceived by a small group of young men without prior experience in extension work. The people who work with the farmers are also young and without extension experience, fourth and fifth-year students from the colleges of agriculture. Yet the programmer has been very successful in encouraging farmers to increase their production. [33] (p. 60).

The program experienced the usual problems during its initial years, including the failure of cooperatives in many BIMAS areas; inadequacies of credit supply and delays in delivery; delayed delivery of agricultural inputs; and "partial failure of communication to the farmers as to the real benefits and how to realize them by use of particular quantities of nitrogen and phosphate fertilizers and related pesticides" [32] (p. 37). By 1968, the program had managed to stem the decline in Java's rice production, but was still unable to produce an increase in production needed to match or exceed population growth on the island [7].

Many of the initial problems were resolved by the mid-1970s, and credit disbursements increased by a factor of seven between 1971 and 1976, from Rp. 8.5 billion to 55.1 billion. Repayment rates were abysmal, with BIMAS arrears reaching as high as 64 per cent of total loans outstanding in the late 1970s [34]. The program did, however, succeed in disseminating HYV seeds and accompanying agricultural inputs and instructions to large numbers of Indonesian rice farmers. As the BIMAS program's early problems were overcome, Indonesia's rice intensification program grew to become the most successful in the world.

With the OPEC price increases of the late 1970s, the Indonesian government was flush with oil revenues, and thus able to aggressively pursue its rice intensification agenda. Various authors ascribe different facets of the BIMAS program with primary causation for the country's impressive increases in rice production. Fox, for instance, points to the massive investments in irrigation infrastructure and HYV seed, chemical fertilizer and pesticide production and distribution [4,5], while most economists argue that price policy was the primary driving force [7,34,35].

As previously mentioned, Indonesia did reach its goal of self-sufficiency in rice in 1984, some fifteen years after launching the national rice intensification program. In recognition of this achievement, the Director General of the United Nations Food and Agriculture Organization (FAO) presented President Suharto a gold medal bearing his image at the organization's biennial meeting in November 1985 [36]. At home, Suharto had taken on the title of "*Bapak Pembangunan*"—the "Father of Development" [31].

4.1. Train and Visit, Spray and Pray

Central to the success of the BIMAS program was the tremendous effort put into developing a national agricultural extension infrastructure, to deliver information and instructions to farmers. Indonesia was a primary site of development of the Training and Visit (T&V) agricultural extension methodology, later promoted by The World Bank in over 70 countries worldwide [37].

Indonesia's New Order government enjoyed a special relationship with the World Bank [38]. Indeed, according to a recent history of the World Bank, "Indonesia was the presidentially designated crown jewel in the Bank's operational crown" [39] (p. 493). Surely, Indonesia's agricultural intensification program was one of the most sparkling gems in the array. According to World Bank annual reports from the years 1976 to 1987, the Bank loaned Indonesia $134 million to support T&V extension services, roughly five per cent of the $3 billion invested in rice intensification during that period (the majority being for irrigation ($2.1 billion) and nitrogen fertilizer production and distribution ($460 million)).

The T&V system stressed exclusive dedication to technical information dissemination through a single hierarchical line of command; in-house technical specialists to provide training to staff and tackle issues reported by field staff; a strict and predetermined schedule of village visits over a two-week cycle to disseminate the key set of messages for the coming two-week cycle [40]. By 1984, at the height of the BIMAS program in Indonesia, some 14,000 fieldworkers paid fortnightly visits to groups of 16 "contact farmers" who were in turn responsible to spread the BIMAS message another 160 to 320 other farmers [6]. These extension workers were also responsible for organizing the distribution of BIMAS packages of seeds and chemicals. By the mid-1980s, over half the harvested area of *Padi sawah* (irrigated rice) in Indonesia was under the BIMAS program.

The system included means of exerting "gentle pressure", if not outright coercion, on farmers in order to meet targets for the use of improved seeds and other inputs. Agricultural inputs were distributed through the village administration. Farmers purchasing seed on credit through the Village Cooperative (KUD) were required to accept the entire package of seeds, fertilizers and pesticides prescribed as part of the blanket recommendations covering the entire rice-producing area. In some areas, crops of farmers who were not planting the latest HYVs were cut down by village officials, or planting of HYVs and fertilizer use were enforced by the army [6].

The cultivation of HYVs over an extensive area increases the new rice's susceptibility to pest infestations. High fertilizer use, close spacing, the dense canopy produced by heavy tillering, creates a favorable situation for the multiplication of certain kinds of pest [41]. The BIMAS approach mandated calendar-based applications of insecticides as the primary means of addressing potential pest problems [16].

Chemical pesticides are called "*obat*" in Indonesian, which means "medicine". The implication being that these various types of "medicine" will help the rice plant stay healthy, in the same way that immunizations protect humans [42]. As with urea and superphosphate fertilizers, the Indonesian government heavily subsidized agricultural pesticides. National expenditures on pesticide subsidies rose from US$50 million per year in the 1970s to over $150 million per year in the mid-1980s [16]. For a period from the late 1960s to the mid-1970s, the government contracted with various international insecticide firms to conduct aerial spraying of large areas of rice land, in an attempt to curb stem-borer infestations [16].

A growing body of scientific research indicates that, while doubling or trebling nitrogen fertilizer application leads to significantly increased production of HYV rice, there is no parallel evidence to show that increasing insecticide use leads to measurable increases in yields (nor, conversely, that reduced pesticide use results in reduced rice yields) [43,44]. In spite of this, 12 per cent of pesticides sold worldwide are applied to rice crops, more than any other crop [45]. Approximately 80 per cent of this amount used in Asia [46].

This combination of factors, including aggressive promotion of HYV rice varieties as part of a Green Revolution package that included heavy-handed extension by trained experts in conjunction with local officials and the military, massive government subsidies on chemical fertilizers and pesticides, and prophylactic use of insecticides to control rice pests, together created a perfect storm for major outbreaks of insecticide-induced resurgent pest outbreaks, specifically, for a rice brown planthopper plague of unprecedented proportions.

4.2. BPH: An Archetypical Insecticide-Induced Resurgent Pest

The rice brown planthopper, *Nilaparvata lugens* Stål (Hemiptera: Delphacidae)—henceforth BPH—is native to all rice-growing areas of Asia, but was hardly heeded until the height of the Green Revolution, when it suddenly became a pest. The two major international texts on Asian rice pests published in the 1960s [47,48] do not even contain pictures of this insect.

The BPH is a small (2–3.5 mm in body length), brownish, sucking insect, belonging to the suborder Delphacidea. Like other homopterous insects, it has a set of four stylets in its mouth which serve as a piercing and sucking organ [49]. The BPH is a typical vascular feeder; it primarily sucks the phloem sap from fresh rice plants. Female adults of BPH spend about 60%–90% of their time sucking on rice plants, with repeated sustained sucking lasting one to seven hours. In addition to removing photosynthates from the plant, the sucking can block the vascular bundles in the rice leaves, causing toxic build-ups of free amino acids, leading to a phenomenon called "hopperburn". Hopperburn appears initially as a yellowing of older leaves, but soon extends to the whole plant, which turns brown and dies. Non-fatal infestations generally result in fewer panicles per plant, fewer grains per panicle, and lower percentages of ripened grain and lighter grain weight, depending on what stage of the rice plant's growth the infestation occurred [50].

Female BPH embed their eggs in the leaf sheaths of rice, which thereby escape insecticide sprays [1]. The lifecycle of the BPH is a mere 22 days, and a single female lays 200 to 400 eggs. In unsprayed wet-rice ecosystems, the mirid bug *Cyrtorrhinus lividipennis* Reuter feeds on BPH eggs, which it finds by tapping on the stems with their antennae, then sucking the eggs [50]. Further control comes from the hundreds of species of spiders that prey on surviving nymphs and adults [51].

In the absence of these natural enemies, the BPH population can expand exponentially, quickly reaching levels that can cause hopperburn damage that is "comparable to a locust plague in its magnitude of space and intensity" [41] (pp. 286). Adding insult to injury, the BPH also spreads rice grassy stunt and ragged stunt viruses, which can cause serious harm to rice crops even if the adult BPH population is knocked out by heavy spraying [50].

Small patchy hopperburn first appeared in the IRRI farm in 1964, only two years after the onset of field experiments with HYV varieties. By 1971–1972, IRRI was experiencing serious outbreaks. In the mid-1970s, BPH outbreaks were causing major crop damage in Indonesia, southern India and Sri Lanka [45]. IRRI's response was to undertake a breeding program to develop BPH-resistant rice varieties, crossing the old Peta strand with a Taiwanese variety called "Mudgo" that was distasteful to BPH. Six of these—IR26 and IR30 (released in 1975), and IR24, IR28, IR32 and IR34 (released in 1976), were quickly adopted to replace earlier non-resistant varieties [5]. These strategies were effective for one or two seasons, but by 1977, a new BPH biotype appeared in the Philippines, Indonesia, Vietnam, and Solomon Islands that broke the resistance of the new resistant varieties [41]. IRRI quickly developed a new biotype 2-resistant variety, IR64. But it was becoming clear that new responses were needed [5].

Indonesian scientists also developed and released several new high-quality resistant varieties, by crossing IR36 with local varieties. The BPH problem subsided for several years, and in 1984 Indonesia attained its long-sought self-sufficiency in rice [6]. However, insecticide subsidies continued to increase and ever greater amounts of these chemicals were applied to rice. In 1985–1986, Indonesia witnessed a dramatic and sudden breakdown of resistance to the BPH in all of the IR64-based varieties. In two seasons, an estimated 275,000 hectares of rice were destroyed by BPH, a major threat to the country's food security [16].

Around this time, researchers found that insecticide use can accelerate the adaptation of pest populations to the resistant varieties [43,52,53]. Natural genetic diversity within planthopper populations is sufficient to provide individuals capable of feeding and reproducing well on the new resistant varieties. Due to their high fecundity and short lifecycle, when insecticides release the BPH from natural enemies, they quickly evolve biotypes that can overcome host plant resistance traits [52].

5. THE NATIONAL INTEGRATED PEST MANAGEMENT PROGRAM

There is a popular story among IPM aficionados about the origins of integrated pest management in Indonesia. The Indonesian bureaucracy during the New Order period developed a pervasive culture of "*Asal Bapak Senang*"—meaning "as long as Father (one's superior) is pleased". Within the Ministry of Agriculture, pest and plant disease infestation records are based on sampling reports of pest observers from the Directorate of Crop Protection. These reports are amalgamated stepwise as they move up through the administrative hierarchy, from sub-district level to the district, province and finally national level. Records of severe pest outbreaks occurring in a jurisdiction are considered a potential embarrassment to one's supervisor, so estimates of crop damage were routinely underreported, becoming progressively smaller as the reports moved up through the system.

For this reason, the major BPH outbreaks of 1985–1986—especially embarrassing in the wake of the country's landmark achievement of rice self-sufficiency in 1984—went largely unreported. It was only when farmers from President Suharto's home village of Kemusuk, near Yogyakarta, made a direct appeal to the President's brother for help that an independent investigation was launched by the National Development Planning Agency (BAPPENAS), and the extent of the damage became apparent. A politically dangerous situation had been created, and President Suharto was furious [6].

IPM scientists from the national Agricultural University (IPB) in Bogor and from FAO seized this opportunity to act, and in October 1986 actually took several matchboxes of insects and spiders into the President's office to explain the principles of integrated pest management in rice [54]. President Suharto was so impressed, that he immediately drafted a Presidential Instruction that banned 28 broad-spectrum pesticides in 57 formulations (leaving only ten brands with four active ingredients, all narrow-spectrum insecticides considered effective against BPH), initiated a phasing out of government pesticide subsidies, and mandated integrated pest management (IPM) as the national policy for crop and plant protection. Other features of the instruction included the creation of 1500 new pest observer positions to more than double the number of Directorate of Crop Protection fieldworkers; enforced use of resistant rice varieties; enforced introduction of one secondary food crop after two irrigated rice crops in certain regions; and "crash programs" through "commando posts" (*POSKO*) to deal with pest outbreaks through targeted applications of narrow-spectrum insecticides, if necessary [6].

Initially, IPM was to be promoted through the T&V extension apparatus already in place. The Indonesian government requested permission from the World Bank to divert US$4.19 million remaining from the second phase of the National Agricultural Extension Project (NAEP II) loan to support IPM training. Senior pest observers were trained as "IPM Master Trainers", and the new Pest Observer recruits and selected extension workers were then given a six-day crash training program. These individuals were to then train contact farmers, in the classic T&V manner. Training guides, flip-charts, slide shows, and thousands of leaflets and pamphlets were hurriedly produced and distributed. Additional funds were expended on travel, honoraria, vehicles, and compensation for farmers attending the courses. The entire budget was expended in seven months, reaching less than ten per cent of the targeted 10,300 farmers. Less than a quarter of those actually entered a rice field during the course of the training program [6]. In post-training assessments, farmers reported that they had not learned much that they could use. Indonesia's first attempt at implementing integrated pest management in rice proved to be a dismal failure.

5.1. A New Approach: Farmer Field Schools

In 1989, USAID provided a grant of US$ 4.7 million to the Indonesian Ministry of Finance and the National Planning Agency (BAPPENAS) to initiate a new two-year pilot IPM Program. Technical assistance was provided by FAO. The decision to house the project in BAPPENAS rather than the Ministry of Agriculture was strategic—it was considered too difficult to devise radical new strategies in an agency so strongly committed to transfer-of-technology T&V extension approaches, and also deeply enmeshed with pesticide company interests. The FAO team established secretariats in Jakarta and Yogyakarta, and worked closely with Directorate of Crop Protection pest observers and selected extension agents in a number of districts in Central and West Java.

The program piloted a new model called Farmer Field School (FFS), which drew from adult non-formal education concepts, based on the assumption that farmers already possess a wealth of experience and knowledge [55]. The other fundamental concept employed was agro-ecological analysis, examining the entire rice paddy agro-ecosystem and relationships between soil, water and weather conditions, rice plants, insect populations and communities, and management activities, rather than focusing just on individual pest problems [56,57,58].

The Farmer Field School takes place over the course of an entire rice growing season, about twelve weeks. The rice field itself becomes a laboratory and a classroom. Often there are two plots: a "non-IPM" plot that is sprayed with pesticides according to Ministry of Agriculture guidelines, and an IPM plot that is managed based on decisions made by the group during their weekly meetings. There are no teachers and students, or trainers and trainees, but rather facilitators and participants (as the program matured, many of the facilitators were themselves FFS farmer-alumni). There are typically about 25 members in an FFS course, who are then divided into groups of five to conduct field observations and analysis. Each session begins with a careful examination of conditions in the fields, observing sample rice hills chosen in a random diagonal transect across each field. The groups make notes of insects, spiders, damage symptoms, weeds and diseases observed on each hill. The stage of plant growth is carefully noted, as are weather and water conditions. Interesting insects and other creatures are captured and placed in small plastic bags. Experiments are set up, to study the effects of plant spacing, water control, varieties, fertilizer and pesticide application, and soil characteristics on plant growth [6,45].

The participants then gather in a nearby home or shed, and prepare drawings of what they observed on large sheets of newsprint. These include the rice plant at its present stage of growth, pests and natural enemies found on the plants, weather, water and soil conditions. Then ensues a lively discussion of the week's findings, using the "*Apa ini?*" ("What is this?") principle. Answering a question directly is considered a lost opportunity for learning. Farmers are encouraged to discuss what they have observed, and come to their own conclusions about the status of the crop and possible control measures (or to set up new experiments to find out the answers). One particularly effective activity is the creation of "insect zoos" by placing muslin netting over a rice plant, for farmers to observe predation and parasitism in action, and thereby learn which organisms are pests and which are biological control agents [6].

Special topics are introduced during most sessions, in modules designed to avoid lecturing. These include things like life cycles of rice field inhabitants; parasitism and parasitoids; the effects of pesticides on natural enemies; or the growth of rat populations. Various "ice-breaker" or other socio-dynamic activities enliven the sessions and foster a sense of belonging among the group. These might include competitions and wagers (for instance, whether a newly discovered bug is a predator or pest), with the "losers" having to carry the "winners" around the village on their backs! [59]

Conventional IPM employs the concept of "economic threshold": *i.e.*, the amount of economic damage that will justify the cost of control. As managers, farmers are expected to weigh expected yield loss (in terms of damage (kg/ha), which they then multiply by price per kg) with estimated pest control costs. For Indonesian rice farmers, this concept seemed overly complex and confusing. In response, the IPM program staff shifted to a concept of "experience threshold", which develops as farmers learn through experience, and focuses more on the procedure of decision-making [6]. Farmers learn to make crop management decisions based on their personal circumstances and the ecological balance in each paddy.

The four guiding principles of the Indonesian IPM Program reflect this holism and the program's overall goal of making farmers confident managers and decision-makers, eager for new ideas and information but free from dependence on directives from above:

1. Grow a healthy crop.
2. Observe fields weekly.
3. Conserve natural enemies.
4. Farmers are IPM experts.

Results were immediate and profound: among FFS farmers, insecticide applications reduced from an average of 2.8 sprays per season to less than one, with most farmers not spraying at all. When they did apply an insecticide, IPM farmers could identify a specific pest [45]. Studies showed that IPM farmers had, on average, slightly higher yields, higher overall returns, and lower economic variance than their non-IPM counterparts [16]. But the most impressive change was not in the rice fields *per se*, but in the character of the IPM rice farmers themselves. Farmers and village officials, the lowliest ranked people in the bureaucratic hierarchy and until recently the passive targets and implementers of government directives, were now actively and intelligently discussing their problems, making considered decisions about pest control, producing complex and accurate drawings and diagrams of insects and food webs in their paddy fields, and speaking frankly and openly in front of others—including visiting dignitaries such as the Minister of Agriculture [6].

IPM farmer groups began participating in local Independence Day parades and other festivals, waving banners festooned with integrated pest management slogans, and carrying giant anatomically-correct models of predacious lycosid and linyphiid spiders. From the outset, IPM in Indonesia took on the hallmarks of a movement.

To prepare a core group of trainers for scaling up the program, the program developed a special 15-month Training of Trainers program for 2300 pest observers from the Directorate of Crop Protection. Agricultural extension workers operating under the Agriculture Ministry's powerful Agency for Agricultural Education and Training and steeped in T&V methodologies, were deemed not particularly suitable candidates for introducing IPM. Ag extension workers have many tasks, among which pest control is a relatively minor role. They are heavily involved—indeed, likely have a vested interest—in input distribution activities, which can conflict with the nature of IPM [6].

The FFS and pest observer training programs were supplemented by a number of research and other activities, including a field laboratory in West Java focusing on the white stemborer (*Scirpophaga innotata* Walker), health impact studies in pesticide-intensive areas in West and Central Java, insect habitat studies, educational and public information materials development, and training evaluation studies. For these research activities, the IPM Secretariat chose to work closely with a few Indonesian and international universities, rather than with the Ministry's own Research and Development Unit (*Badan Litbang*). Their reasons were the same as for the choice of working with pest observers rather than extension workers: *Badan Litbang* was (and is) highly invested in an input-based agricultural modernization paradigm.

The training evaluation studies were used to constantly refine and improve the FFS model, and devise strategies for "scaling up" [60]. The pesticide studies highlighted the vulnerability of rice farmers to pesticide poisoning and the impact and cost for farmers and the

Indonesian people more broadly [61]. Results were shared with farmers during FFS sessions, and participants put dyes in knapsack sprayers to see how much insecticide spray contaminates them when they sprayed their fields.

The IPM program strove to influence national crop management policy and practice. An important opportunity arose in 1990, during a serious outbreak of white stemborer in the key rice producing region of Indramayu, West Java. Indramayu is one of the country's major rice bowls, with extensive irrigation allowing multiple cropping over a vast area.

Stemborers are particularly injurious pests, and difficult to knock out once a population is established. White stemborer eggs can survive fallow periods in the stubble and stems from the previous harvest [62]. The Ministry's proposed response to the 1989–1990 outbreak in Indramayu was to reinitiate aerial spraying, a practice that had been discontinued since the 1970s. The IPM Secretariat convinced Ministry officials to hold off while they devised an alternative strategy. Working with local communities and government, the program mobilized hundreds of villagers and school children (who were granted a special school holiday) to descend on rice fields to collect egg masses. This simple action led to a marked reduction in pest pressure, boosting the national standing of the IPM Program considerably [59].

Buoyed by their success, local farmer groups redoubled their efforts to develop strategies to identify more sustainable solutions to this persistent and harmful rice pest. With the assistance of an IPM Field Coordinator, farmers set up a number of experiments to attempt to better understand stemborer biology.

Each year, white stemborer moths would begin appearing soon after the first planting following the annual three-month fallow period. In their first experiment, farmers were shocked to observe that stemborer moths that were collected and sprayed with insecticides still managed to spawn before they died. In another experiment, farmers learned that spraying insecticides directly on egg masses did not prevent healthy larvae from emerging a week later. They then attempted to curb outbreaks by spreading sand coated in insecticide (carbofuran) in seedbeds. Results were variable, depending on the time of application and time of oviposition of stemborer eggs. They had to look for another approach.

The group set up experiments to learn about the origin of the early-season flights. They found live larvae in the stubble of last season's crops, and so tried burning and flooding the stubble. This, however, did not kill many larvae, and the farmers concluded that physically removing stemborer egg masses was still the most effective method to reduce infestation levels. This is a difficult and time-consuming task. There had to be another way.

The Field Coordinator helped farmers to gain an understanding of aestivation (a form of hibernation during adverse weather conditions), and to design new studies using light traps, in order to more accurately forecast peak flight periods. Stemborer moths have no mouth parts; if they cannot find newly planted rice to lay their eggs within the first week, they die. Using simple rain gauges made from discarded drinking water bottles, the FFS farmers soon learned that the mass flights took place after the first major rainfall of the season. Armed with this knowledge, they were able to organize other farmers in surrounding villages to only plant their nursery beds after peak flights. Once the first flight dies, there are no further generations.

The campaign proved successful, reducing the incidence of white stemborers to less than 5 per cent, compared to 25 per cent in neighboring areas. Through their own experiments, Indramayu farmers were able to devise an effective strategy to solve a problem that has been known by researchers since the 1930s. Coordinated planting of nursery beds has since become the stemborer management strategy for the Indramayu district, a spectacular example of farmer-led agricultural policy [59].

In 1991, the initial two-year program was extended for a third year, with some additional USAID support and some "bridging funds" siphoned from the World Bank National Agricultural Extension (NAEP III) program. The three-year pilot phase ended in 1992. By this time 2300 pest observers had completed the 15-month IPM Training of Trainers course. An additional 6000 agricultural extension workers were also trained in a shorter course. More than 250,000 farmers, in six major rice-producing provinces had participated in FFS courses, corresponding with a 60 per cent reduction in pesticide use in project areas [63].

5.2. Scaling Up: IPM Goes Mainstream

Based on the resounding success of the three-year, six province pilot, in 1992 the decision was made to go national. The World Bank provided a loan of $32 million, supplemented by another $7 million from USAID. The main activity was to be the training of approximately 800,000 farmers (30 per cent of these to be women) to develop appropriate, farmer-responsive technology; supplemented by strengthening the regulatory and environmental management of pesticides.

(The World Bank celebrates its support of the popular IPM Program—a search for "Farmer Field School" on the World Bank website conjures links to 745 documents. However, immediately after the first IPM extension debacle in 1986–1987 using left-over Bank loan funds from the National Agricultural Extension Program (NAEP II), the Bank loaned Indonesia another $70 million for T&V extension (NAEP III), more than twice the amount it eventually committed to the National IPM program. Furthermore, during the six-year lifespan of the Indonesian National IPM program, over $350 million of World Bank loan funds were expended globally on pesticide purchase and distribution. A Pesticide Action Network analysis of World Bank projects between 1999 and 2003 indicated that only 9 per cent of projects worldwide complied with the Bank's own Operational Policy 4.09 on Pest Management, while scores of projects included provisions to "remove bottlenecks" in the procurement and distribution of "agricultural inputs" [64]).

Management of the national IPM program shifted from BAPPENAS to the Ministry of Agriculture. The Indonesian National IPM Program ran from 1993 to 1999. At the same time, the FAO IPM Secretariat in Jakarta became the coordinating center of a 12-country Asia Regional IPM Program.

The program experienced numerous problems related to scaling up. In Indonesia, it took on many of the features of large government projects characteristic of the New Order period, including standardization, emphasis on targets rather than process, lax quality control, leakage (including numerous instances of "ghost" Field Schools), and top-down delivery, as Ministry officials attempted to squeeze and shape the program to fit within pre-existing bureaucratic structures and culture, reverting to T&V command and control models with which they were comfortable [63]. As well, there was constant pressure from both the Ministry of Industry and sections within the Ministry of Agriculture to continue supporting pesticide use in government-led agricultural intensification programs. Meanwhile, the FFS brand became very popular with Indonesian government officials and donors alike (in name if not necessarily in form and spirit), and soon there were "Farmer Field School" projects for artificial insemination and veterinary services, smallholder tree crops, cotton and vegetables [65].

By the time the Indonesian National IPM Program ended in 1999, over a million Indonesian rice farmers had participated in IPM Farmer Field Schools. In combination with somewhat disciplined enforcement of pesticide regulations (the 1986 Presidential Instruction was followed by a 1992 national law on crop management plus a number of other ministerial decrees, until by 2000 the number of banned pesticides formulations reached 120), pesticide misuse in rice was significantly curtailed in Indonesia, and BPH infestation ceased to be a serious problem in the country [63].

At a national workshop held in Yogyakarta in 1999 to conclude a participatory evaluation of the Indonesian IPM Program, FFS alumni formed the National Association of IPM Farmers (IPPHTI, *Ikatan Petani PHT Indonesia*). The FAO Asia Regional IPM Secretariat continued to provide limited support to IPPHTI, primarily for meetings, workshops and organizational development, and to publish a monthly bulletin. FAO also helped secure financial and technical assistance from other funders such as VSO, HIVOS, Diakonia and Norske Foklehjelp—at much reduced levels. The Indonesian Ministry of Agriculture continued to run IPM Field Schools, though the numbers quickly dropped to a trickle.

6. BACK TO THE FUTURE: IPM IN INDONESIA TODAY

It has now been over fifteen years since the conclusion of the Indonesian National IPM Program in 1999. Indonesia has undergone profound changes during the intervening period. Indonesia was particularly hard hit by the Asian Financial Crisis of 1997–1998. President Suharto resigned in May 1998 after months of political and economic turmoil. The country has since undergone sweeping political and institutional changes, significantly altering the national government's ability to implement initiatives such as the 1993–1999 National IPM Program.

It is a truism regarding large government development projects that once funding dries up, much of what was constructed begins immediately to degrade. The National IPM Program was no exception to this rule.

Since 2009, after several years of relatively BPH-free rice crops, the insect has returned with a vengeance. This correlates directly with increased insecticide use; Indonesian rice farmers are now spraying more than ever [66,67]. Severe BPH outbreaks have taken place in key rice producing regions of East, Central, and West Java, with both hopperburn and widespread virus infestation occurring. Crop losses in these high-production areas have been significant, approaching or even surpassing the 1985–1986 outbreak that led to the establishment of the national IPM policy and program [67]. The following sections explore the reasons leading to this recurrence.

6.1. Reformasi: Democratization and Decentralization Indonesian Style

The end of the National IPM Program coincided with a period of major social and political upheaval in Indonesia. After President Suharto stepped down in May 1998, there ensued a period of rapid political and institutional reconfiguration as Indonesia scrambled to recover from the worst political and economic crisis it had faced since 1965. Democratic reforms introduced by Suharto's hand-picked successor B.J. Habibie included the repeal of anti-subversion laws, release of political prisoners, freedom of the press, a new law allowing the formation of political parties, Indonesia's first free legislative elections since 1955, a referendum on the future of East Timor, and a new regional autonomy law that set in motion a rapid and radical devolution of government power and responsibility to regional governments [68,69]. Other changes included significant deregulation of trade and investment [70,71].

With these changes, the context for IPM in Indonesia changed dramatically. Hastily-prepared decentralization laws passed in 1999 fundamentally altered the relationship between Jakarta and regional governments, delegating significant decision-making and implementing powers—and budget resources—to district and municipal governments. Government service delivery initially suffered, as the hierarchical chain of command meticulously constructed over decades of military rule came unraveled [72].

In the agriculture sector, Food and Horticultural Crop Protection Centers and staff came under Provincial government control, while Regional Extension Centers and extension workers became the responsibility of District governments. Interaction and communication between agricultural research and extension services, always weak in Indonesia, virtually ceased [73].

Meanwhile, World Bank and other donor support for agricultural extension in Indonesia and developing countries worldwide had diminished considerably. Emphasis shifted to issues of cost recovery and fiscal (un-)sustainability [40]. Without major donor support, national and local politicians are less likely to invest in extension than in more highly visible projects such as irrigation or road construction. The inherent difficulties in attributing impact make extension an easy target for lower budgetary support [74]. As well, around the world, agricultural research priorities had been shifting for some time from staple foods to export crops [75].

The demise of President Suharto meant that IPM in Indonesia lost a major patron at the highest level of government. Ministry of Agriculture support for IPM, always ambivalent, quickly eroded. The first *Reformasi*-era Minister quietly set about dismantling departmental controls on pesticide distribution and marketing [76].

Decentralization also had the effect of rendering it much more difficult to initiate and coordinate large national programs such as the National IPM Program of 1993–1999. National ministries lost considerable control over sectoral budgets and programs. In principal, ministries' roles are now largely confined to setting policy and minimum standards for government services, and monitoring of policy implementation by regional governments. While some of the more service delivery orientated Ministries such as Health and Education seriously pursued decentralization agendas, some of the other ministries—notably those engaged in natural resource management such as Forestry, Mining and Agriculture—resisted the process [77]. Structures remained largely unchanged, and these ministries strove to maintain large development budgets—euphemistically intended to support the development agendas of regional governments. These budget allocations generally take the form of *projects* and *programs*. In the realm of agriculture, this translates into *inputs*. In essence, the Ministry of Agriculture's tactics to retain relevance (and budget share) in the era of regional autonomy often resembles a reincarnation of the BIMAS program; *i.e.*, delivery of packages—of credit, seed, fertilizer, chemicals—to farmer groups. This is inimical to IPM approaches.

As the regional autonomy law came into effect in January 2001, the exuberance over Indonesia's new political freedom quickly degenerated into tussles over control of natural and fiscal resources. Indonesia's reforms since 1998 ushered in a new era of "money politics" leading to some distinctly non-democratic outcomes. Despite wide-spread recognition of the phenomenon, there is no clearly delineated meaning for the term "money politics", nor is it our intention to provide one here. Following McVey on Thailand [78] (pp 16), it is sufficient to note that in post-New Order Indonesia, "money has come to dominate politics at all levels: one must have money to run and one must make money from office too".

Successfully competing for the office of District Head (*Bupati*) allegedly costs between Rp. 20 to 40 billion (US$ 200–400 thousand) [79]. As public campaign financing is virtually non-existent, the sources of such funds are relatively opaque. Needless to say, politicians accrue considerable debts, which must be repaid. Such repayment often takes the form of *contracts*. Infrastructure projects account for much of this, along with concessions in resource-rich provinces for forest industries, plantations, and other extractive industries. In rice-growing districts, options are generally more limited. Hence, district governments in rice and other food crop producing regions tend to promote input packages (that are then contracted out), similar to those descending from the central government discussed above. This is further compounded by the tendency of governments to pursue populist programs, which again often take the form of subsidies or aid packages.

Another consequence of Indonesia's decentralization and democratization processes has been the politicization of the bureaucracy [80]. Incoming *Bupatis* award supporters with lucrative portfolios, and sideline those who did not back them in the election [81]. Very few supporters of IPM have flourished in this political milieu. This practice is also particularly corrosive of institutional memory.

6.2. The Magic of the Marketplace

Partly as a consequence of the US$ 43 billion rescue package provided by the IMF in the wake of the Asian financial crisis, and with the encouragement of other international donors, since 1998 Indonesia has pursued an aggressive policy of deregulation, privatization and trade liberalization. This initially led to a significant spike in the price of agricultural inputs, placing a significant burden on farmers [82]. One response by the Ministry of Agriculture was to deregulate pesticide imports and distribution. In a 2006 interview, the Minister of Agriculture noted that

(In 2001) the pesticide business was controlled by a small number of multinational companies and just a few local companies. As a result, the price of pesticides to farmers was relatively high. They were cautious, in fact, reluctant, to use pesticides. [83]

He went on to explain that Indonesia could confidently deregulate pesticide imports and sales because Indonesia has already promoted Integrated Pest Management through Farmer Field Schools. Indeed, the country's achievements had been praised by FAO and WHO. Even though prices had dropped, Indonesian farmers adhered to standards of appropriate timing, precise targeting, and correct dosages.

The interview neglected to mention that government support for IPM Field Schools and IPM more broadly had reduced to a trickle, or that only about two per cent of Indonesia's estimated 44 million rice farmers and agricultural workers had ever participated in IPM Field Schools.

Local and international companies wasted no time taking advantage of the new (de-)regulations. Within a few years, the number of pesticide brands being sold in Indonesia skyrocketed to over 2700—more than 250 of these for use on paddy [84]. The majority of these are combinations of generic ingredients imported from China; Indonesian companies wishing to market them simply register a trademark with the Ministry of Agriculture's Pesticide Commission. By 2012, 384 companies were importing more than 50,000 tons of 366 active ingredients to produce the dizzying array of products marketed to farmers and government agencies. Between 2000 and 2012, the total value of pesticide imports to Indonesia increased by a factor of six, from US$ 50 million to just under 300 million [66].

The increased competition touted by the Minister above has given rise to aggressive and sophisticated marketing in the countryside. Villages across Java are festooned with pesticide banners, posters and leaflets. Most villages have several kiosks and shops selling agricultural chemicals. Vendors are offered generous bonuses and prizes for meeting sales targets. They organize seminars, workshops, study tours, gala entertainment events with extravagant door prizes, even one-day pesticide "Farmer Field Schools". Farmers purchasing Rp. 100,000 (about USD $10) worth of pesticide receive a free tee shirt or cap—thus becoming advertisements themselves. Turn in ten empty pesticide containers for recycling, receive a free tee shirt. Farmers are offered an amount of pesticide free of charge in return for putting up a poster in their field (if the results appear positive). Pesticide companies repair broken backpack sprayers for free. Other promotions include lotteries to win televisions, refrigerators, even Hajj trips to Mecca! [85]

Pesticide companies employ their own field workers, known as "formulators". These individuals are better educated, provisioned, and remunerated than government extension workers. As well, the companies take advantage of the dire economic situation of most government extension programs and workers, offering incentives to those who successfully promote their products [76]. Agricultural extension workers' primary responsibility is to collect "Definitive Group Work Plans" from farmer groups; basically a shopping list of inputs requested from government agencies and partner suppliers. This presents a convenient opportunity for close collaboration between extension workers and pesticide distributors.

As a result, more Indonesian farmers are using more pesticides than ever before. If pest problems persist, the general response is to try something more expensive, or to mix different formulations together to create a hopefully more powerful chemical cocktail [86].

Surveys conducted by students from the University of Indonesia and Bogor Agricultural Institute in several rice growing districts in Central

Java in 2011 found a dizzying array of pesticide brands being used, comprised mostly of the active ingredients shown in Table 1 [67,87]:

Table 1. Partial list of pesticides being used by Rice Farmers in Central Java in 2011.

bisultap	etofenprox
buprofezin	fenobucarb
carbosulfan	fipronil
chlorpyrifos	imidacloprid
cyfluthrin	lambda-cyhalothrin
cypermethrin	metomil
deltamethrin	proproxur
diazinon	thiamethoxam
dimehypo	thiocyclam hydrogen oxalate
dinotefuran	

This list includes at least one active ingredient (the organophosphate diazinon) that was banned for use in rice by the 1986 Presidential Decree that launched Indonesia's IPM program, plus a number of other broad-spectrum insecticides identified by Bogor Agricultural Institute crop scientists as prone to promote planthopper outbreaks including chlorpyrifos, (organophospate), fipronil (phenylpyrazole), deltamethrin (pyrethroid), and imidacloprid (neonicotinoid) [67]. Some of the other compounds on the above list appear on the Pesticide Commission's 672-page list of approved ingredients, but are not indicated for use on rice [84]. On the list also is the juvenile retardant buprofezin which is often inappropriately combined with other insecticides. Clearly, the information being provided to Indonesia's rice farmers has more to do with selling chemicals than with controlling planthoppers.

6.3. Masterplanning and Modernizing (Again)

Beginning in about 2004, some seven years and three presidents after the economic and political crises that brought an end to President Suharto's New Order regime, Indonesia entered a new phase of robust economic growth, driven largely by exports of primary commodities (coal, ore and palm oil) to China, combined with increased domestic consumption. National and many local governments are becoming increasingly confident, and have initiated a number of ambitious modernization programs, under the overarching framework of the *Masterplan for the Acceleration and Expansion of Indonesia Economic Development* (MP3EI). This technology and infrastructure-based Masterplan intends to shift Indonesia from the category of emerging to advanced economy within the next 15 years. The Masterplan sets ambitious targets for each of eight primary programs (agriculture, mining, energy, industrial, marine, tourism, telecommunication, and the development of strategic areas). Food agriculture is one of 22 main economic activities targeted for acceleration and expansion [88].

Within the agriculture sector, the Masterplan is further elaborated in a national strategy known as the Revitalization of Agriculture, Fisheries and Forestry (RPPK), and for rice, a program entitled National Rice Production Increase (P2BN), which targets annual production increases of around six per cent, intended to reach a national rice surplus of ten million tons by 2014.

This discourse of acceleration and expansion, and ambitious production targets, are fodder for the classically trained agronomists and agricultural economists steeped in an input/output-based agricultural modernization paradigm, who dominate Indonesia's agriculture bureaucracy. Targets justify budgets; budgets purchase and deliver inputs. Then it is simply a matter of getting farmers to do the rest. For the Indonesian Ministry of Agriculture, this marks a return to the halcyon days of the BIMAS program. Tactics—and results—hearken back to the period before the onset of the BPH crisis of the mid-1980s.

The primary vehicle for the delivery of inputs and instructions to farmers is called a Field School, though it bears little resemblance to the original FFS concept established during the National IPM Program. So-called Integrated Crop Management (ICM) Field Schools consist of four two-hour sessions where farmers are introduced to particular techniques and inputs—quite different from the twelve or more half-day meetings comprising the original IPM Field Schools. The *participatory* aspect of ICM Field Schools appears limited to farmers choosing which *packet* they wish to accept. There is no longer any pre-FFS assessment or research by participants, nor—as IPM Field School proponents are quick to point out—no experimentation, experiential learning, laughing or clapping.

The ICM Field School model was developed by the Ministry of Agriculture's Research and Development Unit (*Badan Litbang*— which had been intentionally circumvented during the original IPM program). Implementation is managed by the powerful Directorate of Cereals in the Directorate General of Food Crops with additional support from the

Directorate General of Agricultural Facilities and Provisions. These two agencies have reaped the greatest benefit from the target-based P2BN program, whilst the Directorate of Crop Protection—the institutional home of the IPM Field School—faces a much more difficult task establishing causal links between its work and production increases [89]. Any protests from this group that IPM Field Schools remain fundamentally important to the health and productivity of national agriculture are met with the response that the ICM Field Schools already incorporate IPM principles; that ICM is in effect "*IPM plus*".

The Directorate of Crop Protection continues to conduct IPM Field Schools—though at a much reduced level than during the National IPM Program. Since 2006 the Directorate has funded about 560 IPM Farmer Field Schools nationally per year, or less than 50 per year in each of the 12 major rice-growing provinces [90]. Compared to the number of ICM Field Schools being supported, this is a mere trickle. In keeping with the production target orientation of the ICM Field Schools, Ministry statistics do not record how many Field Schools were implemented, but rather the number of hectares covered. In 2010, this figure reached 2,393,015 hectares nationally [91], which conservatively equates to about 200,000 Field Schools. To illustrate, in 2012, East Java, Indonesia's main rice-producing province, received budget support from the Ministry of Agriculture to implement just over 200 IPM Farmer Field Schools. The budget for ICM Field Schools, on the other hand, set a target of between 30 and 40 Field Schools per sub-district, or approximately 20,000 ICM Field Schools in one year [73].

Yet another strategy introduced in recent years to achieve ambitious rice production targets is the introduction and dissemination of hybrid rice. Commercial hybrids are a relatively new innovation in rice technology, promising to break the current rice yield ceiling through the effect called *heterosis*, or hybrid vigor. (Although HYV rice is created through the hybridization of different strands, it is then 'bred out' for several generations to achieve a stable genetically fixed, or inbred, variety. Hybrid rice, on the other hand, is the first generation (F1) offspring of crossing two inbred varieties. First generation hybrids express heterosis, often in the form of much higher yields. Second generation (F2) seed produced by the F1 plants is inferior, meaning that farmers need to purchase new F1 seeds every season to achieve the heterosis effect each time).

China has led the way with development and dissemination of hybrid rice seed; presently around 50 per cent of China's paddy land is planted in hybrid varieties. Most estimates suggest that China's hybrid rice yields on average 15 to 20 per cent more than inbred HYV varieties, however, the hybrid crops also require much higher dosages of fertilizer and pesticides.

These hybrid strains appear prone to a range of rice pests, including white stemborer (*Scirpophaga innotata* Walker), whiteback planthopper (*Sogatella furcifera* Horvath), rice leaf roller (*Cnaphalocrocis medinalis*), bacterial blight (*Xanthomonas oryzae pv. oryzae*), sheath blight (*Rhizoctonia solani*), downy mildew (*Sclerospora oryzae, Sclerophthora macospora*), false smut (*Ustilaginoidea virens*), kernel smut (*Tilletia barclayana*) and various viral diseases (tungro, rice ragged stunt, rice grassy stunt) [92]. Perhaps most significantly, due to a combination of plant physiology, high fertilizer use and overuse of chemical agents to combat the array of pests and diseases above, hybrid rice is extremely susceptible to BPH infestation.

Nonetheless, since 1998, the Asian Development Bank, together with FAO, IRRI and the Asia Pacific Seeds Association (APSA) have been aggressively promoting the introduction of hybrid seeds in Bangladesh, India, Indonesia, the Philippines, Sri Lanka and Vietnam. Initially, seeds were distributed with heavy subsidies to offset the purchase cost (usually about ten times the cost of HYV seed) [92]. In each of these countries, the initial euphoria over yield increases rapidly gave way to farmer resistance, as millers and consumers shied away from purchasing hybrid rice, which they claim is too sticky, too mild, and also has a much lower milled rice to husk ratio [93].

The national government in Indonesia has been distributing hybrid seeds every year since 2004—often through the mechanism of ICM Field Schools. Former President Susilo Bambang Yudhoyono appeared on national television harvesting hybrid rice in several provinces each year between 2007 and 2012. The hybrids have not been popular with farmers, due to high cost, the fact that they cannot keep seed for replanting, susceptibility to BPH and other rice pests, and low consumer demand for the product. Nonetheless, the government continued to promote the technology, targeting an area of over 300,000 hectares by 2014 [94]. After serious BPH outbreaks in 2010–2011 (see below), the Ministry and provincial government of Central Java abandoned these ambitious targets—for the time being, at least.

6.4. Return of the Planthopper

Under these conditions, it is hardly surprising that BPH infestations began to manifest in many of Indonesia's major rice-producing areas. Due to the rather chaotic administrative situation created by the ongoing decentralization process, combined with reluctance to report bad news that still persists in Indonesia despite more than a decade of democratic reforms and press freedom, accurate figures on the scale of the problem are unavailable. Throughout the early years of the 2000s, BPH infestation figures ranged in the low thousands of hectares, but between 2008 and 2011 shot up dramatically. The figures from the Directorate of Crop Protection shown in Table 2 most likely significantly understate the magnitude of the problem [95]:

Table 2. Rice brown planthopper (BPH) Infestation 2008–2013.

Year	Rice Paddy Planted (Java)	Infested (ha)	Totally Destroyed (ha)
2008	5,742,270	24,152	608
2009	6,093,303	47,473	1237
2010	6,358,521	137,768	4602
2011	6,165,079	223,656	36,065
2012	6,185,521	30,174	242
2013	6,467,073	61,630	2763

The sharp decline in 2012 can be explained by two factors. First, in 2011 the government (temporarily, at least) abandoned its policy of promoting hybrid rice, as it became increasingly obvious that the hybrid varieties were especially susceptible to BPH outbreaks, and began distributing BPH-resistant inbred varieties Inpari 11 and 13. Also, 2012 was a year of low rainfall in most of Indonesia, compared to the very wet two previous years [76].

It appears that Indonesia has now come nearly full circle since that day in 1986 when IPM scientists took matchboxes of rice field insects into the office of former president Suharto, which led to the establishment of what has been heralded as one of the most successful national integrated pest management programs in the world [96,97].

The foregoing story provides an important cautionary note not about the sustainability of Integrated Pest Management *per se*, but certainly about the sustainability of government-supported IPM programs, and about the self-sustaining and self-replicating nature of sound agronomic practice grounded in farmer empowerment and mobilization. Despite the considerable success of the Indonesian National IPM Program in the 1980s and 1990s, its accomplishments have withered in the subsequent onslaught of bad policy, worse programs and aggressive and relentless marketing by the producers and traders of agricultural chemicals and seeds.

There remains a committed cadre of IPM die-hards scattered across numerous locations where the IPM Program succeeded two decades ago, but they are growing old. Very few of the 14,000 Pest Observers and IPM Master Trainers who joined the program back in 1990 still work for the government; those who do are nearing retirement age, and a new generation has not been prepped to replace them.

At a recent *ad hoc* meeting of national and international IPM supporters to discuss the ongoing BPH epidemic in Indonesia, one observer mooted the rather pessimistic suggestion that the most effective strategy might be to wait until the problem becomes so serious that the national government has no choice but to act—similar to the situation that spawned the original National IPM Program back in 1986 [76]. If and when that does transpire, an eager army of spiders, water striders, dragonflies, damselflies, beetles, wasps and fungi are waiting in the wings to play their crucial role in supporting Indonesia's food security and sovereignty. They will be helped along by a much smaller, older, but no less purposeful and dedicated corps of IPM Field School alumni and their dwindling numbers of committed allies still in government service.

ACKNOWLEDGMENTS

Travel to conduct interviews and collect information used in the preparation of this manuscript was supported by an OSP travel grant from the Faculty of Arts, Monash University, Australia. This support does not include funds for covering the costs to publish in open access.

CONFLICTS OF INTEREST

The author declares no conflict of interest.

REFERENCES

1. Heong, K.L.; Teng, P.S.; Moody, K. Managing rice pests with less chemicals. GeoJournal 1995, 35, 337–349.
2. Kalpe, C. Rice in world trade. Sustainable Rice Production for Food Security, Proceedings of the 20th Session of the International Rice Commission; Tran, D.V., Ed.; Food and Agriculture Organization of the United Nations: Rome, Italy, 2003. Available online: http://www.fao.org/docrep/006/y4751e/y4751e00.HTM (accessed on 5 December 2014).

3. Simatupang, P.; Timmer, C.P. Indonesian rice production: Policies and realities. Bull. Indones. Econ. Stud. 2008, 44, 65–79.
4. Fox, J. Managing the ecology of rice production in Indonesia. In Indonesia: Resources, Ecology, and Environment; Hardjono, J., Ed.; Oxford University Press: Singapore, Singapore, 1991; pp. 61–84.
5. Fox, J. Ecological policies for sustaining high production in rice: Observations on rice intensification in Indonesia. In South-East Asia's Environmental Future; Brookfield, H., Byron, Y., Eds.; Oxford University Press: Kuala Lumpur, Malaysia, 1993; pp. 211–224.
6. Röling, N.; van de Fliert, E. Transforming extension for sustainable agriculture: The case of integrated pest management in Indonesia. Agric. Hum. Values 1994, 11, 96–108.
7. Booth, A. Agricultural Development in Indonesia; Allen and Unwin: Sydney, Australia, 1988.
8. Resosudarmo, B.P.; Alisjahbana, A.; Brodjonegoro, B.P.S. Indonesia's Sustainable Development in a Decentralization Era; Indonesian Regional Science Association: Jakarta, Indonesia, 2002.
9. Lowe, C. Wild Profusion: Biodiversity Conservation in an Indonesian Archipelago; Princeton University Press: Princeton, NJ, USA, 2006.
10. GRiSP (Global Rice Science Partnership). Rice Almanac, 4th ed.; International Rice Research Institute (IRRI): Los Baños, Philippines, 2013.
11. Chang, T.T. The rice cultures. In The Early History of Agriculture; Hutchison, J.G.G., Clark, E., Jope, M., Riley, R., Eds.; Oxford University Press: Oxford, UK, 1977; pp. 143–155.
12. Huggan, R.D. Co-evolution of rice and humans. GeoJournal 1995, 35, 262–365.
13. Visser, M. Rice: The tyrant with a soul. In Much Depends on Dinner: The Extraordinary History and Mythology, Allure and Obsessions, Perils and Taboos of an Ordinary Meal; Visser, M., Ed.; Grove Press: New York, NY, USA, 1986; pp. 155–191.
14. Table of Harvested Area–Productivity–Production Paddy in All Provinces. Available online: http://www.bps.go.id/eng/tnmn_pgn.php?kat=3 (accessed on 5 December 2014).
15. Population of Indonesia by Province 1971, 1980, 1990, 1995 2000, and 2010. Available online: http://www.bps.go.id/eng/tab_sub/view.php?kat=1&tabel=1&daftar=1&id_subyek=12&notab=1 (accessed on 5 December 2014).
16. Settle, W.H.; Ariawan, A.; Astuti, E.T.; Cahyana, W.; Hakim, A.L.; Hindayana, D.; Lestari, A.S.; Pajarningsih. Managing tropical rice pests through conservation of generalist natural enemies and alternative prey. Ecology 1996, 77, 1975–1988.
17. Chiu, S.C. Biological control of the brown planthopper. In Brown Planthopper: Threat to Rice Production in Asia; International Rice Research Institute (IRRI): Los Baños, Philippines, 1979.
18. Heong, K.L.; Aquino, G.B.; Barrion, A.T. Arthropod community structures of rice ecosystems in the Philippines. Bull. Entomol. Res. 1991, 81, 407–416.
19. Reissig, W.H.; Heinrichs, E.A.; Litsinger, J.A.; Moody, K.; Fiedler, L.; Mer, T.W.; Barrion, A.T. Illustrated Guide to Integrated Pest Management in Rice in Tropical Asia; International Rice Research Institute (IRRI): Los Baños, Philippines, 1986.
20. The Green Revolution: Accomplishments and Apprehensions. Available online: http://www.agbioworld.org/biotech-info/topics/borlaug/borlaug-green.html (accessed on 5 December 2014).
21. Easterbrook, G. Forgotten benefactor of humanity. Atl. Mon. 1997, 279, 75–82.
22. Shiva, V. The Violence of the Green Revolution: Ecological Degradation and Political Conflict in Punjab; Zed Press: New Delhi, India, 1992.
23. Yapa, L. What are improved seeds? An epistemology of the green revolution. Econ. Geog. 1993, 69, 254–257.
24. Dalrymple, D.G. The development and adoption of high-yielding varieties of wheat and rice in developing countries. Am. J. Agric. Econ. 1985, 67, 1067–1073.

25. INDONESIA: Stagnating Rice Production Ensures Continued Need for Imports. Available online: http://www.pecad.fas.usda.gov/highlights/2012/03/Indonesia_rice_Mar2012/ (accessed on 5 December 2014).
26. Dove, M.R.; Kammen, D.M. The epistemology of sustainable resource use: Managing forest products, swiddens, and high-yielding variety crops. Hum. Organ. 1997, 56, 91–101.
27. Chang, T.T. Genetics and evolution of the Green Revolution. In Replies from Biological Research; de Vincente, R., Ed.; Consejo Superior de Investigaciones Cientificas: Madrid, Spain, 1979; pp. 187–209.
28. Heinrichs, E.A. Impact of insecticides on the resistance and resurgence of rice planthoppers. In Planthoppers, Their Ecology and Management; Denno, R.F., Perfect, T.J., Eds.; Chapman and Hall: New York, NY, USA, 1994; pp. 571–598.
29. Anderson, B.R.O'G. Language and Power: Exploring Political Cultures in Indonesia; Cornell University Press: Ithaca, NY, USA, 1990.
30. Moertopo, A. The Acceleration and Modernization of 25 Years' Development; Centre for Strategic and International Studies (CSIS): Jakarta, Indonesia, 1973.
31. Elson, R.E. Suharto: A Political Biography; Cambridge University Press: Cambridge, UK, 2001.
32. Mears, L.A.; Afiff, S. A new look at the BIMAS program and rice production. Bull. Indones. Econ. Stud. 1968, 4, 29–47.
33. Roekasah, E.A.; Penny, D.H. BIMAS: A new approach to agricultural extension in Indonesia. Bull. Indones. Econ. Stud.1967, 3, 60–69.
34. Hill, H. The economy. In Indonesia's New Order; Hill, H., Ed.; University of Hawaii Press: Honolulu, HI, USA, 1994; pp. 54–122.
35. Timmer, C.P. The Role of Price Policy in Increasing Rice Production in Indonesia, 1968–1982; Harvard Institute for International Development (HIID) Development Discussion Paper No. 196; HIID: Cambridge, MA, USA, 1985.
36. Bujono, B. Kita dan beras (Rice and us). Tempo 1985, 15, 12–15.
37. Umali-Deininger, D.; Schwart, L.A. Public and Private Agricultural Extension: Beyond Traditional Frontiers; The World Bank: Washington, DC, USA, 1994.
38. Seymore, F.J.; Kartodihardjo, H. Indonesia. In Right Conditions: The World Bank, Structural Adjustment, and Forest Policy Reform; Seymour, F.J., Dubash, N., Eds.; World Resources Institute: Washington, DC, USA, 2004; pp. 83–111.
39. Kapur, D.; Lewis, J.P.; Webb, R. The World Bank: Its First Half Century; Brookings Institution Press: Washington, DC, USA, 1997; Volume I.
40. Anderson, J.R.; Feder, G. Agricultural extension: Good intentions and hard realities. World Bank Res. Obs. 2004, 19, 41–60.
41. Kiritani, K. Pest management in rice. Annu. Rev. Entomol. 1979, 24, 279–312.
42. Winarto, Y.T. The evolutionary changes in rice-crop farming: Integrated pest management in Indonesia, Cambodia, and Vietnam. Southeast Asian Stud. 2004, 42, 241–272.
43. Gallagher, K.D.; Kenmore, P.E.; Sōgawa, K. Judicial use of insecticides deter planthopper outbreaks and extend the life of resistant varieties in Southeast Asian rice. In Planthoppers: Their Ecology and Management; Denno, R.F., Perfects, T.J., Eds.; Chapman and Hall: New York, NY, USA, 1994; pp. 599–614.
44. Pingali, P.L.; Gerpachio, R.V. Living with reduced pesticide use in tropical rice. Food Policy 1997, 22, 107–118.
45. Matteson, P.C. Insect Pest Management in Tropical Asian Irrigated Rice. Annu. Rev. Entomol. 2000, 45, 549–574.
46. Woodburn, A. The current rice agrochemicals market. In Pest Management in Rice, Proceedings of the Conference on Pest Management in Rice, London, UK, 4–7 June 1990; Grayson, B.T., Greene, M.B., Cropping, L.G., Eds.; Elsevier Applied Science: London, UK, 1990; pp. 15–30.
47. Grist, D.H.; Lever, R.J.A.W. Pests of Rice; Longmans, Green and Co: London, UK, 1969.
48. International Rice Research Institute (IRRI). The Major Insect Pests of the Rice Plant; Johns Hopkins: Baltimore, MD, USA, 1967.

49. Sōgawa, K. The rice brown planthopper: Feeding physiology and host plant interactions. Annu. Rev. Entomol. 1982, 27, 49–73.
50. Bae, S.H.; Pathak, M.D. Life history of Nilaparvata lugens (Homoptera: Delphacidae) and susceptibility of rice varieties to its attacks. Ann. Entomol. Soc. Am. 1969, 63, 149–155.
51. Sigsgaard, L. Early season natural biological control of insect pests in rice by spiders–and some factors in the management of the cropping system that may affect this control. In Proceedings of the 19th European Colloquium of Arachnology, Århus, Denmark, 17–22 July 2000; Toft, S., Scharff, N., Eds.; Aarhus University Press: Aarhus, Denmark, 2000; pp. 57–64.
52. Gallagher, K.D. Effects of Host Plant Resistance on the Microevolution of the Rice Brown Planthopper, Nilaparvata Lugens Stål (Homoptera: Delphacidae). Ph.D. Dissertation, University of California, Berkeley, CA, USA, 1984.
53. Gould, F.; Kennedy, G.G.; Johnson, M.T. Effects of natural enemies on the rate of herbivore adaptation to resistant host plants. Entomol. Exp. Appl. 1991, 58, 1–4.
54. Kenmore, P.E.; Head (at that time) of the FAO Asian Regional IPM Programme: FAO Regional Office for Asia and the Pacific, Bangkok, Thailand. Personal Communication, 1989.
55. Dilts, D.R. Training: Reschooling society? Prisma 1984, 38, 78–90.
56. Conway, G.R. Agroecosystem analysis. Agric. Admin. 1985, 20, 31–55.
57. Conway, G.R. The properties of agroecosystems. Agric. Syst. 1987, 24, 95–117.
58. Altieri, M. Agroecology: The Scientific Basis of Alternative Agriculture; Westview Press: Boulder, CO, USA, 1987.
59. Ooi, P.A.C. Beyond the Farmer Field School: IPM and empowerment in Indonesia. Proceedings of the International Conference of IPM-Theory and Practice, Developing Sustainable Agriculture, Guangzhou, China, 15–20 June 1998; Available online: http://pubs.iied.org/pdfs/6145IIED.pdf (accessed on 10 December 2014).
60. Van de Fliert, E.; Pontius, J.; Röling, N. Searching for strategies to replicate a successful extension approach: Training of IPM trainers in Indonesia. Eur. J. Agric. Educ. Ext. 1995, 1, 41–63.
61. Kishi, M.; Hirschorn, N.; Djajadisastra, M.; Saterlee, L.N.; Strowman, S.; Dilts, R. Relationship of pesticide spraying to signs and symptoms in Indonesian farmers. Scand. J. Workplace Environ. Health 1995, 21, 124–133.
62. Hattori, I.; Siwi, S.S. Rice stemborers in Indonesia. Jpn. Agric. Res. Q. 1985, 20, 25–30.
63. Van den Berg, H. IPM Farmer Field Schools: A Synthesis of 25 Impact Evaluation Reports Prepared for the FAO Global IPM Facility; Global IPM Facility, Food and Agriculture Organization of the United Nations (FAO): Rome, Italy, 2001.
64. Karel, B. Pesticide dependence persists at the World Bank. Glob. Pestic. Campaign. 2003, 13, 6–10.
65. Thorburn, C.; Kusumohadi, M. Experiences in Institutionalizing Participatory Approaches in GoI Departments; DELIVERI Project and Directorate General of Livestock and Animal Husbandry: Ujung Pandang, Indonesia, 1986.
66. FAOSTAT. Available online: http://faostat3.fao.org/download/R/RT/E (accessed on 15 December 2014).
67. Triwidodo, H. Jawa (Terancam) Krisis Pangan Jika Kita Tidak Mampu Mengelola Serangan Wereng Secara Benar. Available online: http://cybex.ipb.ac.id/index.php/artikel/detail/topik/247 (accessed on 15 December 2014).
68. Indonesia in Transition: Social Aspects of Reformasi and Crisis; Manning, C., van Diermen, P., Eds.; Zed Books: London, UK, 2000.
69. Aspinal, E., Fealy, G., Eds.; Local Power and Politics in Indonesia: Decentralization and Democratization; Institute of South East Asian Studies (ISEAS): Singapore, Singapore, 2003.
70. Montgomery, R.; Sunarto, S.; Mawardi, S.; Usman, S.; Toyomah, N.; Febriani, V.; Strain, J. Deregulation of Indonesia's interregional agricultural trade. Bull. Indones. Econ. Stud. 2002, 38, 98–117.

71. Iqbal, F., James, W.E., Eds.; Deregulation and Development in Indonesia; The World Bank: Washington, DC, USA, 2002.
72. Van Dijk, K. A Country in Despair: Indonesia between 1997 and 2000; KITLV Press: Leiden, The Netherlands, 2001.
73. Hartono; Food Crop Pest and Disease Observation Laboratory (LPHP): Pandaan, Indonesia. Personal Communication, 2012.
74. Feder, G.; Willett, A.; Zijp, W. Agricultural extension: Generic challenges and the ingredients for solutions. In Knowledge Generation and Technical Change: Institutional Innovation in Agriculture; Wolf, S., Zilberman, D., Eds.; Kluwer Academic Press: Norwell, MA, USA, 2001; pp. 313–353.
75. Pardy, P.G.; Beintema, N.; Dehmer, S.; Wood, S. Agricultural Research: A Growing Global Divide? International Food Policy Research Institute (IFPRI): Washington, DC, USA, 2006.
76. Wienarto, N.; Center for Public Policy Transformation: Jakarta, Indonesia. Personal Communication, 2014.
77. World Bank. Decentralizing Indonesia: A Regional Public Expenditure Review Overview Report; The World Bank: Washington, DC, USA, 2003. Available online: https://openknowledge.worldbank.org/handle/10986/14632 (accessed on 12 December 2012).
78. McVey, R.T., Ed.; Money and Power in Provincial Thailand; Institute for Southeast Asian Studies (ISEAS): Singapore, Singapore, 2000.
79. Hidayat, S. Local power and politics in Indonesia: Decentralization and democratization. In Deepening Democracy in Indonesia? Direct Elections for Local Leaders (Pilkada); Erb, M., Suliyistiyanto, P., Eds.; Institute for Southeast Asian Studies (ISEAS): Singapore, Singapore, 2009; pp. 125–146.
80. Rohdewohld, R. Decentralization and the Indonesian bureaucracy: Major changes, minor impact? In Local Power and Politics in Indonesia: Decentralization and Democratization; Aspinal, E., Fealy, G., Eds.; Institute for Southeast Asian Studies (ISEAS): Singapore, Singapore, 2003; pp. 259–274.
81. Kristiansen, S.; Ramli, M. Buying an income: The market for civil service positions in Indonesia. Contemp. Southeast Asia 2006, 28, 207–233.
82. McKinley, T. The Macroeconomics of Poverty Reduction: The Case Study of Indonesia; United Nations Development Programme (UNDP): New York, NY, USA, 2003.
83. Siragih, B. Keberhasilan deregulasi pestisida. Agrina, 19 June 2006. Available online: http://www.agrina-online.com/show_article.php?rid=10&aid=67 (accessed on 12 December 2014).
84. Ministry of Agriculture, Republic of Indonesia. Nama formulasi, bahan aktif, jenis pestisida dan penggunaan yang diizinkan (Formulation, active ingredient, type of pesticide and permitted usage). Available online: http://psp.deptan.go.id./file/PESTISIDA%20TERDAFTAR%20DAN%20DIIZINKAN%20-%202012.pdf (accessed on 15 December 2014).
85. Thorburn, C. Empire strikes back: The making and unmaking of Indonesia's national integrated pest management program. Agroecol. Sustain. Food Syst. 2104, 38, 3–24.
86. Winarto, Y.; Department of Anthropology: University of Indonesia, Jakarta, Indonesia. Personal Communication, 2012.
87. Ricehoppers.net. Brown Planthopper in Klaten–Boyolali–Sukoharjo, Central Java. Available online: http://ricehoppers.net/2011/05/brown-planthopper-in-klaten%E2%80% 94boyolali%E2%80%94sukoharjo-central-java/ (accessed on 15 December 2014).
88. Government of Indonesia. Masterplan: Acceleration and Expansion of Indonesia Economic Development 2011–2015; Coordinating Ministry for Economic Affairs: Jakarta, Indonesia, 2011.
89. Budianto, E.; Food Crop Plant Protection, Ministry of Agriculture: Jakarta, Indonesia. Personal Communication, 2012.

90. Direktorat Perlindungan Tanaman Pangan. Buku Saku Perlindungan Tanaman Pangan 2011; Ministry of Agriculture: Jakarta, Indonesia, 2011.
91. Direktorat Jenderal Tanaman Pangan. Laporan Tahunan 2010; Ministry of Agriculture: Jakarta, Indonesia, 2010.
92. Kuyek, D.; Zamora, O.B.; Quijano, R. Hybrid Rice in Asia: An Unfolding Threat. Available online: http://www.grain.org/article/entries/34-hybrid-rice-in-asia-an-unfolding-threat (accessed on 15 December 2104).
93. Financial Express. Hybrid Rice Suffers Severe Blow As "Choosy" Consumers Prefer HYV. Available online: http://www.grain.org/bulletin_board/entries/4280-hybrid-rice-suffers-severe-blow-as-choosy-consumers-prefer-hyv (accessed on 15 December 2014).
94. Riyadi, M.A.; Prihatnala, S.; Firdaus, H.; Hernawan, A.K. Permainan Benih Impor Sengsarakan Petani. Gatra, 18 January 2012, pp. 8–14. Available online: http://www.gatra.com/politik-1/7170-permainan-benih-impor-sengsarakan-petani.html (accessed on 15 December 2014).
95. Directorate of Crop Protection. Data OPT Padi. Available online: http://tanamanpangan.pertanian.go.id/ditlintp/statis-17-dataoptpadi.html (accessed on 15 December 2014).
96. Thiers, P. Successful pesticide reduction politics: Learning from Indonesia. Soc. Nat. Resour. 1997, 10, 319–328.
97. Bottrell, D.G.; Schoenly, K.G. Resurrecting the ghost of green revolutions past: The brown planthopper as a recurring threat to high-yielding rice production in tropical Asia. J. Asia-Pac. Entomol. 2012, 15, 122–140.

CHAPTER 12

Development of a Microbial-Based Integrated Pest Management Program for *Helicoverpa* spp. (Lepidoptera: Noctuidae) and Beneficial Insects on Conventional Cotton Crops in Australia

Robert K. Mensah *, Alison Young and Leah Rood-England

Australian Cotton Research Institute, NSW Department of Primary Industries, Locked Bag 1000, Narrabri, NSW 2390, Australia

ABSTRACT

Entomopathogenic fungi, when used as a microbial control agent against cotton pests, such as *Helicoverpa* spp., may have the potential to establish and spread in the environment and to have an impact on both pests and beneficial insects. Information on the effect of entomopathogenic fungi on pests and beneficial insects is crucial for a product to be registered as a biopesticide. The effect of the entomopathogenic fungus BC 639 (*Aspergillus* sp.) against *Helicoverpa* spp. and beneficial insects (mostly predatory insects) was studied in the laboratory and in cotton field trials. The results show that when *Helicoverpa* spp. second instar larvae were exposed to increasing concentrations (from $10^2$to 10^9) of the entomopathogenic fungus BC 639, the optimum dose required to kill over 50% of the insects was 1.0×10^7spores/mL. In the field trials, the number of *Helicoverpa* spp. per metre on plots treated with 1.0 or 0.50 L/ha of BC 639 was the same as on plots treated with the recommended rate of the commercial insecticide, Indoxacarb. However, when plots were treated with 0.25 L/ha of BC 639, this was not as effective at controlling *Helicoverpa* spp. as 1.0 or 0.5 L/ha BC 639 or Indoxacarb. BC 639 had less effect on predatory insects when applied at lower rates (0.50 and 0.25 L/ha) than at higher rates (1.0 L/ha). Thus, BC 639 was more selective against predators when applied at lower rates than at the higher rate, but was also more selective than Indoxacarb. Thus, the ability of BC 639 to control *Helicoverpa* spp. effectively with a minimal effect on predatory insects indicates its potential for enhancing integrated pest management programs and to sustain cotton production.

Keywords: *Aspergillus* sp.; *Helicoverpa* spp.; entomopathogenic fungus; integrated pest management (IPM); transgenic cotton; *Bacillus thuringiensis*

1. INTRODUCTION

Cotton crops in Australia are attacked by a wide range of pests, the major ones being the two lepidopteran pests, *Helicoverpa armigera* (Hübner) and *Helicoverpa punctigera* (Wallengren) [1]. These two pests are responsible for the use of synthetic insecticides in the Australian cotton industry from the 1960s until 1996–1997 [2]. In 1996, transgenic cotton containing genes from *Bacillus thuringiensis* (Bt) were introduced and then taken up by 90% of Australian cotton farmers. This cotton expresses the toxins Cry1Ac (Ingard®) (in transgenic cotton from 1996) and Cry 1Ac + Cry2Ab (Bollgard II®) (in transgenic cotton from 2004). The use of transgenic cotton has reduced the impact of the major pest (*Helicoverpa* spp.) on cotton crops in Australia [3,4,5,6]. These transgenic cotton crops are toxic to *Helicoverpa* spp. and other lepidopteran pests when ingested [7]. Thus, the introduction of transgenic cotton crops has decreased the use of synthetic insecticides against *Helicoverpa* spp. by 75%–80% [6]. In contrast, 10% of cotton farmers in Australia still grow non-transgenic cotton and rely exclusively on the use of synthetic insecticides to manage *Helicoverpa* spp. For these non-transgenic crops, there is a need to control *Helicoverpa* spp. with synthetic insecticides early in the season, which inadvertently suppresses populations of beneficial insects and potentially causes outbreaks of secondary pests, such as green mirids (*Creontiades dilutus*), cotton aphids (*Aphis gossypii*) and green vegetable bugs (*Nezara viridula*). The negative impact of this practice was of great concern to the cotton industry [6,7,8] until the adoption of transgenic (Bt) cotton crops.

Many beneficial insects, particularly predatory insects and specialist parasitoids, have been recorded in cotton crops worldwide. The role of beneficial insects in regulating pests in cotton agro-ecosystems has become increasingly important [9,10,11,12,13]. Although these beneficial insects do not show tightly coupled dynamics with their prey, because their population dynamics are not solely dependent on target pests, they often provide effective control, particularly in agro-ecosystems with non-chemical insecticide regimes [10,11,14,15,16].

In Australia, cotton crops are often grown in remote areas that have very sparse natural vegetation [11,14] and that lay fallow for most of the year. With no natural refuges and food sources for the adult natural enemies of pests, beneficial insect populations decline quickly, thereby limiting natural biological control [11,14,17]. Presently, cotton farmers in Australia are inadvertently discriminating against the natural enemies of *Helicoverpa* spp. and other pests by exclusively using synthetic insecticides on monocultured crops to control *Helicoverpa* spp. on conventional cotton and to control sucking pests on transgenic (Bt) cotton.

Helicoverpa spp. are difficult to manage on conventional cotton crops: first, because of the lack of diversity and stability in the cotton agro-ecosystem; second, because *Helicoverpa* spp. are highly mobile and highly migratory and can rapidly infest cotton crops; and third, because their eggs can escape natural enemy attack, as the natural enemies may not be present or established in sufficiently high numbers [2,11,14,18,19]. These factors have encouraged the use of synthetic broad-spectrum insecticides.

As a result, cotton farmers require new pest control methods to manage *Helicoverpa* spp. and to complement integrated pest management (IPM) on conventional cotton crops. Entomopathogenic fungi are known to be important natural enemies of many pests of agricultural crops [20,21]. Over 750 different species of fungi have been identified to date; these are cosmopolitan organisms that have been isolated from soils and have infected insects around the world

[22,23,24]. All stages of insects, from the egg, larva or nymph to adult, can be killed by fungal pathogens. Fungi, unlike other pathogens, such as viruses and bacteria, do not need to be ingested to infect and kill the host. The advantage of using fungi for biological control is that the fungi are self-generating within or on the surface of the plant and can potentially provide ongoing protection as the plant grows, with little effect on non-herbivorous predators [5]. Several species of entomopathogenic fungi have been developed as alternatives to synthetic insecticides because of the low risk to humans, low risk to the environment and low pest resistance [22]. However, in Australia, only two isolates of *Metarhizium anisopliae* are commercially available, for the control of plague locusts and sugar cane grubs.

The present study evaluated the efficacy of a naturally occurring entomopathogenic fungus, BC 639 (*Aspergillus* sp.), against *Helicoverpa* spp. populations and on retaining populations of beneficial insects (e.g., predatory beetles, predatory bugs, predatory lacewings and spiders) on commercial conventional cotton crops. The study objectives were: (1) to assess the dose-response of the BC 639 fungus on *Helicoverpa* spp. larvae; (2) to assess the efficacy of different application rates of the BC 639 formulated product; (3) to identify the optimum rate of the BC 639 formulated product needed to control *Helicoverpa* spp. populations; and (4) to assess the impact of the BC 639 formulated product on the natural enemies of the target pests.

2. MATERIALS AND METHODS

2.1. Isolation and Formulation of the Entomopathogenic Fungus BC 639

The BC 639 fungus was originally isolated from an infected insect in a commercial cotton farm [13]. The methods of Bahar *et al.* [25] and Mensah and Austin [13] were followed for isolating the spores for culturing and for preparing the formulation. Using a haemocytometer, the spore concentration was estimated to be 1.0×10^7. The conidial germination was determined from 100 spore counts, and 100% germination was observed [5].

2.2. Experiment 1: Dose-Response of *Helicoverpa* spp. Second Instar Larvae in the Laboratory to BC 639

The experiment was conducted in the laboratory at the Australian Cotton Research Institute in Narrabri, NSW, Australia, from 10 to 28 February 2006. *Helicoverpa armigera* larvae were cultured in the laboratory and used in the study at the second instar stage.

BC 639 was tested at 10^1, 10^2, 10^3, 10^4, 10^5, 10^6, 10^8 and 10^9 spores/mL. For each treatment dose, 10 *H. armigera* second instar larvae were placed in a 90-mm Petri dish lined with filter paper and sprayed separately. The larvae from each treatment were then each transferred into separate 45-mm Petri dishes containing a fresh 19-mm conventional cotton leaf disc on filter paper. The Petri dishes were sealed and placed in an incubator set at 26–28 °C. Therefore, each treatment had 10 larvae placed individually in 10 Petri dishes.

The larvae were checked on 1, 2 and 3 days after treatment (DAT), and the number of dead and alive larvae were recorded. The mortality for each treatment was calculated, and a dose-response curve was developed for the concentration of BC 639 (log spores/mL) per mortality (%) of *Helicoverpa* spp. using GraphPad Prism 6 (GraphPad Software, Inc. San Diego, CA, USA). The number of spores/mL required to cause 50% mortality was extrapolated from the dose-response curve.

2.3. Experiment 2: Efficacy of BC 639 against *Helicoverpa* spp. First and Second Instar Larvae in the Laboratory

The experiment was conducted in the laboratory at the Australian Cotton Research Institute in Narrabri, NSW, Australia. *Helicoverpa armigera* first and second instar larvae were used for the experiment. The spore concentration used for this study was 1.0×10^7. The conidial germination was 100%.

BC 639 was formulated into an oil-based product by Becker Underwood Pty. Ltd., Australia. The product was evaluated at 1% v/v (1.0 L/ha), and water was used as the control treatment. First instar larvae were used in the first experiment, and second instar larvae were used in the second experiment. For each experiment and each treatment, one larva per plant was placed on a cotton leaf and sprayed until run-off. Each treatment used 40 cotton plants infested with 40 larvae. After spraying, each larvae was transferred into a 35-mL clear plastic container (P10M; Solo, Urbana, IL, USA) containing a soybean-based artificial diet. The number of dead larvae were counted and recorded daily until all of the larvae had pupated. The mortality (%) was calculated for each treatment.

2.4. Experiment 3: Efficacy of Different Rates of BC 639 on the Survival of *Helicoverpa* spp. and Beneficial Insects on Commercial Cotton Crops

The trial was conducted on a dryland, commercial, conventional cotton farm at Getta Getta near Goondiwindi in the Macintyre Valley in Queensland in the 2006–2007 growing season.

The following treatments were evaluated for their effect on *Helicoverpa* spp., predatory beetles, predatory bugs, predatory lacewings and spiders: (1) 1.0 L/ha BC 639; (2) 0.50 L/ha BC 639; (3) 0.25 L/ha BC 639; (4) 0.85 L/ha Indoxacarb; and (5) unsprayed (untreated) control. The treatment plots were arranged in a randomized complete block design with six replicates per treatment. Each replicated plot measured 8-m wide and 220-m long.

The first foliar treatment was applied on 31 January 2007 and the second on 25 February 2007. Foliar treatments were applied using a ground rig sprayer fitted with flat fan nozzles (3 nozzles/row of cotton) to achieve a droplet size of 200 μm and an application volume of 100 L/ha. The treatment was applied in the morning when the temperature was between 20 °C and 28 °C. The decision for when to apply the treatment was based on the IPM Guidelines and the economic threshold of 2.0 larvae per metre as recommended by CottonLogic [1].

Visual counts of *Helicoverpa* spp. eggs and larvae, predatory beetles, bugs, lacewings and spiders on cotton plants for each treatment were made at approximately weekly intervals. For each of the six treatment replicates, a randomly selected 1-m length of planted row was examined. The number of *Helicoverpa* spp., predatory beetles, predatory bugs, predatory lacewings and spiders per metre for each treatment was determined.

2.5. Experiment 4: Efficacy of Different Rates of BC 639 against *Helicoverpa* spp. and Beneficial Insects on Conventional Cotton Crops at Norwood in 2007–2008

The trial was conducted on a commercial conventional cotton farm at Norwood, near Moree. The trial was conducted from 16 November 2007 to 11 March 2008.

The following treatments were evaluated for their effect on *Helicoverpa* spp. eggs, very small and small, medium and large larvae and beneficial insects: (1) 1.0 L/ha BC 639; (2) 0.50 L/ha BC 639; (3) 0.25 L/ha BC 639 (4) 0.85 L/ha Indoxacarb; and (5) unsprayed (untreated) control. The treatment plots were arranged in a randomized complete block design with four replicates per treatment. Each replicated plot measured 40-m wide (equivalent to the number of rows) and 90-m long. A 40-m wide buffer separated each of the Indoxacarb-treated plots, fungus plots and unsprayed plots.

The number of insects were counted 24 h before treatment and then 3, 7 and 14 DAT for each spray application. Foliar treatments were applied on 30 January 2008 (first spray) and 25 February 2008 (second spray) using a ground rig sprayer fitted with flat fan nozzles (3 nozzles/row of cotton plants) to achieve a droplet size of 200 μm and an application volume of 100 L/ha. The treatments were applied in the morning when the temperature was between 20 °C and 28 °C. On each occasion, the treatments were applied using 100 L of water per hectare. The control plot was unsprayed.

Visual counts of *Helicoverpa* spp. eggs and larvae, predatory beetles, bugs, lacewings and spiders for each treatment were made at approximately weekly intervals. For each of the four treatment replicates, a randomly selected 1-m length of planted row was examined. The number of *Helicoverpa* spp., predatory beetles, predatory bugs, predatory lacewings and spiders per metre for each treatment was determined.

2.6. Data Analysis

All experimental data were analysed using repeated measures analysis of variance (v. 2.03; Graphpad Instat and Prism Software Inc., San Diego, CA, USA). Treatments and sample dates were the independent variables. Tukey-Kramer multiple comparison tests were used to compare the treatment means.

3. RESULTS

3.1. Experiment 1: Dose-Response of *Helicoverpa* spp. Second Instar Larvae to BC 639 in the Laboratory at the Australian Cotton Research Institute

Exposure to progressively increasing concentrations of BC 639 from 10^2 to 10^9 spores/mL resulted in the deaths of *Helicoverpa* spp. second instar larvae (Figure 1). The optimum dose required to kill over 50% of the insects intercepts the curve at 10^7 and was estimated to be 1.0×10^7 spores/mL (Figure 1).

3.2. Experiment 2: Efficacy of BC 639 against Helicoverpa spp. First and Second Instar Larvae in the Laboratory

BC 639 caused 80% and 82% mortality of *H. armigera* first and second instar larvae, respectively, and was significantly different from the control (water)-treated larvae with 5% and 7.5% mortality, respectively (Table 1).

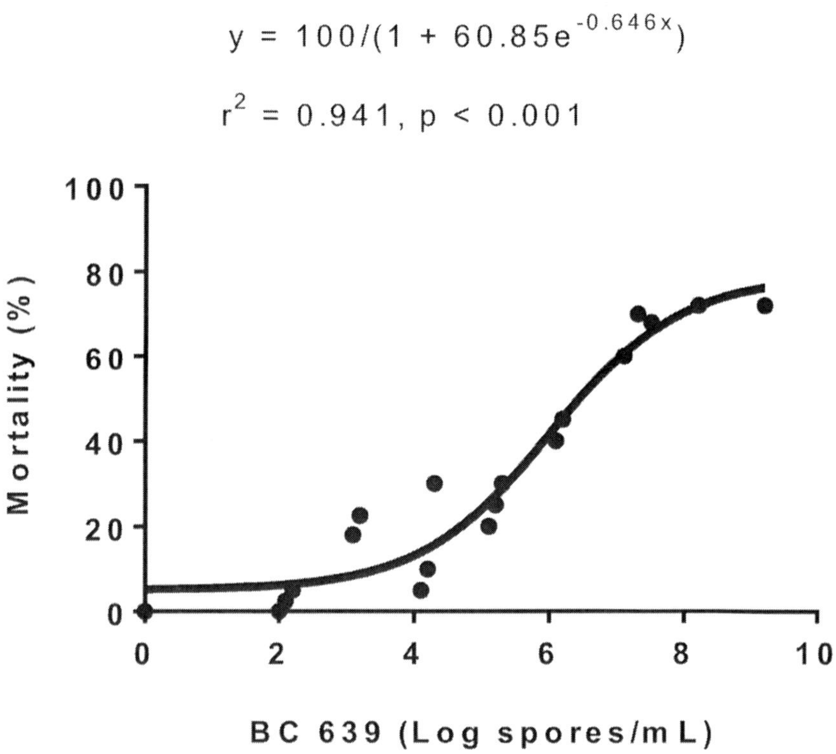

Figure 1. Dose-response curve for BC 639 and mortalities of *H. armigera* second instar larvae in the laboratory, 10–28 February 2006.

Table 1. Effect of BC 639 on *H. armigera* first and second instar larvae (n = 40 larvae/treatment) in the laboratory at the Australian Cotton Research Institute, Narrabri, Australia.

Treatments	% Mortality First Instar ± SE	% Mortality Of Second Instar ± SE
1% v/v BC 639 (*Aspergillus* sp.)	80.50 ± 5.62 [a]	82.50 ± 5.47 [a]
Water (control)	5.00 ± 3.44 [b]	7.50 ± 4.10 [b]

Means within columns followed by the same letters are not significantly different ($p > 0.05$; Tukey–Kramer multiple comparison test).

3.3. Experiment 3: Efficacy of Different Rates of BC 639 on the Survival of Helicoverpa spp. Eggs and Larvae on Conventional Cotton Crops at Getta Getta in 2006–2007

3.3.1. *Helicoverpa* spp. Eggs

After the first spraying, oviposition by *Helicoverpa* spp. was the same among treated and control plots at 3 DAT, but was significantly lower ($p < 0.003$) on the BC 639- and Indoxacarb-treated plots than the unsprayed plots at seven and 14 DAT (Figure 2A). No significant difference ($p > 0.05$) was detected after the first spraying among cotton plants treated with different rates of BC 639 and Indoxacarb (Figure 2A).

After the second spraying, the number of *Helicoverpa* spp. eggs per metre on plots treated with BC 639 and 0.85 L/ha Indoxacarb was the same at 3, 7 and 14 DAT, but they were significantly lower ($p < 0.02$) than the unsprayed plots at three DAT (Figure 2A). Overall, there was no strong evidence that application of the fungus deterred egg laying by *Helicoverpa* spp. on treated cotton plants (Figure 2A).

3.3.2. *Helicoverpa* spp. Very Small + Small (First to Third) Instar Larvae

Significant differences ($p < 0.05$) in the number of first and second instar larvae per metre were found among treatments at seven DAT (Figure 2B). Plots treated with 1.0 L/ha, 0.50 L/ha and 0.85 L/ha Indoxacarb recorded the lowest number of very small and small (VS + S) (first to third) instar larvae per metre after the first spraying (Figure 2B). No significant differences were detected between plots treated with 0.25 L/ha BC 639 and the control (unsprayed) plots after the first spraying (Figure 2B). The numbers of VS + S instar larvae per metre recorded on plots treated with 1.0 L/ha and 0.50 L/ha BC 639 was the same as plots treated with Indoxacarb after the first spraying. At 14 days after the first spraying, the number of VS + S larvae per metre was the same among treated and control plots (Figure 2B). In contrast, no significant differences were detected between the BC 639-treated plots and the control (unsprayed) plots (Figure 2B).

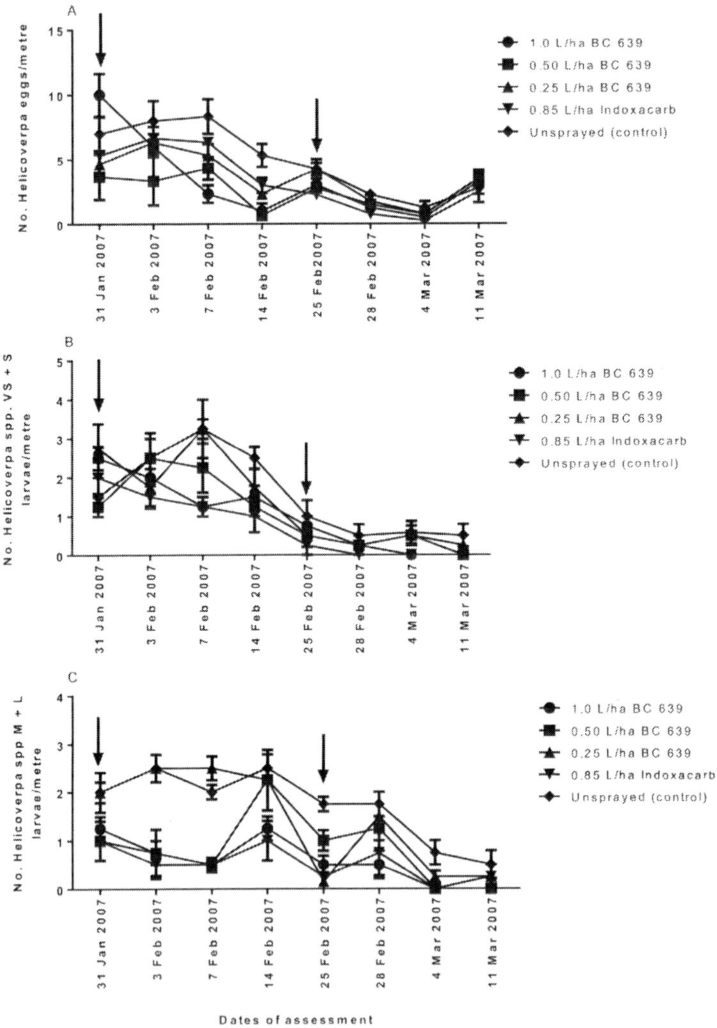

Figure 2. The efficacies of different rates of application of BC 639 and Indoxacarb in reducing numbers of *Helicoverpa* spp. (**A**) eggs, (**B**) very small + small (VS + S) larvae (*i.e.*, instars 1–3) and (**C**) medium + large (M + L) larvae (*i.e.*, instars 4–6) per metre on commercial conventional cotton crops at Getta Getta near Goondiwindi, 2006–2007. The arrows indicate the date of treatment.

3.3.3. *Helicoverpa* spp. Medium + Large (Fourth to Sixth) Instar Larvae

Following the first spraying, the number of medium + large (M + L) larvae was significantly lower ($p < 0.01$) on 1.0 L/ha and 0.50 L/ha BC 639- and 0.85 L/ha Indoxacarb-treated plots than plots treated with 0.25 L/ha BC 639 or the control (unsprayed) plots (Figure 2C). The number of M + L larvae per metre recorded on plots treated with 0.25 L/ha BC 639 was the same ($p > 0.05$) as the control (unsprayed) plots (Figure 2C), indicating that the 0.25 L/ha rate of BC 639 was not effective against *Helicoverpa* spp.

Following the second spraying, the same number of M + L larvae per metre ($p > 0.05$) was found among treatments at three and 14 DAT (Figure 2C). At seven DAT, however, the number of M + L recorded on the unsprayed plots was significantly higher ($p < 0.03$) than the BC 639- and Indoxacarb-treated plots. In contrast, no significant difference ($p > 0.05$) in the number of M + L larvae was found on 1.0 L/ha 0.50 and 0.25 L/ha BC 639-treated plots, Indoxacarb-treated plots and the unsprayed plots at 14 DAT (Figure 2C).

3.3.4. Beneficial Insects

Beneficial insects found at the study site were mainly predatory insects (Table 2). Predators of *Helicoverpa* spp. identified from the treated plots were predatory beetles (only three species were predominant in the study site), bugs, lacewings and spiders (Table 2).

Table 2. Predators of cotton pests sampled and identified from the study sites in 2006–2008.

Order	Family	Species	Group
Coleoptera	Coccinellidae	*Coccinella transversalis* (Fabricius) *Diomus notescens* (Blackburn)	Predatory beetles
	Melyridae	*Dicranolaius bellulus* (Guerin-Meneville)	
Hemiptera	Nabidae	*Nabis capsiformis* (Germar)	Predatory bugs
	Lygaeidae	*Geocoris lubra* (Kirkaldy)	
	Pentatomidae	*Cermatulus nasalis* (Westwood) *Ochelia schellenbergii* (Guerin-Meneville) *Coranus triabeatus* (Horvath)	
Neuroptera	Chrysopidae	*Chrysopa* spp.	Predatory lacewings
	Hemerobiidae	*Micromus tasmaniae* (Walker)	
Araneidae	Lycosidae	*Lycosa* spp.	Spiders
	Oxyopidae	*Oxyopes* spp.	
	Salticidae	*Salticidae* spp.	
	Araneidae	*Araneus* spp.	

3.3.4.1. Predatory Beetles

Following the first spraying on 31 January 2007, the number of predatory beetles per metre on the plots treated with 0.5 and 0.25 L/ha BC 639 (lower rates) was not significantly different ($p > 0.05$) from the unsprayed (control) plots, although plots treated with 1.0 L/ha BC 639 (higher rate) and Indoxacarb were significantly lower ($p < 0.01$) than the unsprayed plots at 3, 7 and 14 DAT (Figure 3A). The number of predatory beetles per metre on plots treated with 1.0 L/ha BC 639 was less than that on unsprayed plots (control), there were significantly more ($p < 0.01$) than on the Indoxacarb-treated plots (Figure 3A).

After the second spray, the number of predatory beetles per metre on the BC 639-treated plots was identical ($p > 0.05$) to the unsprayed plots at three and seven DAT, but was significantly lower ($p < 0.01$) than the unsprayed plots at 14 DAT (Figure 3A). In contrast, the number of predatory beetles per metre on the Indoxacarb-treated plots was significantly lower ($p < 0.001$) than the unsprayed plots at 3, 7 and 14 DAT (Figure 3A).

3.3.4.2. Predatory Bugs

Following the first spraying, the number of predatory bugs per metre on 1.0 and 0.5 L/ha BC 639-treated plots and Indoxacarb-treated plots was the same, but it was significantly lower ($p < 0.01$) than the 0.25 L/ha BC 639 and unsprayed (control) plots at three DAT (Figure 3B). However, at seven and 14 DAT, the number of predatory bugs per metre on the BC 639-treated plots (all rates) was the same as the unsprayed plots, but significantly higher ($p < 0.001$) than the Indoxacarb-treated plots (Figure 3B). The Indoxacarb-treated plots consistently showed fewer predatory bugs per metre after the first spraying (Figure 3B).

Following the second spraying, the number of predatory bugs per metre was not significantly different ($p > 0.05$) between the BC 639-treated plots, Indoxacarb-treated plots and the unsprayed plots (Figure 3B). There could have been an influx of bugs from the unsprayed and BC 639-treated plots into the Indoxacarb-treated plots or the effect could have been due to spray drift.

3.3.4.3. Predatory Lacewings

The predominant predatory lacewings identified from the study plots were *Chrysopa* spp. and *Micromus tasmaniae* (Table 2). Following the first spraying, the number of predatory lacewings per metre on plots treated with 1.0 L/ha BC 639 and Indoxacarb was the same, but it was lower than 0.50 and 0.25 L/ha BC 639 and the unsprayed plots at 3 DAT ($p < 0.0001$), 7 DAT ($p < 0.001$), 14 DAT ($p < 0.01$) and 25 DAT ($p < 0.001$ (Figure 3C)). In contrast, plots treated with 0.50 and 0.25 L/ha BC 639 had the same number of predatory lacewings as the unsprayed plots (Figure 3C).

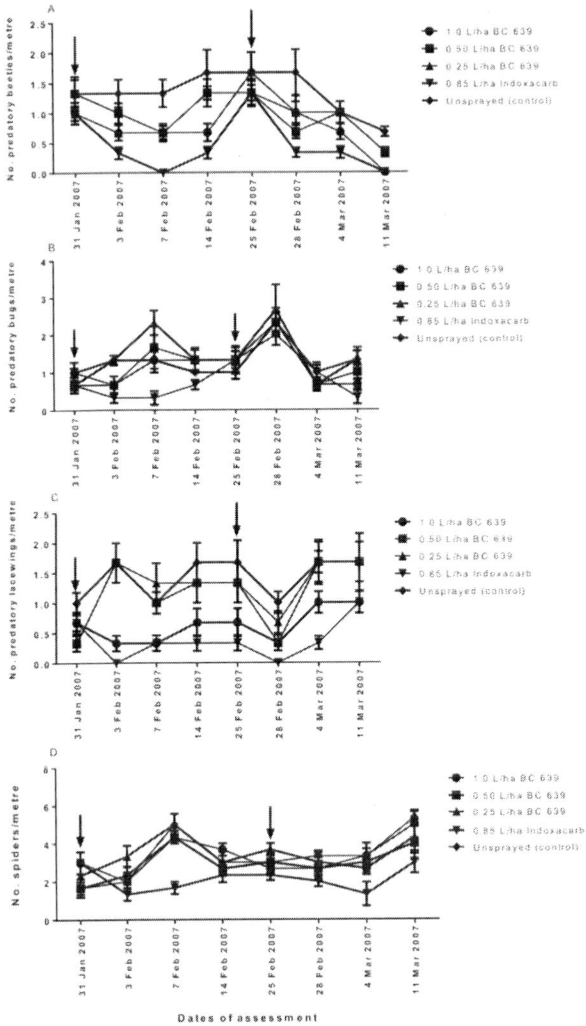

Figure 3. Efficacy of different rates of application of BC 639 and synthetic insecticide (Indoxacarb) on the number of (**A**) predatory beetles, (**B**) predatory bugs, (**C**) predatory lacewings and (**D**) spiders on commercial conventional cotton crops at Getta Getta near Goondiwindi, 2006–2007. The arrows indicate the date of treatment.

Following the second spraying, the number of lacewings per metre in plots treated with all rates of BC 639 and the unsprayed plots was the same at three DAT, but was significantly different ($p < 0.01$) from the Indoxacarb-treated plots (Figure 3C). At seven DAT, the number of predatory lacewings found on 0.5 and 0.25 L/ha and the unsprayed plots was the same ($p > 0.05$), and this was significantly different from the 1.0 L/ha ($p < 0.01$) and Indoxacarb-treated plots ($p < 0.001$) (Figure 3C). No significant difference ($p > 0.05$) was recorded among the BC 639 rates, Indoxacarb and the unsprayed plots at 14 DAT (Figure 3C).

3.3.4.4. Spiders

The spiders identified from the study plots were *Lycosa* spp., *Oxyopes* spp., *Salticidae* spp. and *Araneus* spp. (Table 2). The number of spiders per metre on plots treated with the BC 639 rates and the unsprayed plots was the same following the first (31 January 2007) and second (25 February 2007) treatments (Figure 3D). However, the number of spiders per metre on the Indoxacarb-treated plots following the first and second treatments was significantly lower ($p < 0.01$) than the BC 639 and unsprayed plots at seven DAT after the first and second spraying (Figure 3D).

3.4. Experiment 4: Efficacy of Different Rates of BC 639 against Helicoverpa spp. and Beneficial Insects on Conventional Cotton Crops at Norwood in 2007–2008

3.4.1. *Helicoverpa* spp. Eggs

The first spraying was applied when the density of *Helicoverpa* spp. eggs ranged from 18 to 28 eggs per metre (Figure 4A). After the first spraying, the number of eggs recorded on plots treated with BC 639 rates, Indoxacarb and the unsprayed plots was the same at 3, 14 and 25 DAT ($p > 0.05$) (Figure 4A). However, the number of *Helicoverpa* spp. eggs per metre recorded on plots treated with 1.0 and 0.5 L/ha BC 639 and Indoxacarb was significantly lower ($p < 0.01$) at seven DAT than plots treated with 0.25 L/ha and unsprayed plots (Figure 4A). However, after the second spraying, the number of eggs per metre was the same among treated and unsprayed plots (Figure 4A).

3.4.2. *Helicoverpa* Very Small and Small Larvae

After the first spraying, the plots treated with BC 639 rates and Indoxacarb had a significantly lower ($p < 0.01$) number of VS + S larvae at 3 ($p < 0.0001$), 7 ($p < 0.003$) and 14 DAT than the unsprayed plots (Figure 4B). Similar results were found for plots treated with BC 639 rates, Indoxacarb and the unsprayed plots at 3, 7 and 14 DAT (Figure 4B). After the second spray, the number of *Helicoverpa* VS + S larvae recorded on plots treated with BC 639 rates and Indoxacarb was the same and below the recommended threshold (two larvae per metre) used to manage these pests on conventional cotton crops [1].

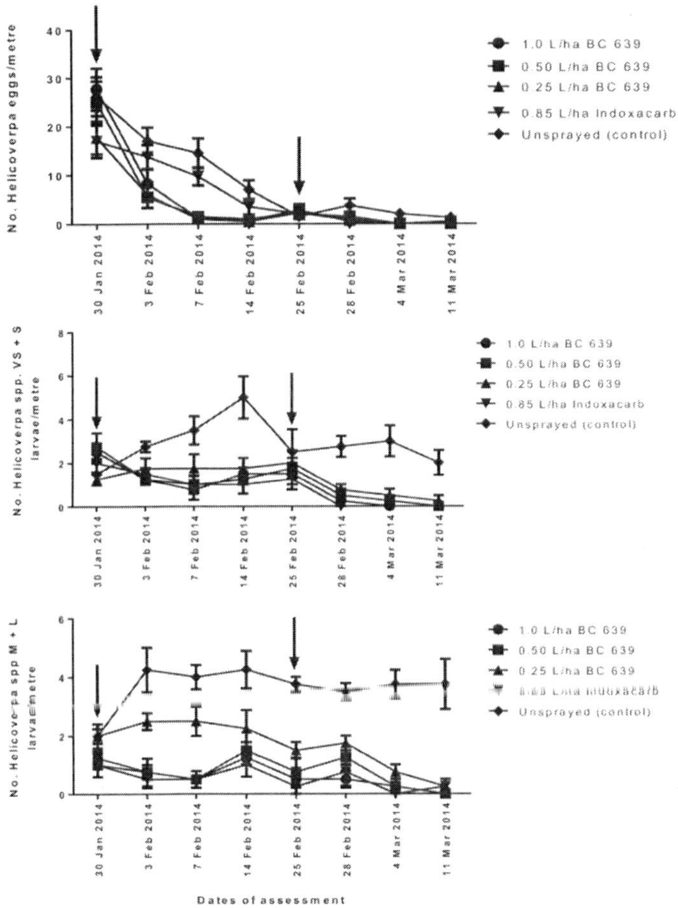

Figure 4. The efficacies of different rates of application of BC 639 and synthetic insecticide (Indoxacarb) on reducing the numbers of *Helicoverpa* spp. (**A**) Eggs, (**B**) very small + small (VS + S) larvae (*i.e.*, Instars 1–3) and (**C**) medium + large (M + L) larvae (*i.e.*, Instars 4–6) per metre on commercial conventional cotton crops at Norwood near Moree, 2007–2008. The arrows indicate the date of treatment.

3.4.3. *Helicoverpa* Medium and Large Larvae

The number of *Helicoverpa* spp. M + L larvae recorded on plots treated with BC 639 rates and Indoxacarb after the first treatment was similar and significantly lower at 3 ($p < 0.0005$), 7 ($p < 0.0001$) and 14 DAT ($p < 0.003$) than the unsprayed plots (Figure 4C). Similarly, the number of M + L larvae per metre on plots treated with BC 639 and Indoxacarb was the same after the second spraying (Figure 4C). The unsprayed plots had the highest number of M + L larvae per metre at 3–14 DAT (3.50–3.75 per metre) (Figure 4C).

3.4.4. Predatory Beetles

Predatory insects identified from the study plots are listed in Table 2. The predominant predatory beetles identified from the study plots were *C. transversalis*, *D. notescens* and *D. bellulus* (Table 2), and the number of the predatory beetles per metre is given in Figure 5A. No significant differences ($p > 0.05$) were found among BC 639 rates, Indoxacarb and the unsprayed plots at three and seven DAT after the first spraying (Figure 5A). However, the number of predatory beetles per metre on the unsprayed plot (3.25 per metre) was significantly higher ($p < 0.001$) than that on the plots treated with BC 639 rates and Indoxacarb at 14 DAT (Figure 5A). Similarly, no significant differences were recorded in the number of predatory beetles per metre among treatments at three and seven DAT, but there were significantly different ($p < 0.01$) numbers between the unsprayed plots, the plots treated with BC 639 and Indoxacarb at 14 DAT after the second spraying (Figure 5A). The effect could have been a result of spray drift. The number of predatory beetles per metre on plots treated with Indoxacarb and BC 639 rates at 14 DAT was not significantly different ($p > 0.05$) (Figure 5A).

3.4.5. Predatory Bugs

The predominant predatory bugs identified from the study plots are given (Table 2). No significant differences ($p > 0.05$) were found among the unsprayed plots, BC 639- and Indoxacarb-treated plots after the treatments (Figure 5B).

3.4.6. Predatory Lacewings

The predominant predatory lacewings identified from the study plots were *Chrysopa* spp. and *Micromus tasmaniae* (Table 2). No significant differences were found among BC 639-treated plots and the unsprayed plots after the first and second treatments (Figure 5C). The Indoxacarb-treated plots consistently showed the lowest number of predatory lacewings per metre (about 50% lower than the BC 639-treated plots) throughout the study. However, it was only at three DAT since the first spray that the difference between the Indoxacarb-treated plots was significantly lower ($p < 0.001$) than the BC 639-treated plots and the unsprayed plots (Figure 5C). This indicates that BC 639 had minimal impact on predatory lacewings compared with Indoxacarb (Figure 5C).

3.4.7. Spiders

The predominant spiders identified from the study plots are given (Figure 5D). No significant differences were found among the unsprayed plots, BC 639- and Indoxacarb-treated plots after the first and second treatments (Figure 5D).

4. DISCUSSION

The study has shown that the application of a formulated product of BC 639 entomopathogenic fungus effectively controlled populations of *Helicoverpa* spp. on commercial cotton crops with minimal or no effect against predators of *Helicoverpa* spp. One exception occurred when the product was applied at 1.0 L/ha (the highest rate in this study), and some negative effect was evident on predatory insects at the Getta Getta study site. In our study, the fungus killed *Helicoverpa* spp. larvae within 3–7 DAT. Application of BC 639 at 1.0 and 0.50 L/ha reduced the number of *Helicoverpa* spp. per metre, and these rates were consistently more effective than the lower 0.25 L/ha rate. Thus, the optimum effective rate of application of BC 639 against *Helicoverpa* spp. on cotton crops to reduce pest damage was 0.50 L/ha. Similar results were obtained when BC 639 was applied to *Creontiades dilutus* on commercial conventional cotton crops [15].

According to [5,25,26], entomopathogenic fungi, such as BC 639, can negatively affect the survival of soft-bodied beneficial insects. In the present study, application of BC 639 at 0.5 and 0.25 L/ha did not significantly affect predatory beetle population, but when BC 639 was applied at 1.0 L/ha, it affected the number of predatory beetles per metre, with a similar effect to that of Indoxacarb (see Figure 3A). Although the application of 1.0 L/ha BC 639 reduced the number of predatory beetles per metre, the negative effect of the 1.0 L/ha was significantly less than the negative effect of Indoxacarb. The present

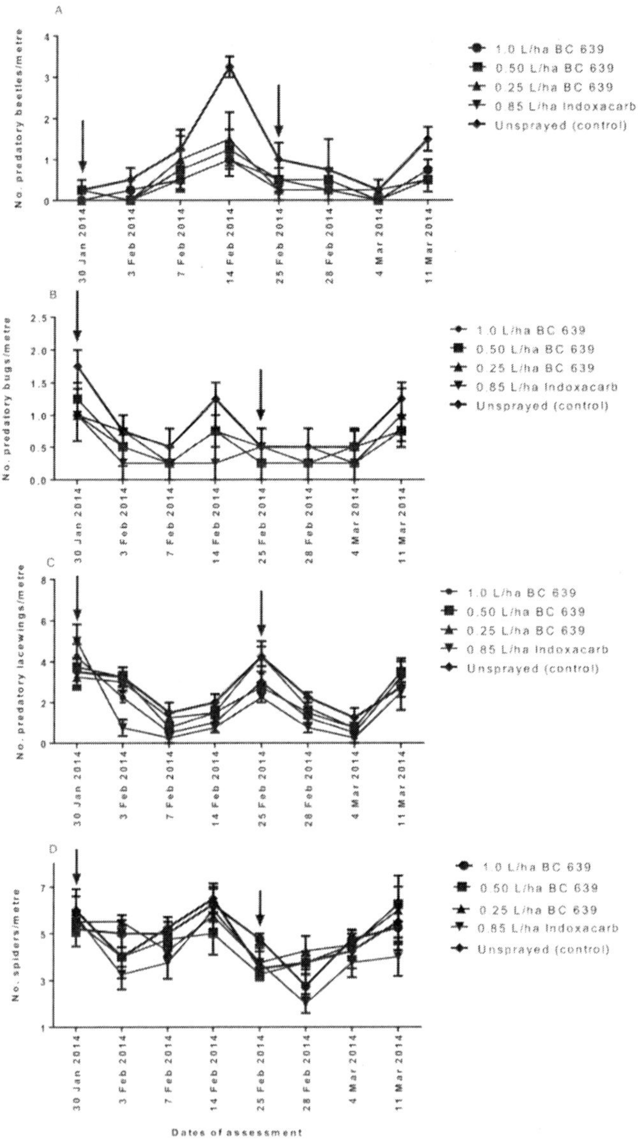

Figure 5. Numbers of (**A**) predatory beetles, (**B**) predatory bugs, (**C**) predatory lacewings and (**D**) spiders per metre on conventional cotton crops at Norwood near Moree in 2007–2008. Arrows indicate the dates of treatment application with BC 639 and Indoxacarb. Error bars represent the standard error of the mean.

results also showed that following the first spraying, 1.0 and 0.50 L/ha BC 639 had a negative effect on the predatory bugs at three DAT, but the population recovered at seven and 14 DAT. In contrast, BC 639 applied at 0.25 L/ha had no effect on the predatory bugs. In the case of predatory lacewings, the 1.0 L/ha BC 639 had a similar negative effect on predatory lacewings as the Indoxacarb treatment. Furthermore, Indoxacarb was consistently disruptive to predatory bugs and spiders, whereas BC 639 had no effect on the spiders. These results suggest that the application of lower rates of BC 639 had a more selective effect on predatory beetles, bugs, lacewings and spiders than the Indoxacarb. Thus, the number of predatory insects recorded for the lower rates of BC 639-treated plots and the unsprayed plots was the same and significantly higher than the 1.0 L/ha BC 639 rate, which prevents us from speculating that BC 639 is essentially benign at all rates on soft-bodied insects. In the case of Indoxacarb, our study found that the selectivity of Indoxacarb only occurred in the treated plots that were adjacent to the unsprayed plots, indicating a possible influx of predators from the unsprayed plots to the Indoxacarb-treated plots.

Entomopathogenic fungi in general are known to be important natural enemies of many pests of agricultural crops [20] and have an advantage over synthetic insecticides in the conservation of beneficial insects [20]. The BC 639 fungus used in this study is self-generating on the surface of the plant and can potentially provide ongoing protection as the plant grows, with little effect on non-herbivorous predators [15,5]. Cotton ecosystems contain many strains of fungi, such as in the soil and on plant debris, but long-term research is required to identify potential fungi and to increase its pathogenicity and adaptability to a particular environment, so as to enhance its ability to infect and control insects on agricultural crops.

The selectivity of BC 639 against predatory insects indicates that this fungus has the potential to enhance conservational biological control, to support IPM and to reduce synthetic insecticide use on conventional cotton crops. IPM involves using all means of managing pest populations with the aim of reducing insecticide use while maintaining profitability, yield and fibre quality [15].

Similar to many other cotton production systems worldwide, the Australian cotton cropping system is strictly monoculture. This practice and the use of synthetic insecticides inadvertently discriminates against beneficial arthropods. Therefore, there is a need to use biological pesticides, such as lower rates of BC 639, to conserve beneficial insects and support IPM in cotton agro-ecosystems. By integrating BC 639 fungal biopesticide with beneficial insects, rather than relying on beneficial insects alone, we can effectively enhance the control of *Helicoverpa* spp. in monoculture cropping systems, such as cotton. This is because *Helicoverpa* spp. are highly migratory and can quickly infest cotton crops and lay their eggs. Such behaviour by *Helicoverpa* spp. means that by the time beneficial insects arrive and establish in the cotton crops, the pest might have developed into medium and large larvae (fourth to sixth instar), which are too large for the beneficial insects (mostly predators) to feed on and, thus, to control. Hence, beneficial insects (particularly predatory insects) need to be well established in the cotton crops prior to the arrival of the pest (*Helicoverpa* spp.) to enable a sufficiently rapid response for population control. Thus, an IPM strategy developed against *Helicoverpa* spp. should involve tools and strategies that enhance the control of these pests; these could include fungal biopesticides, such as BC 639, which is selective against beneficial insects.

An IPM program for large-, medium- and small-scale monoculture crops, such as cotton, should develop in a stepwise progression that initially involves strategies to establish beneficial insects in the cotton field [13], followed by integration with IPM-compatible tools, such as the BC 639 entomopathogenic fungus that controls the target pest, but is selective for beneficial insects. Intervention with broad-spectrum synthetic pesticides should be used as a last resort when pests exceed thresholds and when no other effective selective management options are available.

The success of IPM programs in many cropping systems, especially conventional cotton systems, will depend on the availability of IPM-compatible tools and strategies to farmers. To adopt a true IPM program, farmers will need alternative pest control tools that have a minimal impact on the beneficial insects, whatever crop is grown. For cotton farmers in Australia and other parts of the world, the increase in *Helicoverpa* spp. populations on conventional cotton crops, yield loss, high cost of cotton production, insecticide resistance, disruption of beneficial species and environmental impacts are considered to be the major factors influencing the implementation of a sustainable IPM program for cotton crops. Thus, it is crucial that research continues to identify and develop new IPM-compatible tools and alternative strategies for managing and controlling *Helicoverpa* spp. and other pests [6,12,13]. Hence, the use of biological pesticides, including lower rates of BC 639 entomopathogenic fungus, is very important to support IPM in cotton agro-ecosystems.

The microbial biopesticide BC 639 is selective at lower application rates of 0.5 and 0.25 L/ha against predatory insects and has been used to effectively control *Helicoverpa* spp. on commercial cotton crops.

5. CONCLUSIONS

Application of a formulated entomopathogenic fungus (*Aspergillus* sp. BC 639) effectively controlled populations of *Helicoverpa* spp. on commercial cotton crops with minimal or no effect against predators of *Helicoverpa* spp. when the product was applied at 0.5 and 0.25 L/ha (the lower rates in this study). The fungus killed *Helicoverpa* spp. larvae within 3–7 DAT. Application of BC 639 at 1.0 and 0.50 L/ha was more effective than applications at the lower rate of 0.25 L/ha. The 1.0 and 0.50 L/ha rates of BC 639 had a similar efficacy against *Helicoverpa* spp. larvae as the 0.85 L/ha treatment with the commercial conventional synthetic insecticide Indoxacarb. This result suggests that the optimum effective rate for the application of BC 639 to control *Helicoverpa* spp. infestations on cotton crops was 0.50 L/ha. The microbial biopesticide BC 639 was selective against predatory insects at lower rates, and it can effectively control *Helicoverpa* spp. on commercial cotton crops and, thus, has a great potential to complement IPM programs in cotton crops.

ACKNOWLEDGMENTS

Many cotton farmers and consultants have contributed to the development and evaluation of BC 639 fungal insecticide against *Helicoverpa* spp. in cotton. Special thanks go to Mr. Iain Macpherson and his staff and also the following farmers: Mr. Peter Glennie and Sons (Norwood), David Coulton (Morella), Ben Coulton (Getta Getta), Tony Meppem and Des Magann (Australian Cotton Research Institute, Narrabri). Our special thanks go to the IPM project casual technical staff who supported us in the collection of field data: Ray Morphew, Stacey Cunningham, Carolyne Palmer, Cynthia Wilson, Katinka Atkins, Steve Ryman and Angela Singleton. Our commercial partner, Becker Underwood Pty. Ltd., has been of great help with their technical support. The project is funded by the Cotton Research and Development Corporation (DAN 1101).

AUTHOR CONTRIBUTIONS

Robert K. Mensah was the principal researcher of the project and was involved in the collation, analysis and write up of the laboratory and field data. Technical assistance in the collection of the field and laboratory data was provided by Alison Young and Leah Rood-England.

CONFLICTS OF INTEREST

The authors declare no conflict of interest.

REFERENCES

1. Deutscher, S.; Wilson, L. Insect Thresholds on Cotton, Cottonlogic; Australian Cotton Co-operative Research Centre, Australian Cotton Research Institute: Narrabri, Australia, 1999; p. 25.
2. Fitt, G.P. Cotton pest management: Part 3. An Australian perspective. Annu. Rev. Entomol. 1994, 39, 543–562.
3. Davidson, S. Biotech cotton: A budding field. Ecos 2003, 114, 28–31.
4. Greenplate, J.T.; Mullins, J.W.; Penn, S.R.; Dahm, A.; Reich, B.J. Partial characterisation of cotton plants expressing two toxin proteins from Bacillus thuringiensis: Relative toxin contribution, toxin interaction, and resistance management. Appl. Entomol. 2003, 127, 340–347.
5. Olsen, K.M.; Daly, J.C.; Holt, H.E.; Finnegan, E.J. Season long variation in expression of Cry 1Ac gene and efficacy of Bacillus thuringiensis toxin in transgenic cotton against Helicoverpa armigera. J. Econ. Entomol. 2005, 98, 1007–1017.
6. Wilson, L.J.; Mensah, R.K.; Fitt, G.P. Implementing integrated pest management on Australian cotton. In Insect Pest Management-Field and Protected Crops; Horowitz, A.R., Ishaaya, I., Eds.; Springer-Verlag: Berlin, Germany, 2004; pp. 97–118.
7. Fitt, G.P.; Wilson, L.J. Genetic engineering in IPM: Bt cotton. In Emerging Technologies for Integrated Pest Management: Concepts, Research and Implementation; Kennedy, G.G., Sutton, T.B., Eds.; APS Press: St. Paul, MN, USA, 2000; pp. 108–125.
8. Pyke, B. Adoption of Bollgard II and impact of synthetic insecticide use against Helicoverpa spp. In A Cotton Research and Development Corporation (CRDC) 2007 Report; CRDC: Narrabri, Australia, 2007.
9. Clarke, M.S.; Luna, J.M.; Stone, N.D.; Youngman, R.R. Generalist predator consumption of armyworm and effect of predator removal on damage in no-till corn. Environ. Entomol. 1994, 23, 617–622.
10. Mensah, R.K.; Harris, W.E. Using Envirofeast spray and refugia technology for cotton pest control. Aust. Cotton Grow.1995, 16, 30–33.
11. Mensah, R.K. Habitat diversity: Implications for the conservation and use of predatory insects of Helicoverpa spp. in cotton systems in Australia. Int. J. Pest Manag. 1999, 45, 91–100.
12. Mensah, R.K.; Wilson, L.J. Integrated Management Guidelines for Australian Cotton; Entopak, Cotton Co-operative Research Centre (CRC) Publications, Australian Cotton Research Institute: Narrabri, Australia, 1999.
13. Mensah, R.K.; Austin, L. Microbial control of cotton pests. Part 1: Use of the naturally occurring entomopathogenic fungus Aspergillus sp. (BC 639) in the management of Creontiades dilutus (Stal) (Hemiptera: Miridae) and beneficial insects on transgenic cotton crops. Biocontrol Sci. Technol. 2012, 22, 567–582.
14. Mensah, R.K. Development of an integrated pest management programme for cotton, Part 2: Integration of a Lucerne/cotton interplant system, food supplement sprays with biological and synthetic insecticides. Int. J. Pest Manag. 2002, 48, 95–105.
15. Murdoch, W.W.; Cheeson, J.; Cheeson, P.L. Biological control in theory and practice. Am. Nat. 1985, 125, 344–366.

16. Wratten, S.D. The effectiveness of native natural enemies. In Integrated Pest Management; Burn, A.J., Coaker, T.H., Jepson, P.C., Eds.; Academic Press: London, UK, 1987; pp. 89–111.
17. Hagen, K.S.; Hale, R. Increasing natural enemies through the use of supplementary feeding and non-target prey. In Proceedings Summer Institute for Biological Control of Plant Insects and Diseases; Mississippi University Press: Jackson, MI, USA, 1974; p. 647.
18. Fitt, G.P. The ecology of Heliothis in relation to agroecosystems. Annu. Rev. Entomol. 1989, 34, 17–52.
19. Mensah, R.K. Development of an integrated pest management programme for cotton, Part 1: Establishing and utilizing natural enemies. Int. J. Pest Manag. 2002, 48, 87–94.
20. Hajek, A.E.; Leger, R.J. Interactions between fungal pathogens and insect hosts. Annu. Rev. Entomol. 1994, 39, 293–322.
21. Hajek, A.E.; Delalibera, I. Fungal pathogens as classical biological control agents against arthropods. Biocontrol 2010, 55, 147–148.
22. Butt, T.M.; Jackson, C.W.; Magan, N. Fungi as Biocontrol Agents: Progress, Problems and Potential; Amos, N., Ascher, K.R.S., Eds.; CAB International: Wallingford, UK, 2001.
23. Zimmermann, G. Review on safety of entomopathogenic fungi Beauveria bassiana and Beauveria brongniartii. Biol. Control Technol. 2007, 17, 553–596.
24. Zimmermann, G. Review on safety of entomopathogenic fungi Metarhizium anisopliae. Biol. Control Technol. 2007, 17, 879–920.
25. Bahar, H.M.; Backhouse, D.; Gregg, P.C.; Mensah, R. Efficacy of a Cladosporium sp. fungus against Helicoverpa armigera (Lepidoptera: Noctuidae), other insect pests and beneficial insects of cotton. Biocontrol Sci. Technol. 2011, 21, 1387–1397.
26. Butt, T.M.; Goettel, M. Bioassays of entomogenous fungi. In Bioassays of Entomopathogenic Microbes and Nematodes; Amos, N., Ascher, K.R.S., Eds.; CAB International: Wallingford, UK, 2000.

CHAPTER 13

Improving Strawberry Irrigation with Micro-Sprinklers and Their Impact on Pest Management

Surendra K. Dara[1*], Samuel Sandoval-Solis[1,2], David Peck[3]

[1]*University of California Cooperative Extension, San Luis Obispo, CA, USA*

[2]*Department of Land, Air and Water Resources, Davis, CA, USA*

[3]*Manzanita Berry Farms, Santa Maria, CA, USA*

ABSTRACT

Overhead aluminum sprinklers, which are used during the early stages of strawberry crop development to establish transplants and to leach out salts from the root zone, deliver significant volumes of water. Micro-sprinklers, which are typically used in orchard crops, were evaluated in a commercial strawberry field in California as an alternative to conventional aluminum sprinklers to conserve water without any negative impact on yields. In addition to the water consumption, data were collected from multiple plots within each treatment to determine the impact on plant growth, disease incidence, and seasonal yield. Micro-sprinklers used 32% less water than aluminum sprinklers during a three-week period without affecting fruit yield. They also appeared to lessen the severity of powdery mildew and botrytis fruit rot. This is the first study reporting the use of micro-sprinkler system, which can be a good alternative to the aluminum sprinklers to conserve irrigation water.

Keywords: Strawberry, Irrigation, Micro-Sprinkler, Powdery Mildew, Botrytis Fruit Rot

1. INTRODUCTION

Strawberry is an important commercial crop in California and primarily grown in the Watsonville, Santa Maria, and Oxnard production areas in the Central Coast on about 37,000 acres annually [1] . It is the fifth most important commodity in California with a crop value of $2.5 billion and California produces 90% of fresh market and 96% processing strawberries in the United States [2] .

Strawberry crop requires about 2 acre-feet of irrigation water during a typical production season for fall planting. Fall planting typically starts in October-November and ends in July-August of the following year. Evapotranspiration of strawberry crop varies depending on the micro-climate of the production region, seasonal climatic variations, type of irrigation system used, and general agronomic practices. For example, a two- year study conducted in 2009 and 2011 in the Monterey Bay area showed that the average evapotranspiration varied between 2 and 2.5 acre-feet [3] .

Strawberries are mainly irrigated through a drip system where drip tapes are installed under plastic mulch when beds are prepared. Drip irrigation is supplemented with overhead aluminum sprinkler irrigation during the first few weeks after transplanting to leach out salts from the root zone and to help with transplant establishment. Strawberry crop is sensitive to salinity and overhead aluminum sprinkler irrigation reduces and prevents salt injury [4] . Aluminum sprinklers are also considered very important in the Oxnard area to prevent dry conditions from Santa Ana winds during the early stages of crop establishment. However, supplemental irrigation with these aluminum sprinklers requires a significant amount of water and can be less efficient than other irrigation systems [5] . Some disadvantages of the aluminum sprinklers include increased evaporation, excessive water delivery than what can penetrate into the beds through the openings on the plastic mulch, and potential runoff if managed improperly.

Application efficiency (AE refers to the efficiency of an irrigation system in delivering a specific amount of water) studies showed that micro-sprinklers are more efficient (81% AE) than hand-move sprinkler systems (70% AE) [5] . Water is an important resource in crop production and improved irrigation practices are a critical need in a high value crop like strawberry, especially under the current drought conditions in California. Micro-sprinklers, which are commonly used in orchard systems, could be an efficient alternative to aluminum sprinklers. Installed on strawberry beds, micro- sprinklers can deliver water in a more controlled and targeted manner with minimum or no runoff. Additionally, micro-sprinklers could also contribute to improved humidity in the strawberry microclimate, which discourages twospotted spider mite populations and promotes predatory mites that are sensitive to hot and dry conditions. High humidity in the microclimate is beneficial to entomopathogenic fungi such as Beauveria bassiana, which are applied as biopesticides for controlling twospotted spider mite and insect pests [6] [7] . Some growers have indicated using micro-sprinklers in organically produced strawberries as a means of controlling twospotted spider mites, but there are no data available on its impact.

A study was conducted in Santa Maria during the 2014-2015 production season to assess the potential of micro-sprinklers as an alternative to overhead aluminum sprinklers. Specific objectives of this study included: 1) if micro-sprinklers conserve water without a negative impact on strawberry plant growth and fruit yield; 2) if micro- sprinklers have any impact on pest and predatory mite populations; and 3) if micro- sprinklers have any impact on powdery mildew and botrytis fruit rot.

2. MATERIALS AND METHODS

This study was conducted at Manzanita Berry Farms in Santa Maria on a conventional strawberry field with variety BG-6.3024 planted on 6 November, 2014. A strawberry block with about 135 beds aligned in a north-south direction was divided into two sections. The west half of the block was assigned for the micro-sprinklers and the east half for aluminum sprinklers, which was the grower standard (Figure 1). The length of the beds varied from 306' at the east end to 365' at the west end. Within each section of the treatment, 20' long plots were marked on six random beds as sample plots to measure plant, pest, and disease parameters. Data were collected periodically and analyzed using ANOVA. Significant means were separated using Tukey's HSD test.

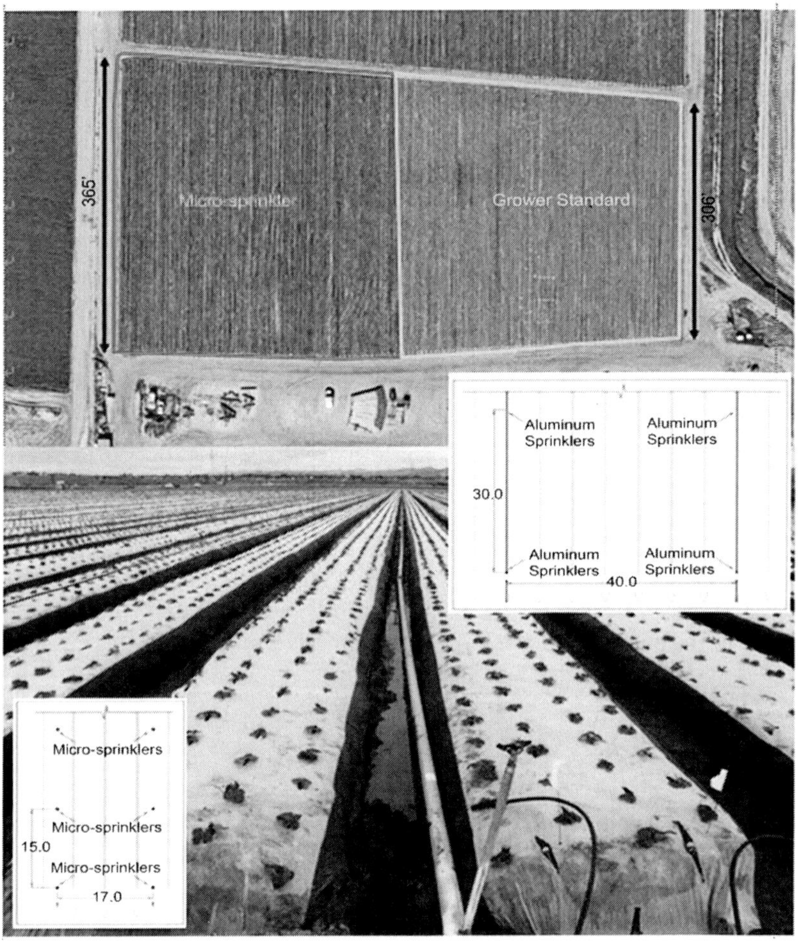

Figure 1. Experimental plots and configuration of micro-sprinklers and aluminum sprinklers.

2.1. Irrigation Management

Micro-sprinklers were installed on every fourth bed (about 16' apart) and had a 15' spacing between two sprinkler heads within a bed. Aluminum sprinklers were installed in furrows every 40' with 7 - 8 beds in between. Sprinkler heads were spaced 30' apart on each pipe. Supplemental irrigation during the transplant establishment period was administered from 6 to 29 November, 2014. During this period, there were 14 irrigation events using aluminum sprinklers where 125 gallons of water was delivered per minute. Micro-sprinklers were used each time for 2 hours at 1 - 3 day intervals where 40 gallons per minute were delivered at 35 PSI pressure.

2.2. Crop Yield

Plots were marked with flags and covered with netting to prevent accidental harvesting by the commercial harvesting crew. Total and marketable strawberry yield data were collected 2 to 3 times every week, following the normal harvest schedule, between 7 February and 12 June, 2015 for a total of 34 sampling dates.

2.3. Plant Canopy and Health

Plant growth was recorded by measuring the width of the plant canopy across and along the length of the bed from 20 random plants within each plot on the 6th day of each month from January to March, 2015. Plant health was monitored at the same time on a scale of 0 to 5 where 0 = dead, 1 = weak, 2 = moderate-low, 3 = moderate-high, 4 = good, and 5 = very good.

2.4. Twospotted Spider Mite and Predatory Mites

One mid-tier leaflet from each of the 10 random plants within each plot was sampled once a month from February to April, 2015. The number of eggs, nymphs, and adult stages of both pest and predatory mites were counted using a mite brushing machine.

2.5. Powdery Mildew

One trifoliate leaf from each of the 20 random plants within each plot were collected on 14 April and 16 and 24 June, 2015. Leaves were examined under a microscope for mycelial growth and powdery mildew severity was recorded on a scale of 0 to 4 where 0 = absent, 1 = 1% - 25%, 2 = 26% - 50%, 3 = 51% - 75%, and 4 = 76% - 100% of leaf area with infection.

2.6. Botrytis Fruit Rot

Strawberries harvested from each plot on 26 March, 13 April, 22 May, and 16 June, 2015, were kept at room temperature in plastic clamshell boxes. Disease severity was measured 3 and 5 days after harvest based on mold growth on the same scale used for powdery mildew.

3. RESULTS AND DISCUSSION

3.1. Irrigation Management

During this period, aluminum sprinklers delivered 120,000 gallons of water over 16 hours of irrigation whereas micro-sprinklers delivered 81,600 gallons over 34 hours (Figure 2). This translates to 38,400 gallons or 32% of water savings in just three weeks for micro-sprinklers. Considering extended periods of overhead irrigation in other field situations, water savings could be significantly higher from micro-sprin- klers.

Micro-sprinkler irrigation events were continued twice a week, 15 min each time, after 29 November for the rest of the production season to discourage twospotted spider mite populations. Distribution uniformity (DU) for the micro-sprinkler system was 74% at 35 PSI when measured on 16 January, 2015. DU could not be measured for aluminum sprinklers, but it is believed to be between 50% - 60% at 70 PSI [5] . Electrical conductivity (EC) was measured for both irrigation systems on 1 January and 1 February, 2015. EC value for micro-sprinklers was 0.54 dS/m, but it varied between 0.47 and 0.49 dS/m for aluminum sprinklers. Although the EC value in micro-sprinkler plots was significantly higher ($P < 0.0007$) than in the aluminum sprinkler plots, it was below the recommended limit of 0.7 dS/m [8] .

3.2. Crop Yield

When seasonal average was considered, there was no significant difference ($P > 0.05$) between the yield of marketable berries in micro-sprinkler and aluminum sprinkler plots (Figure 3). A total of 43,452 gr (95.8 lb) of marketable strawberries were produced per plot in micro-sprinkler treatment and 44,322 gr (97.7 lb) in the aluminum sprinkler treatment (Figure 4).

3.3. Plant Canopy and Health

Plants in the micro-sprinkler plots had significantly smaller canopy in January ($P = 0.004$) and February ($P = 0.0006$), but caught up with the plants in the aluminum sprinkler plots by March ($P = 0.14$) (Figure 5). Plant health rating also followed a similar trend during this period, but significant differences were seen only in February ($P = 0.02$) (Figure 6).

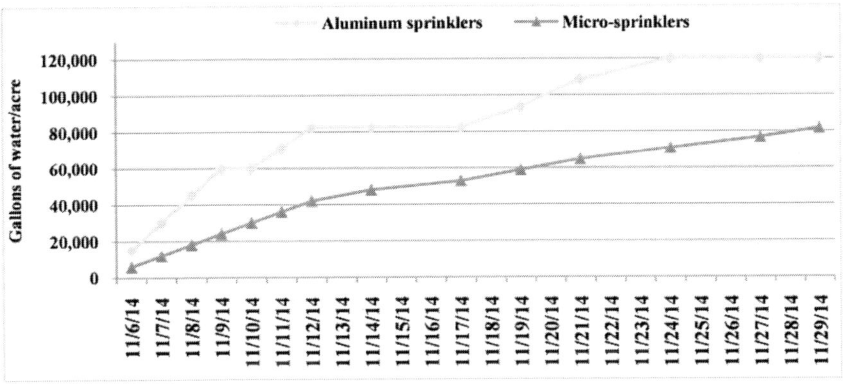

Figure 2. Cumulative irrigation volume delivered by aluminum and micro-sprinkler systems from 6 to 29 November, 2014.

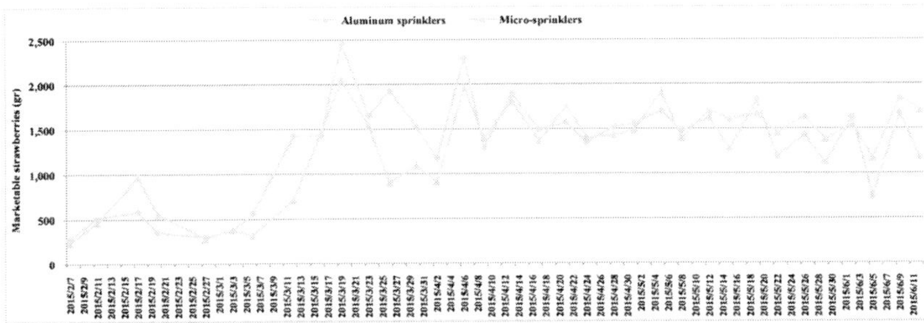

Figure 3. Cumulative marketable strawberry yields at each harvest from aluminum and micro-sprinkler plots.

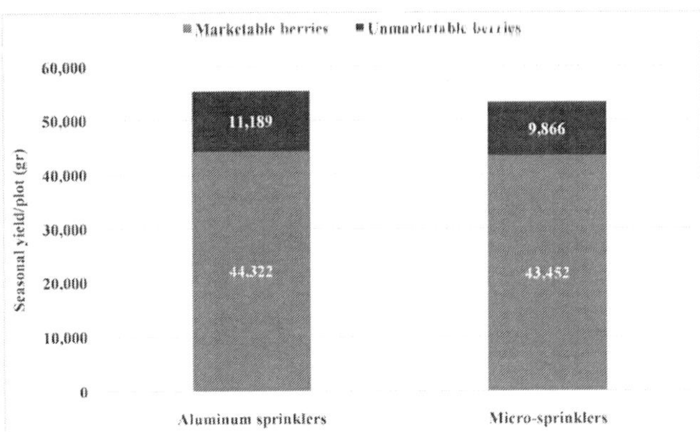

Figure 4. Seasonal totals for marketable and unmarketable strawberries from aluminum and micro-sprinkler plots.

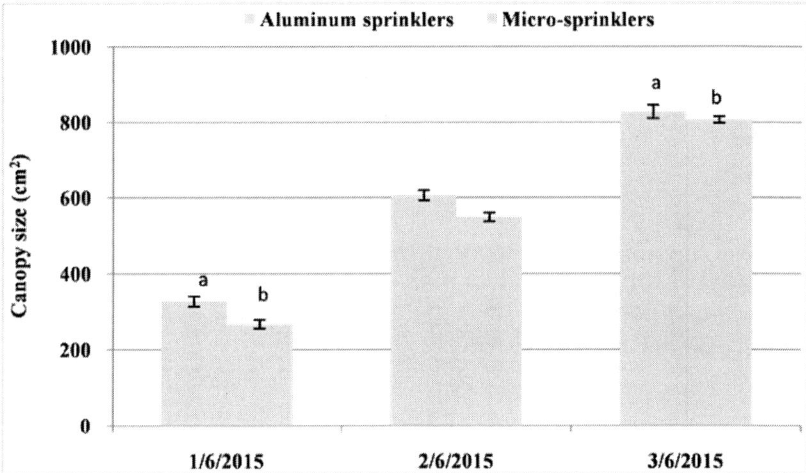

Figure 5. Strawberry plant growth, expressed as canopy size, in aluminum and micro-sprinkler plots.

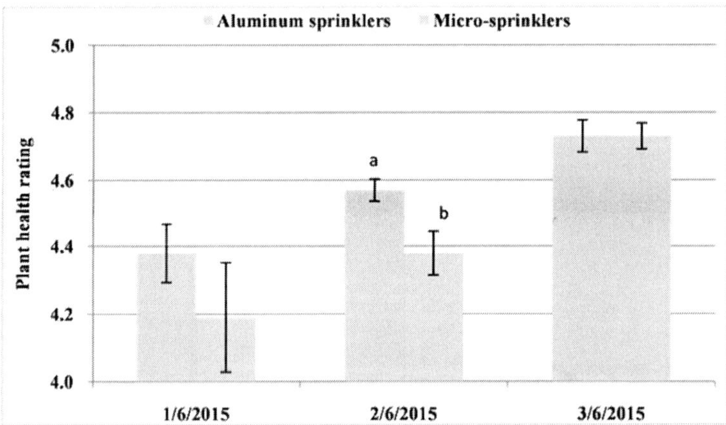

Figure 6. Health of the strawberry plants in aluminum and micro-sprinkler plots.

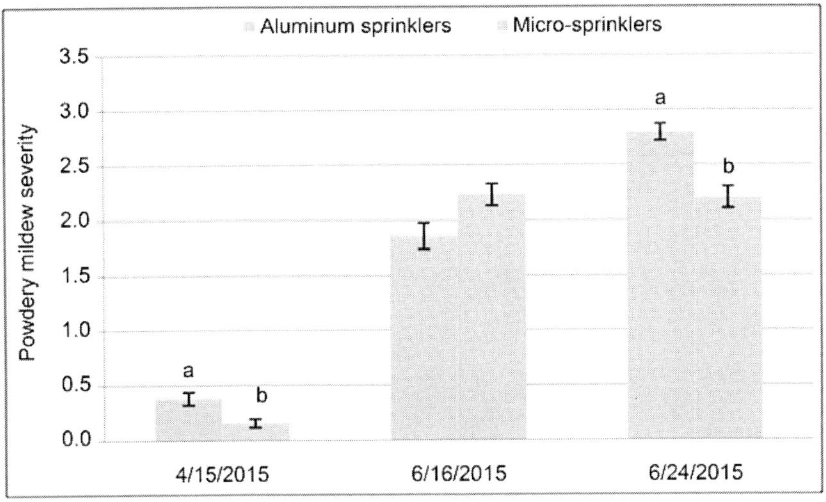

Figure 7. Severity of powdery mildew in aluminum and micro-sprinkler plots.

3.4. Twspotted Spider Mite and Predatory Mites

Due to sparse numbers and uneven distribution of mites, comprehensive data could not be collected.

3.5. Powdery Mildew

Powdery mildew severity was significantly less in micro-sprinkler treatment on 15 April ($P = 0.009$) and June 24 ($P = 0.01$) (Figure 7).

3.6. Botrytis Fruit Rot

Micro-sprinkler plots appeared to have a lower level of botrytis fruit rot, but significant differences were seen only 3 days after harvest for strawberries sampled on 22 May and 16 June ($P = 0.02$) (Figure 8).

The micro-sprinkler system contributed to a significant reduction in supplemental overhead irrigation water without adversely affecting the marketable strawberry yield. Since less pressure is required for micro-sprinklers than aluminum sprinklers to deliver water, energy savings and less carbon emissions are also possible from the micro- sprinkler system. EC value below the recommended level (0.07 dS/m) also suggests that micro-sprinklers were as effective as aluminum sprinklers in leaching out salts from the root zone without excessive water use. Although data on mite

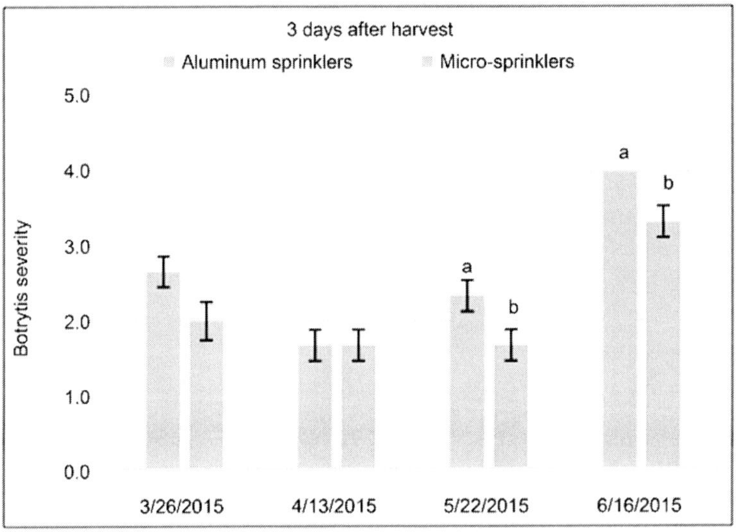

(a)

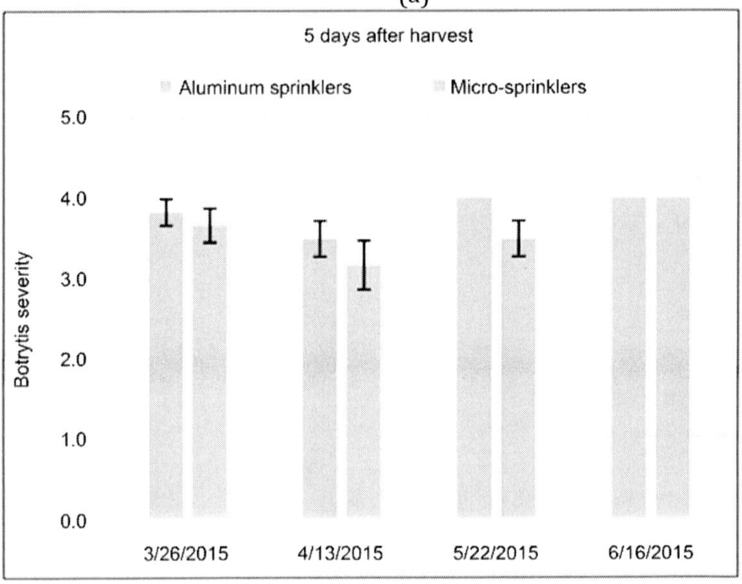

(b)

Figure 8. Severity of botrytis fruit rot 3 (a) and 5 (b) days after harvest in aluminum and micro- sprinkler plots.

populations were insufficient to draw valid conclusions, it should be noted that micro-sprinklers are considered by some growers to assist with twospotted spider mite control throughout the production season. Data also suggest that micro-sprinklers have a positive impact on reducing powdery mildew and botrytis fruit rot.

Micro-sprinkler irrigation over trees improved the color and anthocyanin of apples in a study conducted in Spain [9] . A Polish study in raspberry compared drip and micro-sprinkler irrigation systems to no irrigation [10] . Although not significantly different from drip-irrigated plots, plant height, marketable yields, and berry weight were the highest in micro-sprinkler-irrigated plots. Interestingly, the carotene and vitamin A contents were the highest in unirrigated plots followed by micro-sprinkler, and drip-irrigated plots. Other studies where micro-sprinklers were used in almond in California [11] , onion in India [12] , and mango in Brazil [13] evaluated the application efficiency, influence of the depth of irrigation, and methods of interpreting field measurements of energy fluxes, respectively, but none of them compared micro-sprinklers with conventional irrigation systems. This is the first study in strawberries comparing conventional overhead aluminum sprinklers to micro-sprinklers. This study demonstrated that micro-sprinklers can contribute to a significant reduction in supplemental irrigation water in strawberry production while having a positive impact on disease control. Additional experiments are necessary to strengthen the trends observed in this study and to refine irrigation schedules.

ACKNOWLEDGEMENTS

Thanks to Manzanita Berry Farms' field crew, Chris Martinez, and Fritz Light for their technical assistance and Netafim and RDO Water for their financial support.

CITE THIS PAPER

Dara, S.K., Sandoval-Solis, S. and Peck, D. (2016) Improving Strawberry Irrigation with Micro-Sprinklers and Their Impact on Pest Management. Agricultural Sciences, 7, 859-868.

REFERENCES

1. California Strawberry Commission (CSC) (2015) 2015 Acreage Survey. http://reports.calstrawberry.org/Reports/Acreage%20Survey/2015%20Acreage%20Survey.pdf
2. California Department of Food and Agriculture (CDFA) (2015) California Agricultural Statistics Review, 2014-2015. https://www.cdfa.ca.gov/statistics/PDFs/2015Report.pdf
3. Lin, V., Sandoval-Solis, S., Lane, B.A. and Rodriguez, J.M. (2013) Potential Water Savings through Improved Irrigation Efficiency in Pajaro Valley, California. Final Report, Water Management Research Group, University of California Davis, Davis.
4. Dara, S.K. (2012) Salt Injury in Strawberries. UCCE eNewsletter Strawberries and Vegetables. http://ucanr.edu/blogs/blogcore/postdetail.cfm?postnum=6820 Sandoval-Solis, S., Orang, M., Snyder, R.L., Orloff, S., Williams, K.E. and Rodriguez, J.M. (2013) Spatial Analysis of Application Efficiencies in Irrigation for the State of California. Final Report, Water Management Research Group, University of California Davis, Davis.
5. Shipp, J.L., Zhang, Y., Hunt, D.W.A. and Ferguson, G. (2003) Influence of Humidity and Greenhouse Microclimate on the Efficacy of Beauveria bassiana (Balsamo) for Control of Greenhouse Arthropod Pests. Environmental Entomology, 32, 1154-1163.
6. Dara, S.K. (2016) Managing Strawberry Pests with Chemical Pesticides and Non-Chemical Alternatives. International Journal of Fruit Science, 16, 1-13.
7. Strand, L.L. (2008) Integrated Pest Management for Strawberries. 2nd Edition, UC ANR Publication 3351, Oakland. [Citation Time(s):1]
8. Iglesias, I., Salvia, J., Torguet, L. and Cabús, C. (2002) Orchard Cooling with Overtree Microsprinkler Irrigation to Improve Fruit Colour and Quality of "Topred Delicious" Apples. Scientia Horticulturae, 93, 39-51.
9. Rolbiecki, S., Rolbiecki, R. and Rzekanowski, C. (2002) Effect of Micro-Irrigation on the Growth and Yield of Raspberry (Rubus idaeus L.) cv. "Polana" Grown in Very Light Soil. Acta Horticulturae, 585, 653-657.
10. Koumanov, K.S., Hopmans, J.W., Schwankl, L.J., Andreu, L. and Tuli, A. (1997) Application Efficiency of Micro-Sprinkler Irrigation of Almond Trees. Agricultural Water Management, 34, 247-263.
11. Sarkar, S., Goswami, S.B., Mallick, S. and Nanda, M.K. (2008) Different Indices to Characterize Water Use Pattern of Micro-Sprinkler Irrigated Onion (Allium cepa L.). Agricultural Water Management, 95, 625-632.
12. Teixeira, A.H.C. and Bastiaanssen, W.G.M. (2012) Five Methods to Interpret Field Measurements of Energy Fluxes over a Micro-Sprinkler-Irrigated Mango Orchard. Irrigation Science, 30, 13-28.

CHAPTER 14

Arthropod Pest Control for UK Oilseed Rape – Comparing Insecticide Efficacies, Side Effects and Alternatives

Han Zhang[1] *, Tom Breeze[1] , Alison Bailey[2] , David Garthwaite[3] , Richard Harrington[4] , Simon G. Potts[1]

[1] *Centre for Agri-Environmental Research, School of Agriculture, Policy and Development, University of Reading, Reading, Berkshire, United Kingdom,*

[2] *Land Management and Systems, Faculty of Agribusiness and Commerce, Lincoln University, Christchurch, New Zealand,*

[3] *Pesticide Usage Survey, Fera Science Ltd, Sand Hutton, York, United Kingdom,*

[4] *Rothamsted Insect Survey, Rothamsted Research, Harpenden, United Kingdom*

ABSTRACT

Oilseed rape (*Brassica napus*) is an important combinable break crop in the UK, which is largely protected from arthropod pests by insecticidal chemicals. Despite ongoing debate regarding the use of neonicotinoids, the dominant seed treatment ingredients used for this crop, there is little publicly available data comparing the efficacy of insecticides in controlling key arthropod pests or comparing the impacts on non-target species and the wider environment. To provide an insight into these matters, a UK-wide expert survey targeting agronomists and entomologists was conducted from March to June 2015. Based on the opinions of 90 respondents, an average of 20% yield loss caused by the key arthropod pests was expected to have occurred in the absence of insecticide treatments. Relatively older chemical groups were perceived to have lower efficacy for target pests than newer ones, partly due to the development of insecticide resistance. Without neonicotinoid seed treatments, a lack of good control for cabbage stem flea beetle was perceived. Wide spectrum foliar insecticide sprays were perceived to have significantly greater negative impacts than seed treatments on users' health, natural enemies, pollinators, soil and water, and many foliar active ingredients have had potential risks for non-target arthropod species in UK oilseed rape fields for the past 25 years. Overall, 72% of respondents opposed the neonicotinoid restriction, while 10% supported it. Opposition and support of the restriction were largely based on concerns for pollinators and the wider environment, highlighting the uncertainty over the side effects of neonicotinoid use. More people from the government and research institutes leaned towards neutrality over the issue, compared to those directly involved in growing the crop. Neonicotinoid restriction was expected to result in greater effort and expenditure on pest control and lower production (0–1 t/ha less). Alternatives for future oilseed rape protection were then discussed.

INTRODUCTION

Oilseed rape (*Brassica napus*) production has increased in many European Union (EU) countries since 2005 when the EU set a target for 20% of its energy to be from renewable sources by 2020 [1]. In the UK, oilseed rape production reached 652,000 hectares in 2015 [2], increasing annually by an average 2% over the past ten years. UK oilseed rape is grown for vegetable oil, animal feed and biodiesel. It serves as a break crop with wheat production and accounts for about 80% of the UK combinable break crops [3,4].

Oilseed rape has many arthropod pest and disease problems, prompting widespread use of chemical insecticides [5]. Pesticide usage survey data collected by the UK's Fera Science Ltd. ([6]; methods and figures see S1 Appendix) indicate that, between 1974 and 1988, pest control in oilseed rape in Great Britain was dominated by the use of organophosphate sprays and organochlorines as both foliar sprays and seed treatments. After 1990, pyrethroids quickly replaced these chemicals as foliar sprays, because of their higher efficacy in pest control and lower toxicity to humans, becoming the major insecticide group used from 1995 onwards. On average, pyrethroids represented 80% of total insecticidal weights, and 95% of total area treated for all foliar sprays used in oilseed rape in Great Britain for the past 25 years. Since the ban on organochlorines in 1999 (due to their high toxicity to users, and persistence and bioaccumulation in the environment), neonicotinoids have become the main seed treatments for oilseed rape in Great Britain, with 1,930 kg weight increase every two years. Three neonicotinoid active ingredients, clothianidin, thiacloprid and thiamethoxam, have been restricted in Europe since the end of 2013 [7], and the 2014/15 season was the first without these seed treatments being available.

The neonicotinoid restriction primarily resulted from the perceived negative lethal and sublethal impacts on both managed and wild pollinators [7]. Since the start of the restriction, the evidence base has been strengthened around the effects of neonicotinoids on crop production, human health, pollinators and the wider environment [8–11]. However, major gaps remain (e.g., concentration and toxicity of neonicotinoid metabolites in the environment, see [12]). Strongly contrasting views about the risks of using this chemical group in agriculture exist, and little consensus has been reached [10,11], with an ongoing argument in UK between the National Farmers Union (NFU) and government regarding the lifting of the restriction [13]. The neonicotinoid debate has resulted in uncertainties for the future of oilseed rape production and management for UK farmers, in terms of cropping areas, insecticide choices and impact on profits [14,15].

Two new chemical groups have recently been adopted by UK growers (azomethines and oxadiazines, from around 2012), which could potentially act as alternatives to neonicotinoids. However, as with all currently available chemical groups, there is little publicly available information concerning their relative efficacy in controlling key pests [8], or their side effects on non-target species and the wider environment. Thus it is difficult to assess the overall impact of using a specific insecticide chemical on oilseed rape.

In this study, based on experts' opinions, we provide an insight into these matters by comparing the relative efficacies and side effects of insecticide chemical groups used in UK oilseed rape production. Experts' opinions on the neonicotinoid restriction are also analysed. Because the survey finished in June 2015, the perceived inputs and yields are compared between 2013/14 and 2014/15 season. For future sustainable oilseed rape production in the UK, the potential alternative crop management options are thus discussed.

MATERIALS AND METHODS

Survey structure

An online survey by Qualtrics [16] was conducted to collect UK experts' opinions. The online survey has been approved by the ethics committee of the School of Agriculture, Policy and Development, University of Reading. As stated in the ethical approval (B165) and cover letter, the data can only be shared in summary forms, where no individual will be identified, and raw data will be destroyed in March 2018.

The survey had 28 questions (see S2 Appendix. Survey questionnaire) divided into four parts: (i) general section (6 questions including occupations, experience of giving farmer advice, and knowledge of crops), (ii) wheat section (7 questions, including yields, main pests and damage, insecticide efficacy, and importance of arthropod natural enemies), (iii) oilseed rape section (7 questions, as in the wheat section) and (iv) neonicotinoid section (8 questions, including opinions and reasons for the restriction, expected yields and inputs, and alternative pest management strategies). This paper refers only to the oilseed rape and neonicotinoid sections.

Respondents were also asked to give the regions they refer to when providing the information. To capture the degree of certainty, respondents were asked to give certainty scores on a 1–5 scale (5 being most certain) following several of the questions.

Survey distribution

In January 2015, a pilot study using 15 experts, mainly from universities, was conducted. Because only minor changes to the wording of a small number of questions were made based on their feedback, these responses were included in the analyses. The main nationwide survey was carried out between March and June 2015. Potential respondents were contacted by searching through directories of universities, research institutes, NGOs (non-government organizations), government sectors, consulting firms, and agri-chemical companies. Universities were filtered via the Guardian league table 2015 by subjects of 'Agriculture, forestry & food' and 'Biosciences', and then relevant researchers were selected by scanning through the related department staff pages. Independent consultants were contacted through the Association of Independent Crop Consultants (AICC). Most experts were contacted by email and some through general enquiry web pages. Snowball sampling methods (where a respondent may pass the survey to other related experts) were also utilized to gain more respondents. A reminder with a link to the survey was sent to individuals two weeks after the initial request, and one more reminder two weeks later if no reply was received.

Statistical analyses

Main arthropod pests and related damages in oilseed rape.

Respondents were asked to rank the relative importance of the three most widespread arthropod pests (one being most important of the three) in oilseed rape in the past five years (2009/10 to 2014/15). These scales were used as weights (i.e., weights for the 1st, 2nd and 3rd most important pest were 0.5, 0.33 and 0.17 respectively, with a collective weight of 1) to produce a weighted average response for each pest. Answers were further categorised into different regions and the weighted average for each pest for each region was estimated. See S3 Appendix for the regional distribution.

Respondents were also asked to estimate, without insecticides, the direct feeding damage from each selected arthropod pest and the damage that Turnip yellows virus (TuYV; caused by *Myzus persicae*, peach—potato aphid) would have caused in the past five years, in terms of percentage yield loss (%), followed by the certainty scale (1–5, 5 being most certain). Average damage levels, standard deviations and certainty levels of the three most important pests were estimated.

Efficacy of insecticides and arthropod natural enemies.

Information about the perceived efficacy of both chemical and natural pest control methods was requested in the survey.

The perceived efficacy of each available insecticide chemical group in the 2013/14 season was requested for each of the main pests selected from the previous question, following a 0–6 efficacy scale as below (Table 1):

Table 1. Insecticide efficacy scale.

0	1	2	3	4	5	6
Not sure	0% pest control	1–20% control	21–50% control	51–80% control	81–90% control	91–100% control

Pairwise efficacy comparisons among the available chemical groups were conducted by the Skillings–Mack [17,18] and related post hoc tests [19], using the software R 3.2.5 [20]. Responses were omitted if a respondent answered for only one chemical group for a specific pest. Due to the asymmetrical distribution and the non-parametric analyses for these ordinal data, medians and interquartile ranges were used to summarize the efficacy level.

The same efficacy and certainty scales were used to estimate the importance of arthropod natural enemies on pest control (without insecticides). Qualitative opinions on its importance for oilseed rape production were gathered by selecting from the following options: 'Strongly agree', `Agree', `Neutral',`Disagree', `Strongly disagree' and `Not sure'.

Side effects of insecticides on human health, environment and non-target arthropods.

Perceived negative influences on users' health, natural enemies, pollinators, water and soil were compared between the two application methods for UK oilseed rape(seed treatments and foliar sprays), using a 0–5 scale (0 being no influence, and 5 being greatest influence). Sign tests [21] were used to compare the perceived median differences between seed treatments and foliar sprays on these aspects. Responses were omitted if the respondent only answered for one of the two treatment types. Due to the asymmetrical distribution and the non-parametric analyses for these ordinal data, medians and interquartile ranges were used to summarize the side effect level.

Potential risks of common foliar active ingredients to the non-target arthropods for UK oilseed rape were also analysed by the in-field Hazard Quotient approach ([22]; see S4 Appendix for details). Hazard Quotient is the required assessment for pesticide registration in the European Union. The method combines the laboratory acute test [23] of two sensitive species (cereal aphid parasitoid *Aphidius rhopalosiphi* and predatory mite *Typhlodromus pyri*) and field application rate of the related foliar active ingredient [6] to produce a Hazard Quotient. If the quotient of either species exceeds two, the threshold level, it indicates a potential hazard of the active ingredient towards non-target arthropods. Due to the limited publicly available test results, only one representative active ingredient (the one that was used most often in UK oilseed rape and had test results available) was chosen from each available chemical group. For the same reason, and also that *Typhlodromus pyri* would normally be applicable in orchard fruit crops, only *Aphidius rhopalosiphi* was used as the test species.

Opinions on neonicotinoid seed treatment restrictions.

A five-point Likert scale was used to measure expert opinions on the current neonicotinoid restrictions ('Strongly favour' to 'Strongly oppose' including a 'Not sure' option). Different possible reasons for holding favourable and unfavourable opinions were displayed for respondents to select. Fisher's exact test was used (with Monte Carlo simulated p value, 100,000 replicates) [20] to explore whether or not experiences in farming services and the nature of the employing organization had an impact on respondents' opinions.

RESULTS

Response rates and information about respondents

In total, 455 online surveys were sent out, and 146 responses were received (43% effective response rate accounting for incorrect emails and declines), with 101 completed (70% of total responses). 90 people participated in the oilseed rape and neonicotinoid sections. The following analyses focused on these two sections.

From 2010 to 2014, 43% of respondents (39 out of 90) had worked for independent consultants/ consulting organizations (Fig E.1 in S5 Appendix). University staff accounted for 27%, government 17% and agri-chemical companies 16%. 65% of respondents had provided advice services to farmers.

Main arthropod pests and related damage in oilseed rape (2009/10-2013/14)

Based on a weighted average of expert ranking, the three main arthropod pests in oilseed rape in the past five years were perceived to be cabbage stem flea beetle (*Psylliodes chrysocephalus*) (weighted average response = 28), pollen beetle (*Meligethes aeneus*) (18), and peach—potato aphid (*Myzus persicae*) (8). Different regions varied slightly (e.g., for the North West region, the second most important pest was cabbage seed weevil, and the equal third were aphids and brassica pod midge) (Fig 1; Table E.1 in S5 Appendix).

Due to the limited number of respondents, and the fact that some respondents represented several regions (up to four, S3 Appendix), Fig 1 represents a broad visualization. The following analyses were conducted without categorising into regions.

Without insecticide applications, the perception was that cabbage stem flea beetle would have caused the greatest direct damage, with a mean yield loss of 24% (certainty level 3). However, expert opinion on this also had the largest variation, with a standard deviation of 26%. Second was pollen beetle damage, with a mean value of 19%, and standard deviation of 20% (certainty level 3). Perceived peach—potato aphid direct damage averaged 13%, with a standard deviation of 10% (certainty level 3). TuYV loss without insecticides was perceived to be 15%, with a standard deviation 11%, and a certainty level of 2 (Table 2).

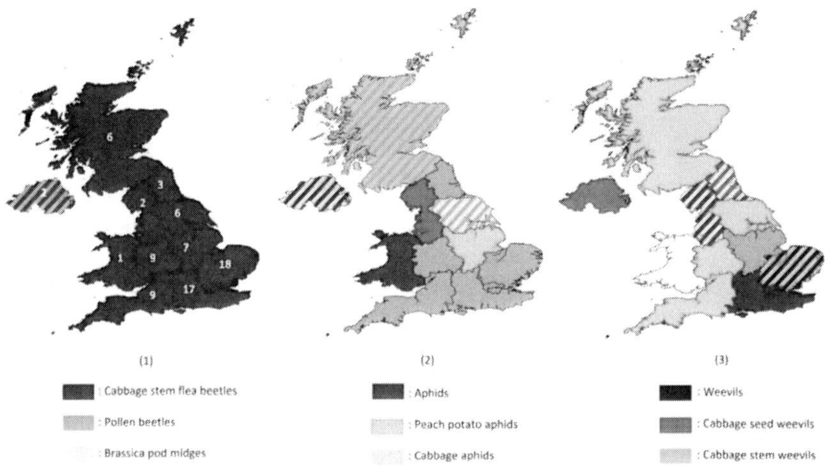

Fig 1. Perceived top three important arthropod pests of oilseed rape for each UK region. (1) Most important arthropod pests; (2) 2nd most important arthropod pests; (3) 3rd most important arthropod pests. Note that different regions are not directly comparable, due to different numbers of respondents for each region (see number in the first graph). The white area for Wales in the third graph denotes no data, since respondents only mentioned two pests. Stripes indicate two pests of equal importance. For the data see Table E.1 in S5 Appendix. Map shapefiles are from the GADM database of Global Administrative Areas and the Office for National Statistics [24,25], and software is QGIS [26].

Table 2. The number of respondents, perceived mean crop yield loss (%) and certainty levels of the main arthropod pests (direct damage), and the turnip yellows virus in UK oilseed rape from 2009/10 to 2013/14 (without insecticides).

	Cabbage stem flea beetles	Pollen beetles	Peach potato aphids	TuYV
Number of respondents	56	41	19	66
Mean damage (s.d.)	24.2 (26.3)	18.9 (20.5)	13.2 (9.9)	15.0 (11.0)
Mean certainty levels (s.d.)	2.8 (1.1)	2.8 (1.0)	2.6 (1.3)	2.3 (1.2)

Note: Numbers in the brackets are the related standard deviations (s.d.). Certainty levels: 1–5, 5 being most certain.

Efficacy of insecticides in oilseed rape (2013/14)

In terms of efficacy of the insecticide groups for oilseed rape protection, because of the potential complex annual variation in efficacy, only the available chemical groups used for the 2013/14 season were studied (Table 3).

Table 3. The perceived median insecticide efficacies (number of missing values; number of 'Not sure' responses; interquartile ranges) for the three main arthropod pests in UK oilseed rape in 2013/14.

Pests (number of respondents)	Carbamates	Pyrethroids	Neonicotinoid seed treatments	Neonicotinoid sprays	Oxadiazines	Azomethines
Peach–potato aphid (16)	2 (1; 1; 1–2) a	2 (0; 1; 1–3) a	5 (0; 0; 4–5.25) b	5 (2; 1; 5–6) bc	4 (6; 3; 4–4) c	4 (0; 1; 4–5.5) b
Pollen beetle(26)	1 (8; 4; 1–2.5) a	3 (2; 0; 2.75–4) a	1 (7; 3; 1–3) ab	4 (2; 2; 4–5) bd	4.5 (5; 8; 4–5) c	4 (5; 3; 3.5–5) d
Cabbage stem flea beetle (44)	1 (16; 4; 1–1.25) a	3 (1; 0; 3–3.5) b	5 (0; 0; 4–5) c	4 (8; 5; 2–5) c	3 (17; 13; 2–4) d	2 (17; 9; 1–3.75) d

Note: efficacy levels: 0 (Not sure), 1 (0% pest control), 2 (1–20%), 3 (21–50%), 4 (51–80%), 5 (81–90%), 6 (91–100%); 'Not sure' answers are omitted; different letters mean significant differences ($\alpha = 0.05$) between two chemical groups (based on Skillings–Mack and related ad-hoc tests).

It should be noted that some chemical groups are not applied against some pests due to certain phenological and pharmacological factors. For example, neonicotinoid seed treatments are not normally used against pollen beetles, because they are primarily applied to autumn sown crops, whereas pollen beetles pose a threat in spring. Neonicotinoid sprays and oxadiazines are mainly used for this pest. Carbamates (mostly pirimicarb) are used mainly for aphid control, and azomethines are applied mainly against aphids and pollen beetles. This information was reflected in Table 3 by the perceived relatively low efficacy levels for the non-targeted pests.

Based on the pairwise comparisons among different chemical groups, neonicotinoids and the two newly introduced chemical groups (oxadiazines and azomethines) were perceived to have higher efficacy than carbamates and pyrethroids for peach–potato aphid (81–90% and 1–20% respectively) and pollen beetle control (51–80% and 21–50% respectively). Neonicotinoid seed treatments were perceived to have more efficacy than pyrethroids against cabbage stem flea beetle (81–90% and 21–50% respectively).

Side effects of insecticides in oilseed rape

Comparing the two main insecticide application methods for UK oilseed rape, foliar sprays were thought to have significantly more negative impacts than seed treatments on users' health, natural enemies, pollinators, water and soil (Table 4). The median influence levels (from 0, no influence, to 5, greatest influence) across all categories were perceived to be 3 for sprays and 1 for seed treatments.

Table 4. Number of pairs, medians (interquartile ranges) of the influence scale (0–5) of the general seed treatments and foliar sprays, and the sign tests of the perceived median differences in side effects between seed treatments and foliar sprays.

Categories	Number of pairs	Seed treatments	Sprays	Sign tests (α = 0.05)
Health	63	1 (0–1)	2 (1.5–3)	<0.00001
Natural enemies	64	1 (1–2)	4 (3–5)	<0.00001
Pollinators	61	1 (1–2)	4 (3–4)	<0.00001
Water	59	1 (1–2)	3 (2–4)	<0.00001
Soil	55	1 (1–2)	2 (1–3)	0.02

Note: influence scale: 0–5, 0 being no influence, 5 being greatest influence.

Focusing on the Hazard Quotients for foliar sprays, different active ingredients have large differences in terms of the potential risks to non-target arthropods. Dimethoate has much higher potential hazard levels, with lambda-cyhalothrin and thiacloprid following. In comparison, pymetrozine, indoxacarb and pirimicarb pose lower potential risks towards non-target arthropods (Fig 2).

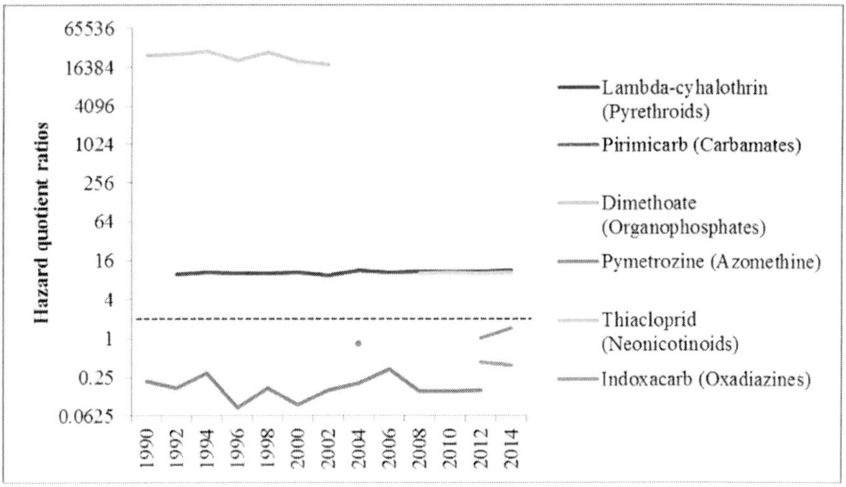

Fig 2. Hazard Quotient ratios of insecticide active ingredients used in oilseed rape in Great Britain (1990–2014). The test species is the aphid parasitoid *Aphidius rhopalosiphi*. Trigger value is 2 (marked with the black dashed horizontal line), above which indicates potential hazard of an active ingredient on non-target arthropod species. Y axis is the log scaled with base 4.

Uncertain future–neonicotinoid debates

From the information gathered, 72% (65 of 90) participants either opposed or strongly opposed the neonicotinoid restrictions in the UK while only 10% (9 of 90) supported or strongly supported this policy. 13 respondents took a neutral stance, and three respondents were not sure about this proposition (Fig E.2 in S5 Appendix).

In terms of organizations (NGOs and food industries were omitted, due to low respondent numbers), an average of 51% respondents from the government, universities, and private research institutes (Group A) opposed the neonicotinoid restriction, compared with 90% of respondents from the agri-chemical companies, commercial/independent consultants, and growers (Group B) (Fig 3). This difference was driven less by respondents who supported the restriction (an average of 20% from Group A compared with 7% from Group B), but more by those who held 'Neutral' and 'Not sure' opinions (30% compared with 3%). To compare this more quantitatively, we regrouped the six options in two ways: one was *oppose* (including 'Oppose' and 'Strongly oppose') versus *favour* (including 'Favour' and 'Strongly favour'); the other was having an opinion (*oppose*/ *favour*) versus 'Neutral'/ 'Not sure'. Fisher's exact test was used to test differences among organizations regarding their opinions on this issue. Results showed that there was no clear difference among organizations for whether to favour or oppose neonicotinoid restriction, but more people from Group A than Group B had a neutral/ not sure proposition (Table E.2 in S5 Appendix).

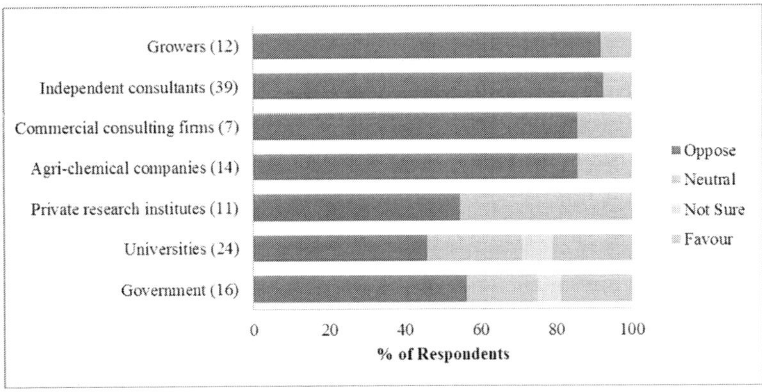

Fig 3. Percentage of respondents' opinions on neonicotinoid restriction by organization type. *Oppose* includes 'Strongly oppose' and 'Oppose' options, *favour* includes 'Strongly favour' and 'Favour' options. Numbers in the brackets are the number of respondents from that organization type who answered the question.

Fisher's exact test also showed a clear division between Group A and B regarding whether a respondent had provided advice services to farmers in the past five years (Table E.3 in S5 Appendix). An average of 41% from Group A and 85% from Group B had done so (Fig 4).

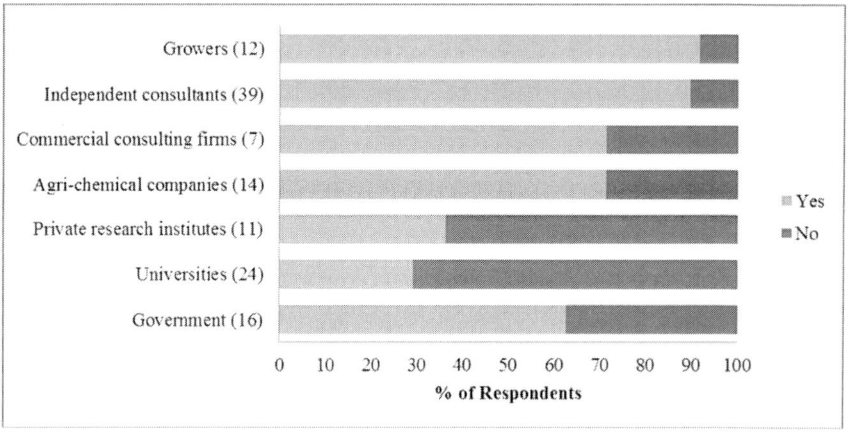

Fig 4. Percentages of respondents who have provided advice services to farmers by organization type (2009/10–2013/14). Numbers in brackets are the number of respondents from that organization type who answered the question.

Although only 10% (9) of respondents favoured the neonicotinoid restriction in the UK, all of them expressed concerns about the negative impacts neonicotinoids could have on pollinators and the wider environment. Four respondents believed that farmers can adjust to the management accordingly, and two suggested that the arthropod pest problems were not severe (Fig 5). Of the 65 respondents that opposed the restriction, the two most widespread reasons were also around environment and pollinators (95% and 89% respectively). 71% of respondents worried that other products are not as efficient as neonicotinoids, and 60% thought that oilseed rape production will be greatly reduced (Fig 6).

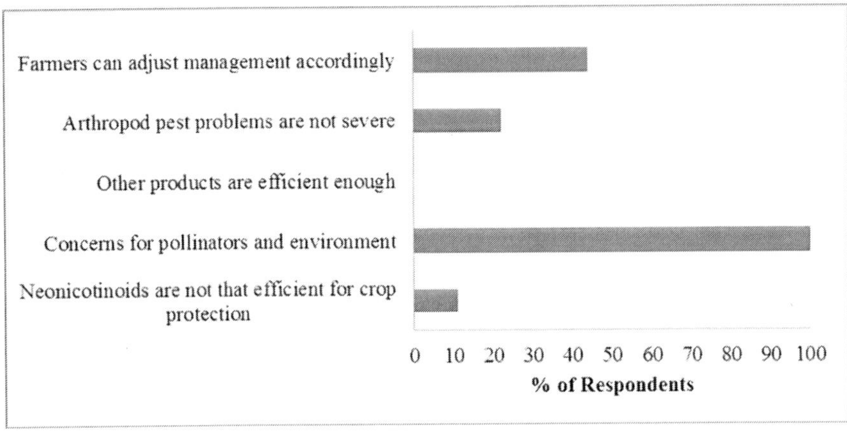

Fig 5. Reasons for favouring neonicotinoid restriction. Total number of respondents is 9.

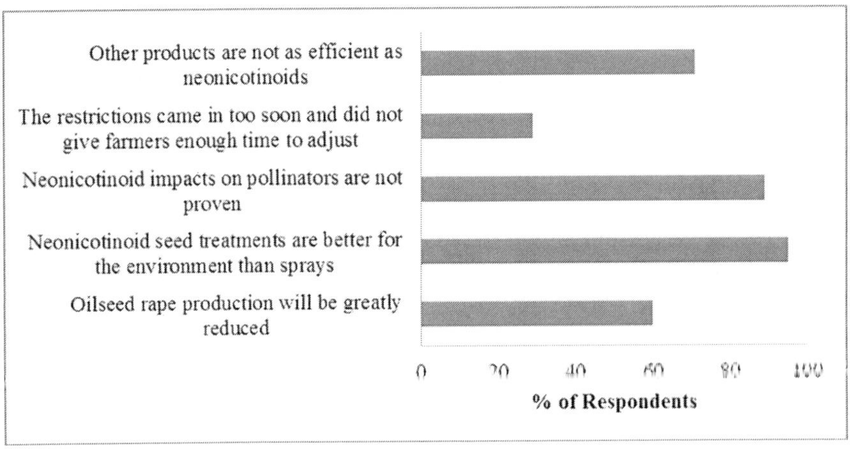

Fig 6. Reasons for opposing neonicotinoid restriction.Total number of respondents is 65.

2014/15 VS 2013/14 oilseed rape growing season

For the 2014/15 season (compared with 2013/14), 64% of respondents (48 out of 75) felt that more time had been spent by agronomists inspecting crops since the restrictions (Fig 7). 70% of respondents (54 out of 78) indicated that oilseed rape farmers spent more money on insecticide products during the 2014/15 season, but 37% (29 out of 78) thought that just a bit more had been spent (Fig 8). Only one respondent thought that farmers spent a bit less than 2013/14, while 15 were unsure. Most of the 76 respondents expected that the winter oilseed rape harvest in 2014/15 season would be between 0–1 t/ha less or about the same as the previous year (36% and 24% respectively). 22% of respondents were not sure about the answer, which was to be expected since yield changes are influenced by many factors in addition to pesticide use. Similar patterns occurred regarding the spring oilseed rape harvest. As the survey finished in June, yields for the year 2014/15 were estimated by the respondents directly (Fig 9).

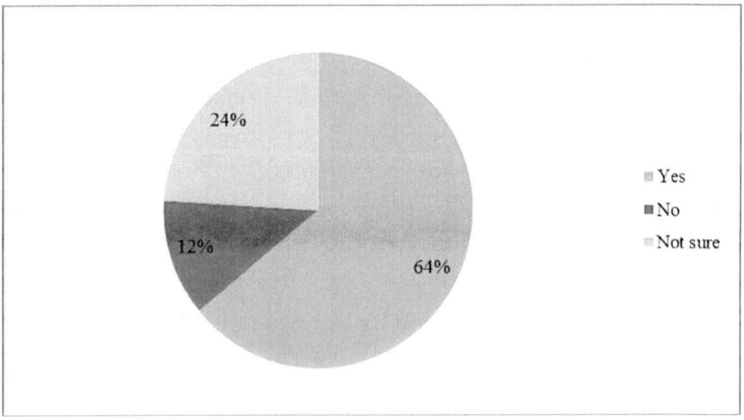

Fig 7. Whether more time was spent by agronomists inspecting oilseed rape fields for arthropod pest abundance. This is a 2014/15 compared with 2013/14 season. Total number of respondents is 75.

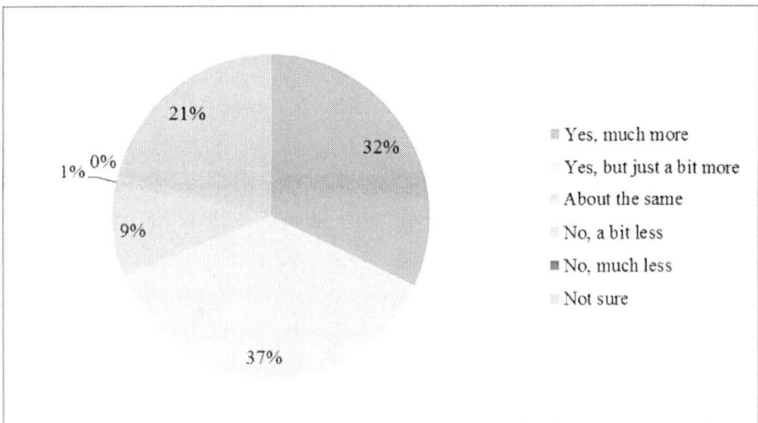

Fig 8. Percentage of respondents regarding whether they thought more money was spent on insecticide products. This is a 2014/15 compared with 2013/14 season. Total number of respondents is 78.

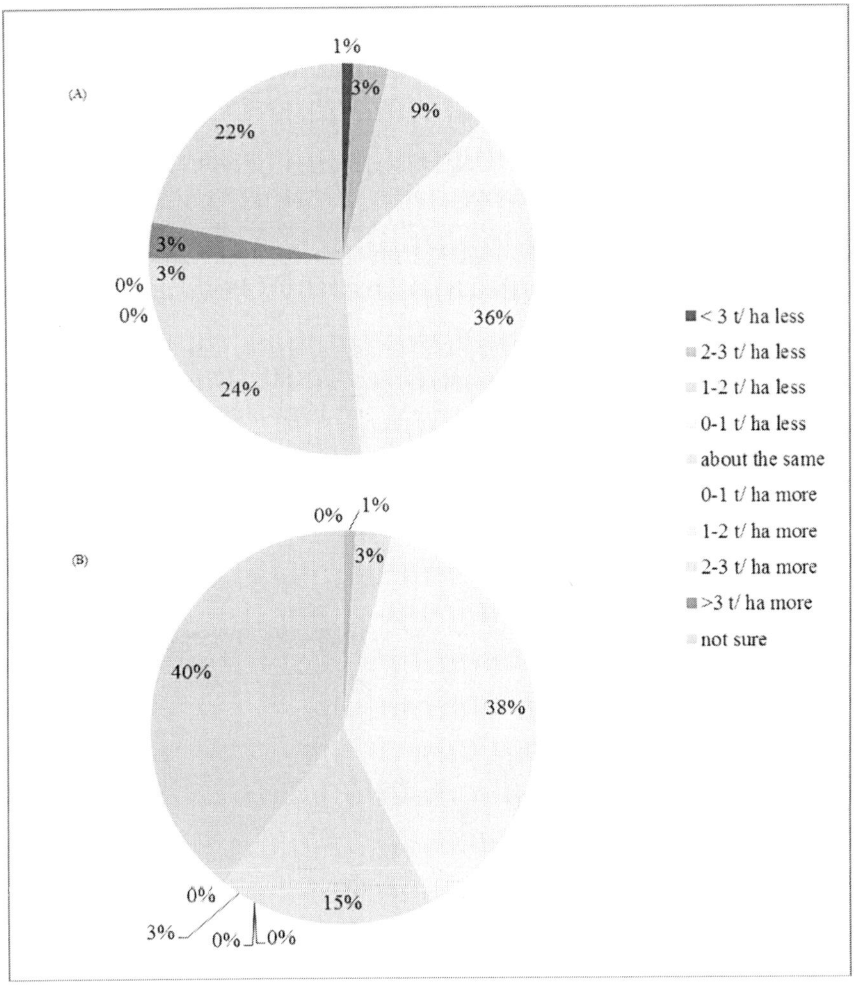

Fig 9. (A) expected winter and (b) spring oilseed rape yield in 2014/15 compared with 2013/14. Total number of respondents is 76 and 71 respectively.

Alternative pest control options

In a hypothetical situation where neonicotinoids would be permanently banned in the UK, respondents' suggestions towards alternative pest management strategies differed: 76% respondents (66 out of 87) would choose to use new insecticides if available. Other options such as 'Grow oilseed rape less often', 'Use new oilseed rape varieties', and 'Grow a smaller area of oilseed rape' were similar in terms of support (46%, 45% and 43% respectively). The use of currently available insecticides was mentioned by 30% of respondents. For the 'Others' option, 65% (13 out of 20) respondents mentioned the use of biocontrol/ IPM (integrated pest management) as an alternative approach (Fig 10). Focusing on natural pest control, 67% (56 out of 83) of respondents agreed that natural enemies are important for oilseed rape production (Fig 11). Without insecticide treatments, 57% of respondents thought that natural enemies could exert 1–20% control on oilseed rape pests, and 32% suggested 21–50% control (Fig 12).

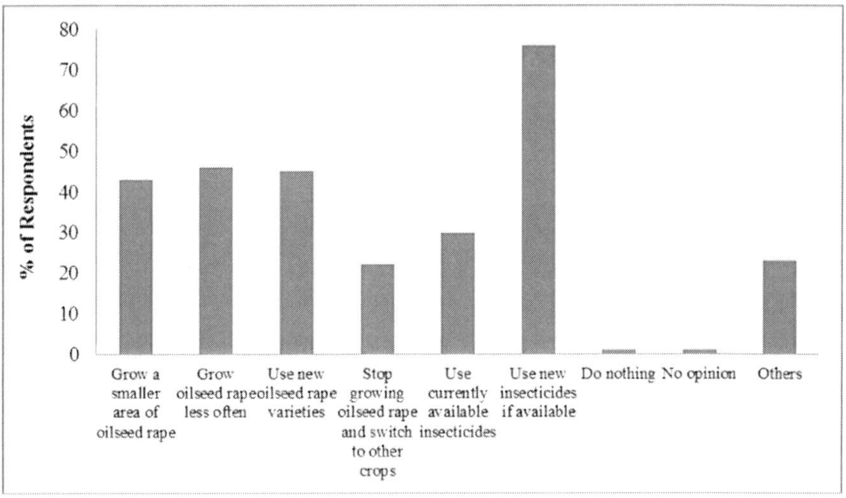

Fig 10. Perceived future management options if neonicotinoid seed treatments were withdrawn completely. Total number of respondents is 87.

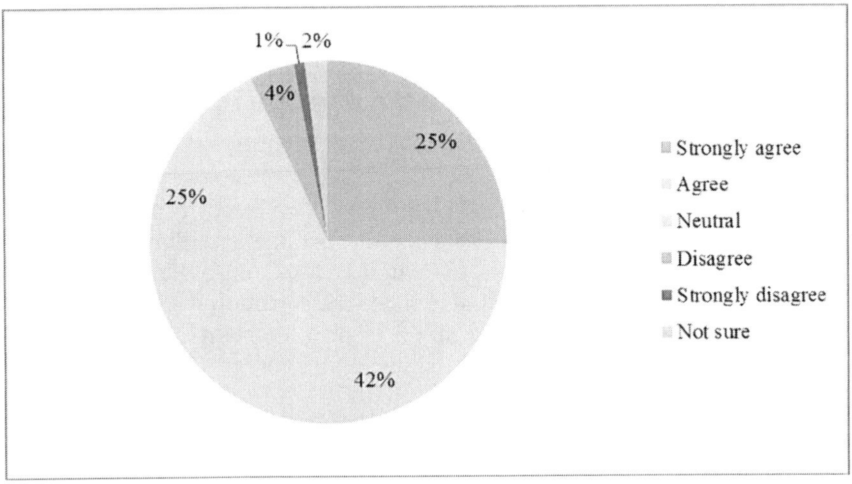

Fig 11. Perceived importance of arthropod natural enemies to oilseed rape production in UK.

Total number of respondents is 83.

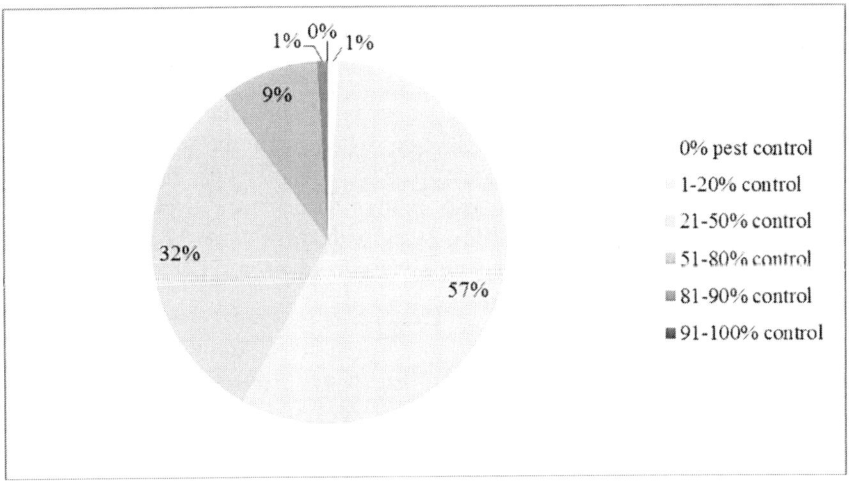

Fig 12. Without insecticides, perceived efficacy of arthropod natural enemies to control key pests in oilseed rape. Total number of respondents is 81.

DISCUSSION

Main arthropod pests and related damage in oilseed rape (2009/10-2013/14)

The most important arthropod pests in UK oilseed rape based on perceptions from the survey (cabbage stem flea beetle, pollen beetle, and peach—potato aphid) were also indicated by [27] on a European scale, except for the peach—potato aphid, which was considered a minor pest in Europe. However, peach—potato aphid is an important pest in the UK, especially as a vector of TuYV ([28] showed an average 15% yield loss by this virus in untreated crops).

The large variation in levels of perceived pest damage suggests that pest damage is likely to vary greatly under different contexts. [29] suggested a 1% untreated yield loss for cabbage stem flea beetle, 0.5% for pollen beetle, and 3% for aphids carrying TuYV. In 2010, [30] found varied pollen beetle damage on different study sites in England (0–6% yield loss). The low certainty levels provided by experts (an average of 3 out of 5) suggest that uncertainties and research gaps remain in this area.

Efficacy of insecticides in oilseed rape (2013/14)

Due to the commercial confidentiality and expensive trials [8], it is difficult to obtain data on insecticide efficacy which can differ among active ingredients within one chemical group [31]. Efficacy levels are also difficult to evaluate and compare because many factors can affect them both temporally and spatially, including insecticide resistance (Table E.4 in S5 Appendix), application methods, plant growth stages, etc. [32,33].

The perceived high efficacy of neonicotinoids against insect pests and related diseases has been demonstrated in several studies ([34]: cabbage stem flea beetle; [35]: pollen beetle; [36]: peach—potato aphid), although others have shown limited efficacy ([37]: peach—potato aphid). For the relatively new groups, oxadiazines and azomethines, the perceived relatively high efficacy against pollen beetle was illustrated in a recent German study [38]. However, limited efficacy by these two groups was also found [39,40]. It should be noted that without careful management (e.g., rotating insecticides with different available modes of action, [41,42]), resistance to the above insecticides could occur for pests in UK oilseed rape, thus reducing the efficacy. Neonicotinoid resistance has already been detected in the peach—potato aphid in Southern Europe [43].

Pyrethroids and carbamates had the lowest expected efficacy against the peach—potato aphid (1–20%), which could be partly due to the insecticidal resistance occurred for this pest in UK (Table E.4). Although in 2010, a HGCA study found that pyrethroids were efficient against pollen beetle (about 90% control, [30]), experts in our study only expressed a median 21–50% control, possibly due to growing resistance since its appearance in 2006 (Table E.4 in S5 Appendix).

The perceived low confidence in cabbage stem flea beetle control is worrying. Apart from neonicotinoid seed treatments, the other available chemical pyrethroids were perceived to exert lower than 50% control, which could partly due to the resistance occurred in this pest (Table E.4 in S5 Appendix) [44,45].

Across all experts, the perceived efficacy rarely exceeded 90% from all available chemical groups used in UK oilseed rape in 2013/14. This may reflect the lack of confidence in assigning efficacy for insecticides used in oilseed rape, but also the fact that many insecticides, even when newly introduced, cannot provide 90% pest control [46]. The perceived low and/or uncertain efficacy in the available chemical groups and the uncertainty in pest control after the neonicotinoid restriction indicate an urgent need for robust, accountable and updated information about efficiency of insecticides for pest control for this crop in the local fields.

Side effects of seed treatments versus sprays

Experts generally perceived seed treatments to be less harmful than sprays. This is to be expected as: (i) compared with sprays, seed treatments have less direct contact with operators and non-target species [36]; (ii) less surface runoff [47]; and (iii) reduced concentration in the environment [48]. However, many counter-arguments have arisen in the past few years against this application method, especially for the neonicotinoid group. Traces of neonicotinoid residues have been detected in humans [49], pollinators and beehives [50,51], soil [52] and water [53]. Negative impacts have also been found on human health [54], pollinators [8,55], natural enemies [10], earthworms [56] and aquatic invertebrates [53]. Possible reasons behind these side effects are the systemic characteristics of neonicotinoids (residues in treated plant tissues) [12], low soil absorption (high leaching potentials) [57], and great toxicity to invertebrates [53].

On the other hand, sprays can also cause side effects for human health and the wider environment. For example, although relatively few human poisonings from pyrethroids have been reported despite their extensive use worldwide [58], sub-lethal reactions have been found, including paresthesia and nausea [58]. Negative impacts have also been detected for pollinators [50,59–61], natural enemies [62–64], soil [65–68] and water systems [69,70].

Since the survey finished in June 2015, more knowledge has been accumulated on the side effects of insecticides toward non-target species and the wider environment. Take the impacts of neonicotinoid seed treatments on pollinators as an example: since the research gaps were identified by [71] (a literature review up to June 2015), many studies have investigated other active ingredients besides imidacloprid [72], pollinator species besides honey bees (*Apis*) [73], and/or impacts on the colony besides individual development [74]. Although more work needs to be done to improve the evidence base on this issue [75], respondents' opinions could be changed by the new evidence.

However, with the currently available evidence, it is still difficult to compare the overall side effects of the two insecticidal application methods, especially between neonicotinoids and pyrethroids, since little research has done so (but see [76] on earthworms).

Hazard Quotient ratios from sprays

By using the in-field Hazard Quotient method, a temporal comparison of hazard levels on non-target arthropods among available foliar active ingredients for UK oilseed rape was presented (Fig 2). However, limitations exist in interpreting the results because this method is based on the laboratory acute toxicity tests, one being that these tests are difficult to account for the influence from the environment (S4 Appendix) [77,78].

The high toxicity of dimethoate to human health and the wider environment has been widely recognized [79]. Although its use in UK oilseed rape was stopped a decade ago, it is still approved for use in wheat. The high threat potential of lambda-cyhalothrin towards non-target arthropods has also been reflected in previous studies [80–82]. Although thiacloprid has been shown to have negative effects on arthropod natural enemies [40], it has been found to have limited side effects on bees by some studies [36,83]. According to [84], pymetrozine has less of an effect than pirimicarb on aphid natural enemies. Indoxacarb has been shown to be less toxic than lambda-cyhalothrin to arthropod predators and parasitoids [85,86], but its potential hazard to honey bees could be high [87]. Relatively low side effects from pirimicarb against natural enemies have also been recorded [88].

Perceptions on neonicotinoid restriction

The tendency for respondents from the universities, private research institutes and government to choose the 'Neutral' opinion on the neonicotinoid restriction debate is worth discussing. These respondents may have had an actual mid-point opinion on this issue, that they neither opposed nor favoured the restriction. It may have been because there was a lack of interest in this topic, or they considered similar overall costs and benefits for either side [89].

Uncertainty could also have been important to this group: they may have had more recent information about the effects of neonicotinoids on crop yields, pollinators and the wider environment, but because of the complexity of this issue and gaps in the current evidence base [8], they could not estimate the net costs and benefits of neonicotinoids. This is also reflected in that the reasons most frequently chosen for both the *oppose* and *favour* groups were all around pollinators and the wider environment (Figs 5 and 6). On the other hand, they may have been less well informed about the field situations or other related risks for farmers than consultants [90], who have provided more advice services to farmers (Fig 4, Table E.3 in S5 Appendix).

It is also possible that some respondents chose the neutral option to avoid the cognitive costs of selecting the most appropriate opinion, even though they may lean towards one side [91], while others chose this midpoint as a 'hidden don't know' [92]. Nevertheless, the proportion of the last reason is estimated to be small in this study, since a 'Not sure' was included as an option to avoid the ambiguity [93], and online surveys potentially have less of this issue since people are more free to express their true opinions [94].

Expected time, money and yield comparisons between 2013/14 and 2014/15

To our knowledge, little publicly available information has been made available on the time and money spent before and after the neonicotinoid restrictions, and limited research has been done regarding the impact of neonicotinoids on crop yield [8], but see [15,95,96]. According to the average response on this issue, compared with 2013/14 season, agronomists spent more time on inspecting oilseed rape crops for pest damage, farmers spent more money for insecticide purchases, and crop yield would be reduced in 2014/15.

Changes in the time spent on pest control activities should be taken into account when considering the pros and cons of neonicotinoid seed treatments, because it represents a hidden benefit if agronomists/ farmers spend less time on pest control, but more time on other activities. As for insecticide purchases, [15] indicates a negative relationships between neonicotinoid seed treatments and foliar sprays for UK oilseed rape, which could potentially lead to more insecticide costs to farmers after the restriction [28].

Although most respondents expected lower oilseed rape yield in 2014/2015 than 2013/2014, this does not reflect the actual average yields (3.9 and 3.6 t/ ha respectively, [2]). It is difficult to assess the impact of neonicotinoid restriction on the yield change based on one year data, since the increased yield may well be due to nicer weather and lower pest pressure during the year [2].

Alternative methods of pest control in oilseed rape

The results from this study suggest a clear preference towards using new insecticides if these become available. This partly reflected a lack of confidence in the old chemical groups, but also an acknowledgment of the importance of insecticides for crop protection [97]. Developing new insecticides (especially with new modes of action) would help current insecticide resistance problems in oilseed rape. However, insecticide discovery has been a challenge, with the shrinking number of agri-chemical companies involved in the research, and the expensive and time-consuming development process [98].

Respondents also advised growers to use new oilseed rape varieties if available. Indeed, crop breeding in the UK has contributed to yield protection by improving crop resistance to pests and diseases [99]. A new oilseed rape variety 'Amalie' has been recommended for use against TuYV in the 2016/17 growing season [100], and a recently completed pre-breeding project has further explored the potential to develop commercial oilseed rape varieties to tolerate this virus infection [101].

With those who advised farmers to grow small areas of oilseed rape in the future, this concern has also been expressed through a farmer survey during 2014/15 by the Farm Business Survey (FBS) team [14], where the most important reasons for a future reduction in area were crop rotations, reduced crop price, and cabbage stem flea beetle damage. When comparing the 2014/15 with 2013/14 [2], the total area has decreased by about 3% (22,000 ha).

In line with the experts' suggestions, the importance of IPM has also been emphasized by EU and UK policymakers [102], and numerous UK organizations (e.g., LEAF -Linking Environment and Farming, Natural England). In order to develop further IPM for oilseed rape production, one of the most crucial aspects is to understand insecticide efficacy on pest control, and its changes over time due to resistance: this will be important for developing action thresholds to use chemicals strategically [32]. Another crucial aspect of IPM, as expressed by experts in this study, is natural pest control. Many studies have been carried out to evaluate the impact of natural enemies on pest suppression in oilseed rape, and to seek methods of conserving them [27]. However, knowledge gaps still exist in this area [27].

A big challenge will be to combine these two aspects when developing IPM strategies, so that the side effects of insecticides on natural enemies could be reduced to a minimum. Indeed, by conducting research among eight EU countries, [103] have found consistent negative effects of insecticides on biological control potential.

Farmers would adopt new strategies only if they work better than current practices [102]. Profit is one fundamental aspect in the judgement. However, to our knowledge, little literature is available which estimates the influence of natural enemies on crop yield or net profits, especially for large scale field crops by conventional or IPM farmers [104,105], and none has focused on oilseed rape, partly because of the difficulty of conducting field experiments. More research is needed to estimate the economic value of this important service provided by natural enemies.

CONCLUSION

Insecticides used in UK oilseed rape production have been designed to be more efficient in controlling pests, and less harmful to non-target species and the wider environment. However, their efficacy levels are not fully understood, and may not be sufficient in the long-term, due to the limited publicly available studies and fast development of insecticide resistance in pest species. Similarly, it is difficult to assess their side effects, partly because little research has comprehensively compared the impacts of different insecticides in a standardised manner. The type and extent of benefits for farmers are also fundamental when assessing insecticides. For these reasons, the decision as to whether further to restrict neonicotinoid seed treatments in oilseed rape needs careful evaluation. It is a challenge to take into account the multi-faceted aspects when assessing an insecticide; one way forward could be to translate each aspect into economic values, and then apply cost benefit analysis. In order to do so, more research is needed regarding the influence of a chemical on crop protection, farm profit, the environment and related ecosystem services. This study provided an insight into these aspects, but limitations exist due to a relatively small sample of expert opinions.

Integrated pest management presents an important potential future strategy for oilseed rape production, and the importance of insecticides and natural pest control should be better recognised and incorporated. Economic valuation of pest control services by natural enemies for oilseed rape needs to be quantified, coupled with improved communication and knowledge exchange between government, researchers, consultants and growers.

ACKNOWLEDGMENTS

We sincerely thank all the experts who participated, distributed, and commented on this survey. We sincerely thank that the Fera Science Ltd. shared the insecticide data for this study.

REFERENCES

1. European Renewable Energy Council (EREC). Renewable Energy Technology Roadmap-20% by 2020. Brussels: EREC; 2008. Available from: https://www.energy-community.org/portal/page/portal/ENC_HOME/DOCS/324184 /0633975AA77A7B9CE053C92FA8C06338.PDF.
2. 2.2.Department for Environment, Food & Rural Affairs (DEFRA). Farming Statistics Final crop areas, yields, livestock populations and agricultural workforce at June 2015—United Kingdom. London: DEFRA; 2015. Available from: https://www.gov.uk/government/statistics/farming-statistics-final-crop-areas-yields-livestock-populations-and-agricultural-workforce-at-1-june-2015-uk.
3. Twining S, Clarke J. Future of UK winter oilseed rape production. Wolverhampton: ADAS UK Ltd; 2009. Available from: http://www.voluntaryinitiative .org.uk/importedmedia/library/1152_s4.pdf.
4. Department for Environment, Food & Rural Affairs (DEFRA). Agriculture in the United Kingdom 2014. London: DEFRA; 2015. Available from: https://www.gov. uk/government/statistics/agriculture-in-the-united-kingdom-2014.
5. Agriculture and Horticulture Development Board (AHDB). Oilseed rape guide. Kenilworth: AHDB; 2015. Aavailable from: https://cereals.ahdb.org.uk/crop-management/growth-guides.aspx.
6. Fera. Pesticide Usage Surveys; 2015 [cited 2015 Aug 26]. Database: Pesticide Usage Surveys [Internet]. Available from: http://pusstats.fera.defra.gov.uk/index.cfm.
7. European Commission. Commission Implementing Regulation (EU) No 485/2013. Off J Eur Union. 2013. Available from: http://eur-lex.europa.eu/LexUriServ /LexUriServ.do?uri=OJ:L:2013:139:0012:0026:en:PDF.
8. Godfray HCJ, Blacquière T, Field LM, Hails RS, Potts SG, Raine NE, et al. A restatement of recent advances in the natural science evidence base concerning neonicotinoid insecticides and insect pollinators. Proc R Soc B Biol Sci. 2015;282(20151821).
9. van der Sluijs JP, Amaral-Rogers V, Belzunces LP, Bijleveld van Lexmond MFIJ, Bonmatin JM, Chagnon M, et al. Conclusions of the Worldwide Integrated Assessment on the risks of neonicotinoids and fipronil to biodiversity and ecosystem functioning. Environ Sci Pollut Res. 2015;22(1):148–154.

10. European Academies Science Advisory Council (EASAC). Ecosystem services, agriculture and neonicotinoids. Brussels: EASAC; 2015. Available from: http://www.easac.eu/home/reports-and-statements/detail-view/article/ecosystem-se.html.
11. Potts SG, Imperatriz-Fonseca VL, Ngo HT, Biesmeijer JC, Breeze TD, Dicks L V., et al., editors. Summary for policymakers of the assessment report of the Intergovernmental Science-Policy Platform on Biodiversity and Ecosystem Services on pollinators, pollination and food production. Bonn: Intergovernmental Science-Policy Platform on Biodiversity and Ecosystem Services (IPBES); 2016. Available from: http://www.ipbes.net/work-programme/pollination.
12. Bonmatin JM, Giorio C, Girolami V, Goulson D, Kreutzweiser DP, Krupke C, et al. Environmental fate and exposure; neonicotinoids and fipronil. Environ Sci Pollut Res. 2015;22:35–67.
13. Case P. NFU vows to continue bid for neonics derogation. Farmers Weekly. 13 May 2016. Available from: http://www.fwi.co.uk/news/nfu-vows-to-continue-bid-for-neonics-derogation.htm. Cited 3 June 2016.
14. Scott C, Bilsborrow P. An interim impact assessment of the neonicotinoid seed treatment ban on oilseed rape production in England. Newcastle: Rural Business Research; 2015. Available from: http://www.ncl.ac.uk/afrd /survey/# farmbenchmarkingservice.
15. Budge GE, Garthwaite D, Crowe A, Boatman ND, Delaplane KS, Brown MA, et al. Evidence for pollinator cost and farming benefits of neonicotinoid seed coatings on oilseed rape. Nature. 2015;5(12574).
16. Qualtrics. Provo, Utah, USA. 2015. Available from: http://www.qualtrics.com.
17. Skillings JH, Mack GA. On the Use of a Friedman-Type Statistic in Balanced and Unbalanced Block Designs. Technometrics. 1981;23(2):171–177.
18. Srisuradetchai P. The Skillings-Mack Test Statistic for Block Designs with Missing Observations. Vienna: R Foundation for Statistical Computing; 2015. Available from: https://cran.r-project.org/package=Skillings.Mack.
19. Pohlert T. The Pairwise Multiple Comparison of Mean Ranks Package (PMCMR). Vienna: R Foundation for Statistical Computing; 2014. Available from: http://cran.r-project.org/package=PMCMR.
20. R Core Team. R: A Language and Environment for Statistical Computing. R Foundation for Statistical Computing. Vienna, Austria. 2016. Available from: https://www.r-project.org/.
21. Millard SP. EnvStats: An R Package for Environmental Statistic. New York: Springer; 2013. Available from: http://www.springer.com.
22. European Standard Characteristics of Non-Target Arthropod Regulatory Testing (ESCORT). Guidance document on regulatory testing and risk assessment procedures for plant protection products with non-target arthropods. Wageningen: ESCORT; 2000.
23. European Food Safety Authority (EFSA). Rapporteur Member State assessment reports submitted for the EU peer review of active substances used in plant protection products. Parma: EFSA; 2015. Available from: http://dar.efsa.europa.eu/dar-web/provision.
24. Open Geography Portal. NUTS Level 1 (January 2015) Generalised Clipped Boundaries in England and Wales; 2016 [cited 2016 Feb 6]. Database: ONS Geography Open Data [Internet]. Available from: http://geoportal.statistics.gov.uk /datasets/87bca93367bd4ad5ad8199960aa3b68c_2.
25. GADM. United Kingdom shapefile; 2015 [cited 2016 Feb 6]. Database: GADM database of Global Administrative Areas [Internet]. Available from: http://www.gadm.org/country.
26. QGIS Development Team. QGIS Geographic Information System. Open Source Geospatial Foundation Project. 2016. Available from: http://qgis.osgeo.org.
27. Williams IH, editor. Biocontrol-Based Integrated Management of Oilseed Rape Pests. 1st ed. Springer Netherlands; 2010.

28. Nicholls CJ. Research Review No.77: Implications of the restriction on the neonicotinoids: imidacloprid,clothianidin and thiamethoxam on crop protection in oilseeds and cereals in the UK. Kenilworth: Agriculture and Horticulture Development Board; 2013. Available from: https://cereals.ahdb.org.uk/media/197715/rr77.pdf.
29. Clarke J, Wynn S, Twining S, Berry P, Cook S, Ellis S, et al. HGCA Research Review No.70: Pesticide availability for cereals and oilseeds following revision of Directive 91/ 414/ EEC; effects of losses and new research priorities. Kenilworth: Agriculture and Horticulture Development Board; 2009. Available from: http://cereals-2.ahdb.org.uk/publications/documents/cropresearch/RR70_Research_ReviewX.pdf.
30. Ellis S, Berry P. Project Report No. 495: Re-evaluating thresholds for pollen beetle in oilseed rape. Kenilworth: Agriculture and Horticulture Development Board; 2012. Available from: https://cereals.ahdb.org.uk/media/200518/pr495.pdf.
31. Šmatas R, Makūnas V, Brazauskienė I, Petraitienė E. Sensitivity of pollen beetle (Meligethes aeneus F.) to insecticides with different modes of action and their efficacy in the field conditions. Zemdirbyste-Agriculture. 2012;99(2):197–202.
32. Nault BA, Anthony SM. Impact of Insecticide Efficacy on Developing Action Thresholds for Pest Management: A Case Study of Onion Thrips (Thysanoptera: Thripidae) on Onion. J Econ Entomol. 2010;103(4):1315–1326. pmid:20857743
33. Slater R, Ellis S, Genay J-P, Heimbach U, Huart G, Sarazin M, et al. Pyrethroid resistance monitoring in European populations of pollen beetle (Meligethes spp.): A coordinated approach through the Insecticide Resistance Action Committee (IRAC). Pest Manag Sci. 2011;67(6):633–638. pmid:21268233
34. Maienfisch P, Angst M, Brandl F, Fischer W, Hofer D, Kayser H, et al. Chemistry and biology of thiamethoxam: a second generation neonicotinoid. Pest Manag Sci. 2001;57(10):906–913. pmid:11695183
35. Zimmer CT, Nauen R. Cytochrome P450 mediated pyrethroid resistance in European populations of Meligethes aeneus (Coleoptera: Nitidulidae). Pestic Biochem Physiol. 2011;100(3):264–272.
36. Elbert A, Haas M, Springer B, Thielert W, Nauen R. Applied aspects of neonicotinoid uses in crop protection. Pest Manag Sci. 2008;64:1099–1105. pmid:18561166
37. Panini M, Dradi D, Marani G, Butturini A, Mazzoni E. Detecting the presence of target-site resistance to neonicotinoids and pyrethroids in Italian populations of Myzus persicae. Pest Manag Sci. 2014;70(6):931–938. pmid:23963887
38. Kupfer S, Schroeder G. Studies on the Selective Use of Insecticides Against the Pollen Beetle (Meligethes aeneus) in Agricultural Practice in Brandenburg from 2006 to 2014. Gesunde Pflanz. 2015;67(2):59–73.
39. Vaitelyte B, Petraitiene E, Smatas R, Brazauskiene I. Control of Meligethes aeneus, Ceutorhynchus assimilis and Dasineura brassicae in winter oilseed rape (Brassica napus L.). Zemdirbyste-Agriculture. 2011;98(2):175–182.
40. Jansen J-P, Gomez GSMY. A large field trial to assess the short-term and long-term effects of 5 insecticides used to control the pollen beetle on parasitic hymenoptera in oilseed rape. IOBC-WPRS Bull. 2014;103:9–16.
41. Sparks TC, Nauen R. IRAC: Mode of action classification and insecticide resistance management. Pestic Biochem Physiol. 2015;121:122–128. pmid:26047120
42. Bass C, Denholm I, Williamson MS, Nauen R. The global status of insect resistance to neonicotinoid insecticides. Pestic Biochem Physiol. 2015;121:78–87. pmid:26047114
43. Slater R, Paul VL, Andrews M, Garbay M, Camblin P. Identifying the presence of neonicotinoid resistant peach-potato aphid (Myzus persicae) in the peach-growing regions of southern France and northern Spain. Pest Manag Sci. 2012;68(4):634–638. pmid:22052736
44. Insecticide Resistance Action Committee (IRAC). IRAC Newsletter Issue 37: Cabbage stem flea beetle (CSFB), Psylliodes chrysocephala-Resistance on the move. IRAC; 2015. Available from: http://www.irac-online.org/content/uploads /econnection37.pdf.

45. Zimmer CT, Muller A, Heimbach U, Nauen R. Target-site resistance to pyrethroid insecticides in German populations of the cabbage stem flea beetle,Psylliodes chrysocephala L. (Coleoptera:Chrysomelidae). Pestic Biochem Physiol. 2014;108:1–7. pmid:24485308
46. Buntin GD. Economic thresholds for insect management. In: Higley LG, Pedigo LP, editors. Economic thresholds for integrated pest management. Lincoln, NE: University of Nebraska Press; 1986. pp. 128–147.
47. Palmquist K, Salatas J, Fairbrother A. Pyrethroid Insecticides: Use, Environmental Fate, and Ecotoxicology. In: Perveen F, editor. Insecticides—Advances in Integrated Pest Management. Rijeka: InTech; 2012. pp. 251–278.
48. Nuyttens D, Devarrewaere W, Verboven P, Foqué D. Pesticide-laden dust emission and drift from treated seeds during seed drilling: A review. Pest Manag Sci. 2013;69:564–575. pmid:23456984
49. Ueyama J, Harada KH, Koizumi A, Sugiura Y, Kondo T, Saito I, et al. Temporal Levels of Urinary Neonicotinoid and Dialkylphosphate Concentrations in Japanese Women between 1994 and 2011. Environ Sci Technol. 2015;49(24): 14522–14528. pmid:26556224
50. Hladik ML, Vandever M, Smalling KL. Exposure of native bees foraging in an agricultural landscape to current-use pesticides. Sci Total Environ. 2016;542:469–477. pmid:26520270
51. Mullin CA, Frazier M, Frazier JL, Ashcraft S, Simonds R, VanEngelsdorp D, et al. High Levels of Miticides and Agrochemicals in North American Apiaries: Implications for Honey Bee Health. PLoS One. 2010;5(3): e9754 pmid:20333298
52. Jones A, Harrington P, Turnbull G. Neonicotinoid concentrations in arable soils after seed treatment applications in preceding years. Pest Manag Sci. 2014;70:1780–1784. pmid:24888990
53. Morrissey CA, Mineau P, Devries JH, Sanchez-Bayo F, Liess M, Cavallaro MC, et al. Neonicotinoid contamination of global surface waters and associated risk to aquatic invertebrates: A review. Environ Int. 2015;74:291–303. pmid:25454246
54. Mason R. Immune Suppression by Neonicotinoid Insecticides at the Root of Global Wildlife Declines. J Enviromental Immunol Toxicol. 2013;1(1):3–12.
55. Godfray HCJ, Blacquière T, Field LM, Hails RS, Petrokofsky G, Potts SG, et al. A restatement of the natural science evidence base concerning neonicotinoid insecticides and insect pollinators. Proc R Soc B Biol Sci. 2014;281(20140558).
56. Pisa LW, Amaral-Rogers V, Belzunces LP, Bonmatin JM, Downs CA, Goulson D, et al. Effects of neonicotinoids and fipronil on non-target invertebrates. Environ Sci Pollut Res. 2014;22(1):68–102.
57. European Food Safety Authority (EFSA). Conclusion regarding the peer review of the pesticide risk assessment of the active substance imidacloprid. EFSA J. 2008;6: 148r.
58. Bradberry SM, Cage SA, Proudfoot AT, Vale JA. Poisoning due to Pyrethroids. Toxicol Rev. 2005;24(2):93–106. pmid:16180929
59. van dame R, Meled M, Colin M-E, Belzunces LP. Alteration of the homing-flight in the honey bee Apis mellifera L. exposed to sublethal dose of deltamethrin. Env Toxicol Chem. 1995;14:855–860.
60. Sanchez-Bayo F, Goka K. Pesticide residues and bees—A risk assessment. PLoS One. 2014;9(4): e94482. pmid:24718419
61. Charreton M, Decourtye A, Henry M, Rodet G, Sandoz J-C, Charnet P, et al. A Locomotor Deficit Induced by Sublethal Doses of Pyrethroid and Neonicotinoid Insecticides in the Honeybee Apis mellifera. PLoS One. 2015;10(12): e0144879. pmid:26659095
62. Holland JM, Luff ML. The effects of agricultural practices on Carabidae in temperate agroecosystems. Integr Pest Manag Rev. 2000;5:109–129.
63. .Desneux N, Pham-Delègue MH, Kaiser L. Effects of sub lethal and lethal doses of lambda-cyhalothrin on oviposition experience and host-searching behaviour of a parasitic wasp, Aphidius ervi. Pest Manag Sci. 2004;60(4):381–389. pmid:15119601

64. Devotto L, Carrillo R, Cisternas E, Gerding M. Effects of lambda-cyhalothrin and Beauveria bassiana spores on abundance of Chilean soil surface predators, especially spiders and carabid beetles. Pedobiologia (Jena). 2007;51(1):65–73.
65. Hartnik T, Sverdrup LE, Jensen J. Toxicity of the pesticide alpha-cypermethrin to four soil nontarget invertebrates and implications for risk assessment. Environ Toxicol Chem. 2008;27(6):1408–1415. pmid:18211124
66. Tripathi G, Kachhwaha N, Dabi I. Ecophysiological category based toxicological responses in metabolism of earthworms: Impact of a pyrethroidal insecticide. Pestic Biochem Physiol. 2010;98(3):333–341.
67. Das Gupta R, Chakravorty PP, Kaviraj A. Susceptibility of epigeic earthworm Eisenia fetida to agricultural application of six insecticides. Chemosphere. 2011;84(5):724–726. pmid:21489602
68. Song Y, Kai J, Song X, Zhang W, Li L. Long-term toxic effects of deltamethrin and fenvalerante in soil. J Hazard Mater. 2015;289:158–164. pmid:25725337
69. .Amweg EL, Weston DP, Ureda NM. Use and toxicity of pyrethroid pesticides in the Central Valley, California, USA. Environ Toxicol Chem. 2005;24:966–972. pmid:15839572
70. Ng CM, Weston DP, You J, Lydy MJ. Patterns of Pyrethroid Contamination and Toxicity in Agricultural and Urban Stream Segments. In: Gan J, Spurlock F, Hendley P, Weston DP, editors. Synthetic Pyrethroids. American Chemical Society; 2008. pp. 355–369.
71. Lundin O, Rundlöf M, Smith HG, Fries I, Bommarco R. Neonicotinoid Insecticides and Their Impacts on Bees: A Systematic Review of Research Approaches and Identification of Knowledge Gaps. PLoS One. 2015;10(8):e0136928. pmid:26313444
72. Mogren CL, Lundgren JG. Neonicotinoid-contaminated pollinator strips adjacent to cropland reduce honey bee nutritional status. Sci Rep. 2016;6:29608. pmid:27412495
73. Woodcock BA, Isaac NJB, Bullock JM, Roy DB, Garthwaite DG, Crowe A, et al. Impacts of neonicotinoid use on long-term population changes in wild bees in England. Nat Commun. 2016;7:12459. pmid:27529661
74. Wu-Smart J, Spivak M. Sub-lethal effects of dietary neonicotinoid insecticide exposure on honey bee queen fecundity and colony development. Sci Rep. 2016;6:32108. pmid:27562025
75. .Walters KFA. Neonicotinoids, bees and opportunity costs for conservation. Insect Conserv Divers. 2016;9(5):375–383.
76. Wang Y, Cang T, Zhao X, Yu R, Chen L, Wu C, et al. Comparative acute toxicity of twenty-four insecticides to earthworm, Eisenia fetida. Ecotoxicol Environ Saf. 2012;79:122–128. pmid:22244824
77. Ellis SR, Hodson ME, Wege P. The influence of different artificial soil types on the acute toxicity of carbendazim to the earthworm Eisenia fetida in laboratory toxicity tests. Eur J Soil Biol. 2007;43:S239–245.
78. Kimball KD, Levin SA. Limitations of Laboratory Bioassays: The Need for Ecosystem-Level Testing. Biosci. 1985;35 (3): 165–171.
79. pan-UK. Dimethoate. Pesticides News. Mar 2002. Available from: http://www.pan-uk.org/pestnews/Actives/dimethoa.htm. Cited 29 March 2016.
80. Langhof M, Gathmann A, Poehling HM, Vogt H, Heimbach U, Vinuela E. Field study on the effects of insecticide drift on nontarget arthropods: residual toxicity, effect on populations and recolonisation processes. Bull OILB/SROP. 2003;26:73–82.
81. Desneux N, Decourtye A, Delpuech JM. The sublethal effects of pesticides on beneficial arthropods. Annu Rev Entomol. 2007;52:81–106. pmid:16842032
82. Ceuppens B, Eeraerts M, Vleugels T, Cnops G, Roldan-Ruiz I, Smagghe G. Effects of dietary lambda-cyhalothrin exposure on bumblebee survival, reproduction, and foraging behavior in laboratory and greenhouse. J Pest Sci. 2015;88(4):777–783.
83. Jeschke P, Nauen R, Schindler M, Elbert A. Overview of the Status and Global Strategy for Neonicotinoids. J Agric Food Chem. 2010;59:2897–2908. pmid:20565065

84. Jansen JP, Defrance T, Warnier AM. Side effects of flonicamide and pymetrozine on five aphid natural enemy species. Biocontrol. 2011;56:759–770.
85. Nowak JT, McCravy KW, Fettig CJ, Berisford CW. Susceptibility of Adult Hymenopteran Parasitoids of the Nantucket Pine Tip Moth (Lepidoptera: Tortricidae) to Broad-Spectrum and Biorational Insecticides in a Laboratory Study. J Econ Entomol. 2001;94(5):1122–1127. pmid:11681674
86. Galvan TL, Koch RL, Hutchison WD. Toxicity of commonly used insecticides in sweet corn and soybean to multicolored Asian lady beetle (Coleoptera: Coccinellidae). J Econ Entomol. 2005;98(3):780–789. pmid:16022306
87. Stoner KA, Eitzer BD. Using a Hazard Quotient to Evaluate Pesticide Residues Detected in Pollen Trapped from Honey Bees (Apis mellifera) in Connecticut. PLoS One. 2013;8(10):e77550. pmid:24143241
88. Bacci L, Rosado JF, Picanço MC, Pereira EJG, Silva G a, Martins JC. Concentration-mortality responses of Myzus persicae and natural enemies to selected insecticides. J Environ Sci Heal A. 2012;47(12):1930–1937.
89. DeMars CE, Erwin TD. Scoring Neutral or Unsure on an Identity Development Instrument for Higher Education. Res High Educ. 2004;45(1):83–95.
90. .Harwood J, Heifner R, Coble K, Perry J, Somwaru A. Managing Risk in Farming: Concepts, Research, and Analysis. Washington (DC): U.S. Department of Agriculture; 1999. Available from: htp://195.43.12.200/mysite6/adobo/080.pdf.
91. Krosnick JA. Response Strategies for Coping with the Demands of Attitude Measures in Surveys. Appl Cogn Psychol. 1991;5:214–236.
92. Sturgis P, Roberts C, Smith P. Middle Alternatives Revisited: How the neither/nor Response Acts as a Way of Saying "I Don"t Know'? Sociol Methods Res. 2014;43(1):15–38.
93. Ryan C, Garland R. The use of a specific non-response option on Likert-type scales. Tour Manag. 1999;20:107–113.
94. Dillman DA. Mail and internet surveys: the tailored design method. 2nd ed. New York: John Wiley & Sons Inc; 2000.
95. Ahmed N, Kanan H, Inanaga S, Ma Y, Sugimoto Y. Impact of pesticide seed treatments on aphid control and yield of wheat in the Sudan. Crop Prot. 2001;20:929–934.
96. Stanley DA, Garratt MPD, Wickens JB, Wickens VJ, Potts SG, Raine NE. Neonicotinoid pesticide exposure impairs crop pollination services provided by bumblebees. Nature. 2015;528(7583):548–550. pmid:26580009
97. Cooper J, Dobson H. The benefits of pesticides to mankind and the environment. Crop Prot. 2007;26(9):1337–1348.
98. Sparks TC. Insecticide discovery: An evaluation and analysis. Pestic Biochem Physiol. 2013;107(1):8–17. pmid:25149229
99. Carter R, Clarke J, Tompkins S. Review of the objectives of modern plant breeding and their relation to agricultural sustainability. Cambridge: ADAS UK Ltd; 2015. Available from: http://www.bspb.co.uk/sg_userfiles/15-06-01_Review_of_objectives_ of_modern_plant_breeding_and_relation_to_agricultural_sustainability-_Report_Update.pdf.
100. Agriculture and Horticulture Development Board (AHDB). AHDB Recommended Lists for cereals and oilseeds 2016/17. Kenilworth: AHDB; 2016. Available from: https://cereals.ahdb.org.uk/media/800462/recl1617-recommended-lists-2016-2017-summer-edition.pdf.
101. Crop Improvement Research Club (CIRC). Summer 2016-Impact summary. Swindon: CIRC; 2016. Available from: http://www.bbsrc.ac.uk/documents/circ-2016-impact-summary/.
102. Hillocks RJ. Farming with fewer pesticides: EU pesticide review and resulting challenges for UK agriculture. Crop Prot. 2012;31(1):85–93.
103. .Geiger F, Bengtsson J, Berendse F, Weisser WW, Emmerson M, Morales MD, et al. Persistent negative effects of pesticides on biodiversity and biological control potential on European farmland. Basic Appl Ecol. 2010;11:97–105.

104. Sandhu H, Wratten S, Costanza R, Pretty J, Porter JR, Reganold J. Significance and value of non-traded ecosystem services on farmland. PeerJ. 2015;3:e762. pmid:25737811

105. .Letourneau DK, Ando AW, Jedlicka JA, Narwani A, Barbier E. Simple-but-sound methods for estimating the value of changes in biodiversity for biological pest control in agriculture. Ecol Econ. 2015;120:215–225.

CHAPTER 15

A Fundamental Step in IPM on Grapevine: Evaluating the Side Effects of Pesticides on Predatory Mites

Alberto Pozzebon, Paola Tirello, Renzo Moret, Marco Pederiva and Carlo Duso *

Department of Agronomy, Food, Animals, Natural Resources and Environment (DAFNAE), University of Padova, Viale dell'Università 16, Legnaro-Padova 35020, Italy

ABSTRACT

Knowledge on side effects of pesticides on non-target beneficial arthropods is a key point in Integrated Pest Management (IPM). Here we present the results of four experiments conducted in vineyards where the effects of chlorpyrifos, thiamethoxam, indoxacarb, flufenoxuron, and tebufenozide were evaluated on the generalist predatory mites *Typhlodromus pyri* Scheuten and *Amblyseius andersoni* (Chant), key biocontrol agents of herbivorous mites on grapevines. Results show that indoxacarb and tebufenozide had a low impact on the predatory mites considered here, while a significant impact was observed for chlorpyrifos, flufenoxuron, and thiamethoxam. The information obtained here should be considered in the design of IPM strategies on grapevine.

Keywords: side-effects; chemical treatments; *Vitis vinifera*; Phytoseiidae; *Typhlodromus pyri*; *Amblyseius andersoni*

1. INTRODUCTION

Knowledge of the impact of pesticides on beneficial arthropods is critical in Integrated Pest Management (IPM). The authorization of a new pesticide in Europe is subjected to the evaluation of its effects on non-target organisms, including a number of beneficials. These evaluations can be made at different levels (e.g., laboratory, semi-field, and field) obtaining complementary information [1,2]. Most of the data in the specific literature come from laboratory and semi-field studies as they are less time-consuming and expensive than field tests. However, field tests are useful for evaluation of the pesticides' impact in realistic use conditions, especially for pesticides having a significant impact in laboratory or semi-field studies [3,4,5]. Predatory mites belonging to the Phytoseiidae family, particularly generalists *sensu* McMurtry and Croft [6], have been widely considered in field evaluations of pesticide side-effects (e.g., [2,7,8,9]). Generalist predatory mites can persist on perennial crops when prey is scarce by feeding on alternative foods and prevent herbivore mite outbreaks for a long time [10,11,12]. Thus, the assessment of pesticide impact on them can be performed excluding potential effects mediated by their prey [13]. At the same time, plant-pathogens, especially those involved in fungal diseases, can represent alternative or supplemental food resources for various predatory mites and their presence can alter the outcome of trials aimed at assessing the impact of pesticides [14,15,16]. Field experiments should therefore be conducted where symptoms caused by fungal diseases are negligible to low.

There is an abundant literature on the side-effects of pesticides on predatory mites but the variability of environmental factors, experimental methods, target species, and pesticide formulates makes comparison of these data difficult. An attempt to summarize the results conducted using standardized methods is reported in the IOBC Pesticide Side Effect Database published by the International Organisation for Biological and Integrated Control/West Palearctic Regional Section (IOBC/WPRS) [17]. Pesticides are classified according their overall effects (Classes 1–4, from harmless to harmful), providing selectivity rates to people involved in IPM. When studied, lethal and sub-lethal effects are combined and, thus, cited publications should be examined by those interested in investigating these aspects. Nevertheless, results presented in reports used in regulatory procedures for pesticides' use authorization are often reported giving useful information to IPM practitioners.

In this paper we summarize the results of a number of experiments conducted in vineyards to evaluate the effects of a number of insecticides on the generalist predatory mites *Typhlodromus pyri* Scheuten and *Amblyseius andersoni* (Chant). These species are key biocontrol agents of herbivorous mites [11,18,19,20,21,22,23]. The choice of insecticides was based on their importance in pest control strategies in viticulture. For some insecticides, it was possible to compare our results with those reported in the IOBC Pesticide Side Effect Database. For other pesticides, records could integrate gaps in the above-mentioned database.

2. EXPERIMENTAL SECTION

The effects of insecticides on predatory mites were investigated in four vineyards located in the Veneto region (Northeastern Italy), two comprising cultivar Glera grapevine (experiments 1 and 2; Miane and San Pietro di Feletto, respectively, province of Treviso; Table 1), one with cultivar Merlot (experiment 3; Spresiano, province of Treviso; Table 1), and one with cultivar Cabernet Sauvignon (experiment 4; Spresiano, province of Treviso; Table 1). These vineyards were inhabited by predatory mites (*T. pyri* and *A. andersoni*) that are considered key biocontrol agents of herbivore mites as well as target beneficials in the evaluation of pesticide side effects. The impact of insecticides was compared in four field experiments where an untreated control treatment was also included (Table 1). Experimental design was a randomized block with four replicates per treatment. Timing of applications was determined by the mode of action of different insecticides and the phenology of main target pests for each vineyard. In particular, from preliminary observations, *Empoasca vitis* (Goethe) (Hemiptera Cicadellidae) was the target pest in experiments 1 and 2, while *Lobesia botrana* (Denis and Schiffermüller) (Lepidoptera Tortricidae) was the target pest in experiments 3 and 4 (Table 1). Preliminary observations revealed that *T. pyri* was the dominant species in the vineyards of experiments 1 and 2, while *A. andersoni* was dominant in the vineyards of experiments 3 and 4. Regarding insecticide use in previous years, all these vineyards had been treated with organophosphates, carbamates, and chitin-inhibitors. Therefore, the dominance of these predatory mite species was likely be due to their adaptation to specific environmental conditions rather than pesticide history: experimental sites 1 and 2 were located in hilly areas whereas experimental sites 3 and 4 were in lowland areas. The experiments were performed in mid-summer and the climatic conditions were typical for the area where the experimental sites were located. Samplings were conducted before and approximately every week after insecticide applications for about one month. A total of 100 leaves per treatment (25 leaves per replicate) were removed and transferred to the laboratory where predatory mite individuals were counted under a dissecting microscope; a number of females were mounted on slides, in Hoyer's medium, and identified under a phase contrast microscope using current keys [24]. Data were analyzed with a restricted maximum likelihood (REML) repeated measures model. Mite densities were considered as response variables with repeated measures made at different times, *i.e.*, sampling dates. Using an F test (α = 0.05), we evaluated the effect of insecticide application, time, and their interaction. Contrasts (α = 0.05) were designed for pairwise comparison between treatments before and after insecticide applications. Degrees of freedom were estimated using the Kenward-Roger method [25]. According to Akaike's Information Criterion, first-order autoregressive was chosen as best fitting covariance structure for correlating different sampling dates [25]. Data were checked for analysis assumptions before analyses and log(x + 1) transformation was applied when necessary. The effect (E) of insecticides was estimated using the Henderson and Tilton's formula [26].

Table 1. Experimental design and insecticides applied in the four experiments.

Treatments	Active Ingredients	Commercial Products (a.i. Concentration)	Dose	Timing of Application (Days from First Application)
Experiment 1				
control	-		-	-
chlorpyrifos	chlorpyrifos	Dursban 75 WG (75%)	70 g/hL	0 days
thiamethoxam	thiamethoxam	Actara 25 WG (25%)	25 g/hL	0 days
indoxacarb	indoxacarb	Steward EC (30%)	15 g/hL	0 days
Experiment 2				
control	-		-	-
chlorpyrifos	chlorpyrifos	Dursban 75 WG (75%)	70 g/hL	0 days
thiamethoxam	thiamethoxam	Actara 25 WG (25%)	25 g/hL	0 days
flufenoxuron	flufenoxuron	Cascade 50 DC (4.7%)	100 mL/hL	0 days
Experiment 3				
control	-		-	-
chlorpyrifos	chlorpyrifos	Dursban 75 WG (75%)	70 g/hL	10 days
flufenoxuron	flufenoxuron	Cascade 50 DC (4.7%)	100 mL/hL	0 days
Experiment 4				
control	-		-	-
chlorpyrifos	chlorpyrifos	Dursban 75 WG (75%)	70 g/hL	10 days
flufenoxuron	flufenoxuron	Cascade 50 DC (4.7%)	150 mL/hL	0 days
indoxacarb	indoxacarb	Steward EC (30%)	15 g/hL	4 days
tebufenozide	tebufenozide	Mimic (23%)	60 mL/hL	0 days

3. RESULTS

3.1. Experiment 1

Typhlodromus pyri was the dominant species in this vineyard, and its density was influenced by treatment and time, while their interaction was not significant ($F_{3, 11.5} = 11.70$; $p = 0.001$; $F_{3, 26.7} = 8.67$; $p < 0.001$; $F_{9, 27.7} = 1.05$; $p = 0.427$; respectively; Figure 1). The occurrence of herbivorous mites (e.g., tetranychids and eriophyids) was negligible. Prior to insecticide applications, *T. pyri* densities were similar among plots ($F_{3, 42.6} = 1.35$; $p = 0.269$). After applications, *T. pyri* densities were lower with thiamethoxam and chlorpyrifos as compared to the untreated control ($F_{1, 13.8} = 25.35$; $p < 0.001$; E = 31.45; $F_{1, 13.8} = 21.72$; $p < 0.001$; E = 44.21; respectively). No effects were observed for indoxacarb ($F_{1, 13.8} = 2.24$; $p = 0.157$; E = 6.58; respectively). Among insecticide treatments, predatory mite densities were lower with thiamethoxam and chlorpyrifos than indoxacarb ($F_{1, 13.8} = 10.02$; $p = 0.007$; $F_{1, 13.8} = 7.94$; $p = 0.013$; respectively). No differences were observed between chlorpyrifos and thiamethoxam ($F_{1, 13.8} = 0.14$; $p = 0.71$).

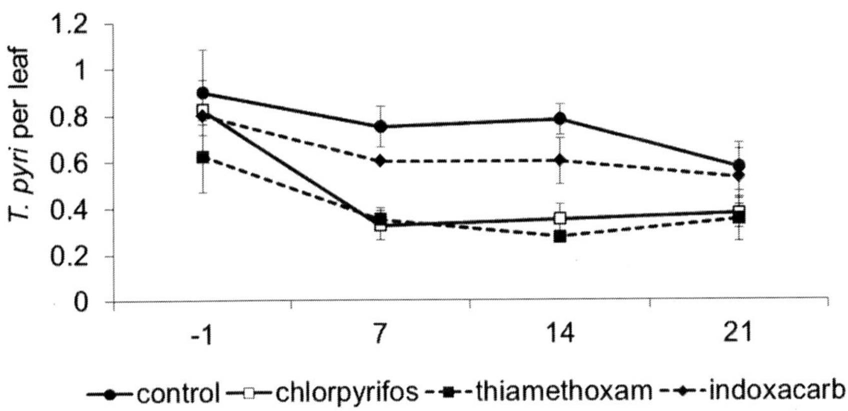

Figure 1. Population dynamics of *Typhlodromus pyri* (mean ± std. err.) observed in different treatments during experiment 1. On the X-axis, days after insecticide application are reported.

3.2. Experiment 2

Typhlodromus pyri was dominant among predatory mites in this vineyard. Its abundance was influenced by treatment, time, and their interaction ($F_{3, 24.3}$ = 9.61; p < 0.001; $F_{4, 46.3}$ = 7.72; p < 0.001; $F_{12, 46.7}$ = 2.40; p = 0.016; respectively; Figure 2). Herbivorous mite populations occurred at negligible levels. In pre-treatment observations, *T. pyri* densities were similar among plots ($F_{3, 57}$ = 0.35; P = 0.789). Following insecticide applications, *T. pyri* numbers were lower where insecticides were applied as compared to the control (thiamethoxam: $F_{1, 28.1}$ = 15.56; p < 0.001; E = 27.20; flufenoxuron: $F_{1, 28.1}$ = 22.29; p < 0.001; E = 34.85; chlorpyrifos: $F_{1, 28.1}$ = 22.94; p < 0.001; E = 34.82). No differences emerged among plots with insecticide applications (thiamethoxam *vs.* flufenoxuron: $F_{1, 28.1}$ = 0.73; p = 0.402; thiamethoxam *vs.* chlorpyrifos: $F_{1, 28.1}$ = 0.51; p = 0.479; chlorpyrifos *vs.* flufenoxuron: $F_{1, 28.1}$ = 0.02; p = 0.893).

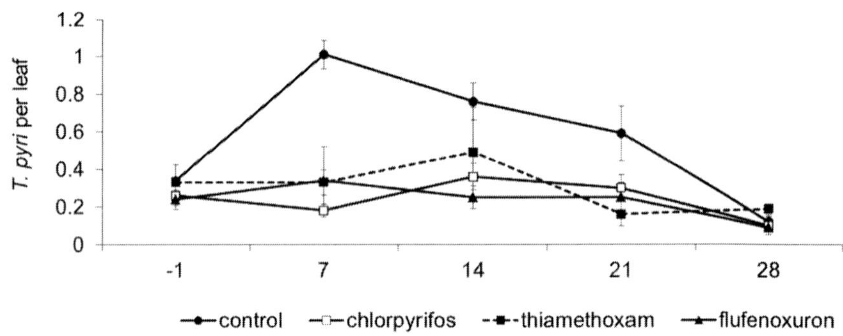

Figure 2. Population dynamics of *Typhlodromus pyri* (mean ± std. err.) observed in different treatments during experiment 2. On the X-axis, days after insecticide application are reported.

3.3. Experiment 3

In this experiment, the predatory mite *Amblyseius andersoni* was commonly observed on grape leaves. Its population dynamics were influenced by treatment and time, while no effect was found of the interaction "treatment × time" ($F_{2, 17.9}$ = 5.94; p = 0.011; $F_{4, 34.2}$ = 5.93; p = 0.001; $F_{8, 34.9}$ =1.02; p = 0.436; respectively; Figure 3). The occurrence of herbivorous mites was very low. Prior to insecticide application, no differences were observed among treatments ($F_{2, 44.8}$ = 0.37; p = 0.691). Later, predatory mite numbers were higher in the control than in flufenoxuron and chlorpyrifos plots (flufenoxuron: $F_{1, 19.3}$= 12.50; p = 0.002; E = 58.95; chlorpyrifos: $F_{1, 22}$ = 5.01; p = 0.003; E = 55.43). No differences emerged between the insecticide-treated plots ($F_{1, 22}$ = 0.20; p = 0.656).

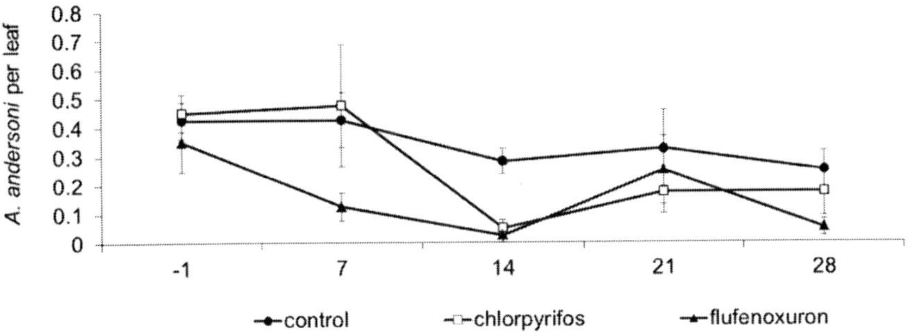

Figure 3. Population dynamics of *Amblyseius andersoni* (mean ± std. err.) observed in different treatments during experiment 3. On the X-axis, days after insecticide application are reported.

3.4. Experiment 4

In this vineyard *A. andersoni* was dominant among predatory mites. Its abundance was influenced by treatment and time, while no effect of their interaction was found ($F_{4,\ 35.7} = 3.36$; $p = 0.012$; $F_{5,\ 65} = 15.64$; $p < 0.001$; $F_{20,\ 68.4} = 0.97$; $p = 0.510$; respectively; Figure 4). The occurrence of herbivorous mites was negligible. Before insecticide application, the abundance of *A. andersoni* was similar among plots ($F_{4,\ 87} = 0.91$; $p = 0.462$). After insecticide applications, lower predatory mite numbers were found in flufenoxuron and chlorpyrifos than in control plots (flufenoxuron: $F_{1,\ 37.5} = 9.46$; $p = 0.004$; E = 41.50; chlorpyrifos: $F_{1,\ 40.5} = 8.58$; $p = 0.006$; E = 55.52). No significant effects were observed for indoxacarb and tebufenozide ($F_{1,\ 37.5} = 3.50$; $p = 0.069$; E = 29.09; $F_{1,\ 37.5} = 1.13$; $p = 0.295$; E = −6.89; respectively). However, it should be noted that predatory mite population densities decreased to zero after 21 days from indoxacarb application and increased thereafter (Figure 4). No differences were observed among insecticides (flufenoxuron *vs.* chlorpyrifos: $F_{1,\ 40.5} = 0.19$; $p = 0.664$; flufenoxuron *vs.* tebufenozide: $F_{1,\ 37.5} = 4.06$; $p = 0.051$; flufenoxuron *vs.* indoxacarb: $F_{1,\ 37.5} = 1.45$; $p = 0.235$; chlorpyrifos *vs.* tebufenozide: $F_{1,\ 40.5} = 2.45$; $p = 0.125$; chlorpyrifos *vs.* indoxacarb: $F_{1,\ 40.5} = 0.70$; $p = 0.408$; tebufenozide *vs.* indoxacarb: $F_{1,\ 37.5} = 0.66$; $p = 0.423$).

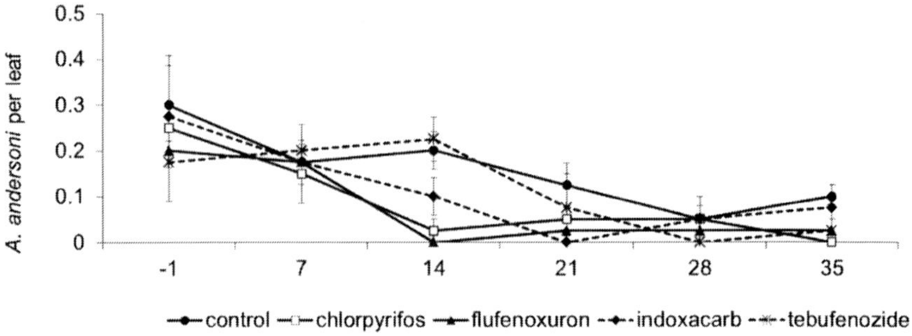

Figure 4. Population dynamics of *Amblyseius andersoni* (mean ± std. err.) observed in different treatments during experiment 4. On the X-axis, days after insecticide application are reported.

4. DISCUSSION AND CONCLUSIONS

The statistical analyses revealed a significant impact of insecticides with E-values exceeding 27%, with the exception of indoxacarb in the fourth experiment (E = 29.1%). In the traditional IOBC classification [2], pesticides that have an overall effect (E) lower than 25% for field trials and 30% for laboratory trials are considered harmless. Pesticides with an effect range from 25% to 49% are considered slightly harmful. Boller *et al.* [27] condensed the first two toxicity classes obtained in field and semi-field tests in one class (N = harmless or slightly harmful), probably to mark the distinction between non-hazardous and hazardous pesticides. Most of the results obtained here fit with records reported in the IOBC Pesticide Side Effect Database while others increase the information available to people involved in IPM strategies.

Chlorpyrifos has been considered harmful for *T. pyri* and *A. andersoni* in both the laboratory and field [28]. Similar effects are reported for other predatory mites such as *Euseius finlandicus* (Oudemans) (IOBC Pesticide Side Effect Database). Chlorpyrifos has been reported as moderately harmful for *Kampimodromus aberrans* (Oudemans) in field conditions [29]. In other trials chlorpyrifos did not significantly affect predatory mite populations (e.g., [30]). Pesticide history and the appearance of resistance can explain the variability in toxicity encountered in other studies. Barbar *et al.* [31] evaluated the susceptibility to chlorpyrifos by predatory mites (*Typhlodromus exhilaratus* Ragusa) occurring on a vine crop exposed to the pesticide or originating from an unsprayed orchard. As expected, the strain of *T. exhilaratus* from the vine was more tolerant to chlorpyrifos as compared to the other strain. Resistance of predatory mites to organophosphates has spread in fruit orchard regions over the world and this provides another explanation for the variability of their impact on target species, including *T. pyri* [32]. *K. aberrans* strains resistant to chlorpyrifos have been detected in vineyards in Northeastern Italy and mechanisms of resistance have recently been studied [33,34]. The reduction in fecundity of *K. aberrans* females exposed to chlorpyrifos can be observed even in resistant strains [35,36]. Such sub-lethal effects have been little explored for other predatory mites. In a field study carried out in North America, chlorpyrifos applications led to higher spider mite densities in vineyards and pesticide impact on generalist phytoseiid mites was probably the key mechanism involved in mite outbreaks [21]. We can conclude that chlorpyrifos should be used with caution in IPM for its negative effects on some predatory mite species. Recent papers have demonstrated that the side effects of chlorpyrifos on predatory mites can be mitigated when they have fresh alternative food sources available [37].

In previous trials (laboratory or field tests) flufenoxuron has been classified as harmless or slightly harmful for *T. pyri* (e.g., [2,38]) and slightly harmful for *K. aberrans* [29]. In our experiments the impact of flufenoxuron on *T. pyri* and especially *A. andersoni* appeared to be more deleterious. Variability in the impact of flufenoxuron could depend on species and strain features as well as pesticide history. A laboratory study conducted on two *K. aberrans* strains showed low effects of flufenoxuron on the survival of predatory mite females but a significant reduction in the fecundity of females belonging to one of these strains [35].

Thiamethoxam showed some detrimental effects on *T. pyri* populations. In other trials where *K. aberrans* was involved, effects on survival were negligible to low, while thiamethoxam reduced predatory mite fecundity [35,36]. Similar effects have been reported in studies conducted with other predatory mite species [39,40]. The functional response to prey in predatory mites can be altered by neonicotinoids including thiamethoxam [41]. Moreover, the total effect of thiamethoxam on predatory mites increased with the routes of exposure involved; in particular, residual and contaminated food exposures increased pesticide effects [13]. The use of neonicotinoid insecticides in IPM programs should therefore be carefully evaluated (e.g., [42,43,44]). Little is reported about the susceptibility of predatory mites to tebufenozide. In the IOBC Pesticide Side Effect Database this insecticide is reported as harmless or slightly harmful to *T. pyri*. Similar effects have been reported for other predatory mites [45]. Our data refer to *A. andersoni* only and confirm the low impact of tebufenozide on predatory mite populations.

Indoxacarb appeared to be the most selective pesticide towards *T. pyri* and *A. andersoni*, confirming data reported for *K. aberrans* [35,36] and other predatory mites [45,46].

The results obtained here provide useful information for the definition of IPM strategies on grapevine. Indoxacarb and tebufenozide had a low impact on the predatory mites considered here, while the highest impact was observed for chlorpyrifos, flufenoxuron, and thiamethoxam. The use of the latter insecticides should be carefully considered in the light of the conservation of predatory mite populations on grapevine.

AUTHOR CONTRIBUTIONS

Alberto Pozzebon and Carlo Duso conceived and designed the experiments; Alberto Pozzebon, Paola Tirello, Renzo Moret, Marco Pederiva performed the experiments; Alberto Pozzebon, and Carlo Duso analyzed the data; Carlo Duso contributed reagents/materials/analysis tools; Alberto Pozzebon and Carlo Duso wrote the paper.

CONFLICTS OF INTEREST

The authors declare no conflict of interest.

REFERENCES

1. Bakker, F.M.; Jacas, J.A. Pesticides and phytoseiid mites: Strategies for risk assessment. Ecotoxicol. Environ. Saf. 1995, 32, 58–67.
2. Sterk, G.; Hassan, S.A.; Baillod, M.; Bakker, F.; Bigler, F.; Blümel, S.; Bogenschütz, H.; Boller, E.; Bromand, B.; Brun, J.; et al. Results of the seventh joint pesticide testing programme carried out by the IOBC/WPRS-Working Group "Pesticides and Beneficial Organisms". BioControl 1999, 44, 99–117.
3. Candolfi, M.P.; Barrett, K.L.; Campbell, P.; Forster, R.; Grandy, N.; Huet, M.C.; Lewis, G.; Oomen, P.A.; Schmuck, R.; Vogt, H. Guidance Document on Regulatory Testing and Risk Assessment Procedures for Plant Protection Products with Non-Target Arthropods; Society of Environmental Toxicology and Chemistry Europe: Brussels, Belgium, 2001.
4. Thomson, L.J.; Hoffmann, A.A. Field validation of laboratory-derived IOBC toxicity ratings for natural enemies in commercial vineyards. Biol. Control 2006, 39, 507–515.

5. Thomson, L.J.; Hoffmann, A.A. Ecologically sustainable chemical recommendations for agricultural pest control? J. Econ. Entomol. 2007, 100, 1741–1750.
6. McMurtry, J.A.; Croft, B.A. Life-styles of phytoseiid mites and their role in biological control. Annu. Rev. Entomol.1997, 42, 291–321.
7. Candolfi, M.P.; Bakker, F.; Canez, V.; Miles, M.; Neumann, C.; Pilling, E.; Primiani, M.; Romijn, K.; Schmuck, R.; Storck-Weyhermüller, S.; et al. Sensitivity of non-target arthropods to plant protection products: Could Typhlodromus pyri and Aphidius spp. be used as indicator species. Chemosphere 1999, 39, 1357–1370.
8. Pozzebon, A.; Duso, C.; Pavanetto, E. Side effects of some fungicides on phytoseiid mites (Acari, Phytoseiidae) in north-Italian vineyards. Anz. Schädlingskunde J. Pest Sci. 2002, 75, 132–136.
9. Hardman, J.M.; Franklin, J.L.; Jensen, K.I.N.; Moreau, D.L. Effects of pesticides on mite predators (Acari: Phytoseiidae) and colonization of apple trees by Tetranychus urticae. Phytoparasitica 2006, 34, 449–462.
10. Bakker, F.M.; Klein, M.E. Transtrophic interactions in cassava. Exp. Appl. Acarol. 1992, 14, 293–311.
11. Duso, C. Role of Amblyseius aberrans (Oud.), Typhlodromus pyri Scheuten and Amblyseius andersoni (Chant) (Acari, Phytoseiidae) in vineyards. I: The effects of single or mixed phytoseiid population releases on spider mite densities (Acari, Tetranychidae). J. Appl. Entomol. 1989, 107, 474–492.
12. Walde, S.J.; Nyrop, J.P.; Hardman, J.M. Dynamics of Panonychus ulmi and Typhlodromus pyri: Factors contributing to persistence. Exp. Appl. Acarol. 1992, 14, 261–291.
13. Pozzebon, A.; Duso, C.; Tirello, P.; Bermudez Ortiz, P. Toxicity of thiamethoxam to Tetranychus urticae Koch and Phytoseiulus persimilis Athias-Henriot (Acari Tetranychidae, Phytoseiidae) through different routes of exposure. Pest Manag. Sci. 2011, 67, 352–359.
14. Pozzebon, A.; Duso, C. Grape downy mildew Plasmopara viticola, an alternative food for generalist predatory mites occurring in vineyards. Biol. Control 2008, 45, 441–449.
15. Pozzebon, A.; Loeb, G.M.; Duso, C. Grape powdery mildew as a food source for generalist predatory mites occurring in vineyards: Effects on life-history traits. Ann. Appl. Biol. 2009, 155, 81–89.
16. Pozzebon, A.; Borgo, M.; Duso, C. The effects of fungicides on non-target mites can be mediated by plant pathogens. Chemosphere 2010, 79, 8–17.
17. IOBC Pesticide Side Effect Database. Available online: http://www.iobc-wprs.org/ip_ipm/IOBC_Pesticide_Side_Effect_Database.html (accessed on 15 September 2015).
18. Schruft, G. Grape. In Spider Mites. Their Biology, Natural Enemies and Control; Helle, W., Sabelis, M.W., Eds.; Elsevier: Amsterdam, The Netherlands, 1985; pp. 359–365.
19. Duso, C.; de Lillo, E. Grape. In Eriophyoid Mites Their Biology, Natural Enemies and Control; Lindquist, E.E., Sabelis, M.W., Bruin, J., Eds.; Elsevier: Amsterdam, The Netherlands, 1996; pp. 571–582.
20. Marshall, D.B.; Lester, P.J. The transfer of Typhlodromus pyri on grape leaves for biological control of Panonychus ulmi (Acari: Phytoseiidae, Tetranychidae) in vineyards in Ontario, Canada. Biol. Control 2001, 20, 228–235.
21. Prischmann, D.A.; James, D.G.; Wright, L.C.; Teneyck, R.D.; Snyder, W.E. Effects of chlorpyrifos and sulfur on spider mites (Acari: Tetranychidae) and their natural enemies. Biol. Control 2005, 33, 324–334.
22. Duso, C.; Pozzebon, A.; Kreiter, S.; Tixier, M.S.; Candolfi, M.P. Management of phytophagous mites in European vineyards. In Arthropod Management in Vineyards: Pests, Approaches, and Future Directions; Bostanian, N.J., Vincent, C., Isaacs, R., Eds.; Springer: New York, NY, USA, 2012; pp. 191–217.
23. Lorenzon, M.; Pozzebon, A.; Duso, C. Effects of potential food sources on biological and demographic parameters of the predatory mites Kampimodromus aberrans,

Typhlodromus pyri and Amblyseius andersoni. Exp. Appl. Acarol. 2012, 58, 259–278.

24. Tixier, M.-S.; Baldassar, A.; Duso, C.; Kreiter, S. Phytoseiidae in European grape (Vitis vinifera L.): Bio-ecological aspects and keys to species (Acari: Mesostigmata). Zootaxa 2013, 3721, 101–142.
25. Littell, R.C.; Milliken, G.A.; Stroup, W.W.; Wolfinger, R.D. SAS System for Mixed Models, 2nd ed.; SAS Institute: Cary, NC, USA, 2006; pp. 159–202.
26. Henderson, C.F.; Tilton, E.W. Tests with acaricides against the brow wheat mite. J. Econ. Entomol. 1955, 48, 157–161.
27. Boller, E.F.; Vogt, H.; Ternes, P.; Malavolta, C. Working Document on Selectivity of Pesticides. IOBC/WPRS, 2005. Available online: http://www.iobc-wprs.org/ip_ipm/IOBC_IP_Tool_Box.html#2 (accessed on 24 August 2015).
28. Hassan, S.A.; Bigler, F.; Bogenschütz, H.; Boller, E.; Brun, J.; Chiverton, P.; Edwards, P.; Mansour, F.; Naton, E.; Oomen, P.A.; et al. Results of the fourth joint pesticide testing programme by the IOBC/WPRS-Working Group "Pesticides and Beneficial Arthropods". J. Appl. Entomol. 1988, 105, 321–329.
29. Tirello, P.; Vettore, S.; Pozzebon, A.; Lorenzon, M.; Duso, C. Effects of some insecticides on Kampimodromus aberrans: Laboratory and fields studies. IOBC/WPRS Bull. 2011, 67, 257–260.
30. Cross, J.V.; Berrie, A.M. Effects of repeated foliar sprays of insecticides or fungicides on organophosphate-resistant strains of the orchard predatory mite Typhlodromus pyri on apple. Crop Prot. 1994, 13, 39–44.
31. Barbar, Z.; Tixier, M.S.; Kreiter, S. Assessment of pesticide susceptibility for Typhlodromus exhilaratus and Typhlodromus phialatus strains (Acari: Phytoseiidae) from vineyards in the south of France. Exp. Appl. Acarol. 2007, 42, 95–105.
32. Wearing, C.H. Historical tests of the toxicity of pesticides to Typhlodromus pyri (Acari: Phytoseiidae) and their relevance to current pest management in New Zealand apple orchards 1. Laboratory tests with eggs and larvae. Biocontrol Sci. Technol. 2014, 24, 780–809.
33. Tirello, P.; Pozzebon, A.; Duso, C. Resistance to chlorpyriphos in the predatory mite Kampimodromus aberrans. Exp. Appl. Acarol. 2012, 56, 1–8.
34. Cassanelli, S.; Ahmad, S.; Duso, C.; Tirello, P.; Pozzebon, A. A single nucleotide polymorphism in the acetylcholinesterase gene of the predatory mite Kampimodromus aberrans (Acari: Phytoseiidae) is associated with chlorpyrifos resistance. Biol. Control 2015, 90, 75–82.
35. Tirello, P.; Pozzebon, A.; Duso, C. The effect of insecticides on the non-target predatory mite Kampimodromus aberrans: Laboratory studies. Chemosphere 2013, 93, 1139–1144.
36. Duso, C.; Ahmad, S.; Tirello, P.; Pozzebon, A.; Klaric, V.; Baldessari, M.; Malagnini, V.; Angeli, G. The impact of insecticides applied in apple orchards on the predatory mite Kampimodromus aberrans (Acari Phytoseiidae). Exp. Appl. Acarol. 2014, 62, 391–414.
37. Pozzebon, A.; Ahmad, S.; Tirello, P.; Lorenzon, M.; Duso, C. Does pollen availability mitigate the impact of pesticides on generalist predatory mites? BioControl 2014, 59, 585–596.
38. Rodrigues, R.; Gonçalves, R.; Silva, C.; Torres, L. Toxicity of five insecticides on predatory mites (Acari: Phytoseiidae) in vineyards in two Portuguese regions. IOBC/WPRS Bull. 2004, 27, 37–44.
39. Duso, C.; Malagnini, V.; Pozzebon, A.; Buzzetti, F.M.; Tirello, P. A method to assess the effects of pesticides on the predatory mite Phytoseiulus persimilis (Acari Phytoseiidae) in the laboratory. Biocontrol Sci. Technol. 2008, 18, 1027–1040.
40. Doker, I.; Pappas, M.L.; Samaras, K.; Triantafyllou, A.; Kazak, C.; Broufas, G.D. Compatibility of reduced-risk insecticides with the non-target predatory mite Iphiseius degenerans (Acari: Phytoseiidae). Pest Manag. Sci. 2015, 71, 1267–1273.

41. Poletti, M.; Maia, A.H.; Omoto, C. Toxicity of neonicotinoid insecticides to Neoseiulus californicus and Phytoseiulus macropilis (Acari: Phytoseiidae) and their impact on functional response to Tetranychus urticae (Acari: Tetranychidae). Biol. Control 2007, 40, 30–36.
42. Beers, E.H.; Brunner, J.F.; Dunley, J.E.; Doerr, M.; Granger, K. Role of neonicotinyl insecticides in Washington apple integrated pest management. Part II. Nontarget effects on integrated mite control. J. Insect Sci. 2005, 5. PMID:16341248.
43. Stavrinides, M.C.; Mills, N.J. Demographic effect of pesticides on biological control of Pacific spider mite (Tetranychus pacificus) by the western predatory mite (Galendromus occidentalis). Biol. Control 2008, 48, 267–273.
44. Beers, E.H.; Schmidt, R.A. Impacts of orchard pesticides on Galendromus occidentalis: Lethal and sublethal effects. Crop Prot. 2014, 56, 16–24.
45. Bernard, M.B.; Cole, P.; Kobelt, A.; Horne, P.A.; Altmann, J.; Wratten, S.D.; Yen, A.L.; Alan, L. Reducing the impact of pesticides on biological control in Australian vineyards: Pesticide mortality and fecundity effects on an indicator species, the predatory mite Euseius victoriensis (Acari: Phytoseiidae). J. Econ. Entomol. 2010, 103, 2061–2071.
46. Bostanian, N.J.; Vincent, C.; Hardman, J.M.; Larocque, N. Toxicity of indoxacarb to two species of predacious mites and a predacious mirid. Pest Manag. Sci. 2004, 60, 483–486.

CHAPTER 16

Pest Control and Resistance Management Through Release of Insects Carrying a Male-Selecting Transgene

Tim Harvey-Samuel[1,2], Neil I. Morrison[2*], Adam S. Walker[2] , Thea Marubbi[2] , Ju Yao[3,4], Hilda L. Collins[3] , Kevin Gorman[2] , T. G. Emyr Davies[5] , Nina Alphey[1,6], Simon Warner[2] , Anthony M. Shelton3 and Luke Alphey[1,2,7]

[1] *Department of Zoology, University of Oxford, South Parks Road, Oxford, Oxfordshire OX1 3PS, UK.*

[2] *Oxitec Ltd, 71 Innovation Drive, Milton Park, Oxford, Oxfordshire OX14 4RQ, UK.*

[3] *Cornell University/NYSAES, Barton Lab 416, 630 W. North Street, Geneva, NY 14456, USA.*

[4] *Institute of Plant Protection, Xinjiang Academy of Agricultural Science, Urumqi, China.*

[5] *Biological Chemistry & Crop Protection Department, Rothamsted Research, Harpenden, Hertfordshire AL5 2JQ, UK.*

[6] *Department of Life Sciences, Imperial College London, Silwood Park Campus, Buckhurst Road, Ascot, Berkshire SL5 7PY, UK.*

[7] *The Pirbright Institute, Woking, Surrey GU24 0NF, UK.*

ABSTRACT

Background

Development and evaluation of new insect pest management tools is critical for overcoming over-reliance upon, and growing resistance to, synthetic, biological and plant-expressed insecticides. For transgenic crops expressing insecticidal proteins from the bacterium *Bacillus thuringiensis* ('*Bt* crops') emergence of resistance is slowed by maintaining a proportion of the crop as non-*Bt* varieties, which produce pest insects unselected for resistance. While this strategy has been largely successful, multiple cases of *Bt* resistance have now been reported.

One new approach to pest management is the use of genetically engineered insects to suppress populations of their own species. Models suggest that released insects carrying male-selecting (MS) transgenes would be effective agents of direct, species-specific pest management by preventing survival of female progeny, and simultaneously provide an alternative insecticide resistance management strategy by introgression of susceptibility alleles into target populations. We developed a MS strain of the diamondback moth, *Plutella xylostella*, a serious global pest of crucifers. MS-strain larvae are reared as normal with dietary tetracycline, but, when reared without tetracycline or on host plants, only males will survive to adulthood. We used this strain in glasshouse-cages to study the effect of MS male *P. xylostella* releases on target pest population size and spread of *Bt* resistance in these populations.

Results

Introductions of MS-engineered *P. xylostella* males into wild-type populations led to rapid pest population decline, and then elimination. In separate experiments on broccoli plants, relatively low-level releases of MS males in combination with broccoli expressing Cry1Ac (*Bt* broccoli) suppressed population growth and delayed the spread of *Bt* resistance. Higher rates of MS male releases in the absence of *Bt* broccoli were also able to suppress *P. xylostella* populations, whereas either low-level MS male releases or *Bt* broccoli alone did not.

Conclusions

These results support theoretical modeling, indicating that MS-engineered insects can provide a powerful pest population suppressing effect, and could effectively augment current *Bt* resistance management strategies. We conclude that, subject to field confirmation, MS insects offer an effective and versatile control option against *P. xylostella* and potentially other pests, and may reduce reliance on and protect insecticide-based approaches, including *Bt* crops.

Keywords: *Bacillus thuringiensis* Diamondback moth Insect Insecticide resistance management Pest *Plutella xylostella* Sterile insect technique Transgenic

BACKGROUND

Pest insects are a major threat to global food production, biodiversity conservation, and human and animal health [1, 2, 3]. Synthetic insecticides are widely used for control; however, potential off-target ecological damage, and the capacity of pest populations to develop resistance, has driven demand for alternative methods of pest control. Integrated pest management (IPM) approaches have been developed, utilizing multiple tools including biological insecticides (applied and expressed in transgenic crops), enhanced biological control, mating disruption, and the release of sterile insects (the sterile insect technique, SIT) to sustainably manage insect pest populations.

The SIT reduces target pest populations through sustained mass-releases of radiation-sterilized insects, thus reducing the frequency of mating between fertile insects [4]. SIT relies on the mate-seeking and mating behavior of released insects, and is therefore species-specific and can be effective against pests that are difficult to control by other methods. SIT has been successful in area-wide eradication and suppression programs against numerous crop pests [5]. Wider applicability of SIT is hindered by several challenges, including the negative effects on insect performance of sterilization by irradiation [6, 7, 8, 9, 10] and difficulty in conducting large-scale sex-sorting for male-only releases [11, 12, 13].

We have previously developed a male-selecting (MS) transgenic system to overcome these obstacles [14, 15]. In this system, pest colonies are engineered with tetracycline-repressible dominant female-specific lethal transgenes. Provision of tetracycline (or suitable analogues) to larval stages suppresses transgene lethality allowing mass rearing. Once released, mating between transgenic and wild insects results in mortality of female progeny (female-specific lethality) due to the absence of suitable quantities of tetracycline in the field, thereby reducing the reproductive potential of the target population [15, 16, 17, 18, 19, 20]. Through targeting female progeny but allowing male transgene heterozygotes to survive to reproduce, the MS system is predicted to be significantly more efficient at suppressing populations than those which target both sexes (such as SIT) [19]. Additionally, this system avoids the negative impacts of irradiation on released insect competitiveness [21] and enables large-scale (off tetracycline) production of single-sex (male) release cohorts. Male-only releases can significantly improve per-male efficiency [22, 23] by concentrating the reproductive effort of released insects on wild females. For the Mediterranean fruit fly (medfly, *Ceratitis capitata*), SIT programs have, in tandem with sterilization by radiation, relied on translocation-based sex-sorting systems in which a dominant marker is translocated to the Y chromosome [24]. However, these traits are difficult to translate to new pest species, are unstable, and compromise insect productivity in mass rearing [25, 26].

The release of male insects carrying MS transgenes ("MS males") has been shown to be effective in suppressing target pest populations in cage experiments against the mosquito *Aedes aegypti*, *C. capitata*, and the olive fly (*Bactrocera oleae*) [18, 20, 27]. However, the pest suppression potential of such a transgenic system has not yet been investigated in lepidopterans, which include many of the most destructive pests of forestry and agriculture worldwide [28]. Beyond this direct population-reducing effect, modeling suggests that releases of MS males into a target population may simultaneously provide an insecticide resistance management benefit. Mating between released males and wild females results in the survival of male transgene heterozygotes and the introgression of their background genetics into the wild pest population [29, 30]. With an insecticide-susceptible genetic background in released insects, this introgression will increase the frequency of susceptibility alleles within the target pest population.

This proposed mechanism of resistance management is analogous to that currently utilized in transgenic crops engineered to express insecticidal Cry toxin proteins from the bacterium *Bacillus thuringiensis*. A major advantage of these '*Bt* crops' is their low environmental impact, with the effects of the toxin limited to target species both taxonomically (due to the high species-specificity of *Bt* toxins) and ecologically (as toxin expression is limited to crop tissue, which needs to be ingested to take effect). Cultivation of *Bt* crops, which primarily target lepidopteran and coleopteran pests, has increased rapidly over the past two decades, reaching 78.8 million hectares in 2014 [31]. Resistance in pest populations is an ongoing threat to transgenic *Bt* crop efficacy [32, 33]. The most widely applied resistance management strategy is known as high dose/refuge [34]. Here, the *Bt* toxin is expressed at sufficiently high levels for resistance to be functionally recessive and a proportion of the crop grown includes non-*Bt* varieties (the refuge). As with modeled MS-based resistance management, refugia therefore act as a source of susceptible alleles which introgress (via mating) into the pest population, reducing the frequency of *Bt*-resistant homozygotes. Although the high-dose/refuge strategy has been largely successful in delaying *Bt* resistance, the development of *Bt*-resistant populations has now been reported in the field, particularly in lepidopteran pests (reviewed in [32]). Additionally, recommendations on refuge size vary considerably depending on the cultivated species and the number of *Bt* transgenes expressed, but may be as high as 50 % of the crop, potentially exposing large areas to economic levels of damage [35]. Novel means of delaying *Bt* resistance, especially those which would function as effective pest control measures in their own right, would therefore be of economic and ecological benefit.

Models of population genetics and dynamics predict that release of MS males carrying insecticide susceptibility alleles could form effective components of insecticide resistance management (IRM) strategies for *Bt* crops [29, 30]. Such releases could substantially reduce refuge size requirements for equivalent levels of resistance management, or make refuges redundant altogether, as well as providing a potential remedial action to reverse the spread of resistance where present. Furthermore, the overall population suppression benefit of an integrated program combining *Bt* plants and MS insects is anticipated to be better than either alone. These effects are predicted at release rates significantly lower than those usually employed for population control in SIT programs.

Testing resistance management systems for *Bt* crops requires an insect that has evolved resistance to *Bt* proteins and plants that express such proteins. Diamondback moth, *Plutella xylostella*, is a major pest of brassica crops costing an estimated US$4-5 billion annually in losses and control costs worldwide [36]. *P. xylostella* was the first agricultural insect pest to have evolved resistance in the field to *Bt* proteins [37, 38]. Of the >500 insecticide-resistant arthropods documented, *P. xylostella* ranks second in the number of cases of resistance and is considered one of the most difficult insect pests to control [39]. Novel means of controlling this highly resistant pest are thus required [40]. Brassica plants have been developed that express *Bt* proteins [41, 42, 43, 44, 45] and the combination of *P. xylostella* and *Bt* brassica plants has served as an effective model system for studying resistance management of *Bt* proteins expressed in plants [46, 47]. We have developed a MS system in *P. xylostella* potentially suitable both as a novel control tool for growers, and as a model system for testing the predicted benefits of MS insect releases for managing resistance to *Bt* and other insecticides [15]. A strain of *P. xylostella* transformed with this system, called OX4319L, shows tightly controlled, highly penetrant female-specific lethality [15] and is sexually competitive against non-transgenic individuals [48]. Herein, we describe experiments that demonstrate the direct population-suppressing and *Bt* resistance management effects of releasing OX4319L males into wild-type populations.

RESULTS

Population suppression

This experiment investigated the direct population suppression potential of the OX4319L MS transgene-carrying *P. xylostella* strain, independent of other control measures. Releases of transgene-homozygous OX4319L males into two experimental cages began 9 weeks after the initial wild-type introductions into the cages. At this point the population size, estimated by weekly consistency of egg production, in each cage was judged to have reached equilibrium (Figure 1a). The first re-introductions of transgenic progeny (as pupae, evidenced by positive screening of the DsRed2 fluorescent transgene marker) into the two treatment cages took place 2 weeks later (Figure 1c), indicating successful mating by OX4319L males. The proportion of re-introduced pupae that were transgenic (fluorescence proportion) increased as releases continued into the treatment cages, eventually reaching 100 % 7 and 9 weeks after OX4319L releases began. Under the restrictive (non-tetracycline) conditions of this experiment, the transgenic (fluorescent) phenotype was restricted to a single genotype (male heterozygotes) as female transgene carriers are unable to survive to adulthood [15]. Transgenic males were therefore only able to mate with wild-type females and the fluorescence proportions recorded here are equal to twice the transgene allele frequency in the population [48]. The increasing introgression of MS transgenes into these treatment populations had a substantial effect on the population sex ratio and reproductive capacity. By week 15 (6 weeks after OX4319L releases began) the number of dead adult females collected in each of the treatment cages had decreased considerably relative to that of control cages (Figure 1b), concurrent with a reduction in the reproductive output in these cages (Figure 1a). As the generation time (egg to egg) of the insects in these experiments was approximately 3 weeks, these time periods (between initiation of MS male releases and suppression, approximately two generations) fit the hypothesis that introductions of the OX4319L transgene into the treatment populations were causing reductions in the number of females reaching adulthood, and thus the number of eggs being laid in the subsequent generation. Ten weeks after OX4319L releases began (week 19) the reproductive output of both treatment cages had dropped to 0, and no dead female moths were collected after this point. In this experimental protocol this equates to approximately three generations. Experiments were continued for another 2 weeks after egg-laying ceased, confirming that their populations were extinct.

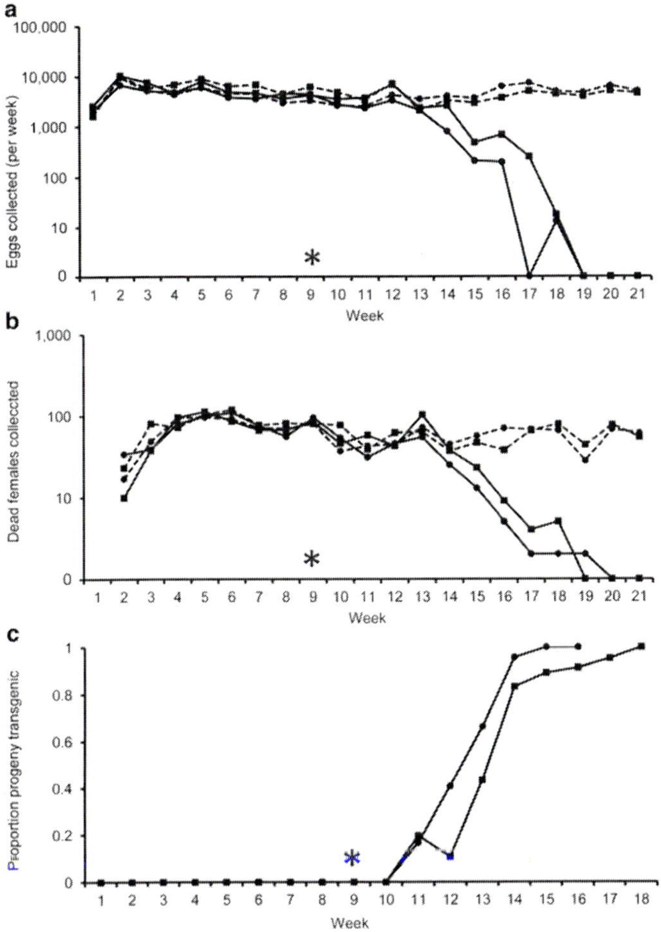

Figure 1. Suppression of caged populations of *Plutella xylostella* by weekly introduction of OX4319L males. Graphs showing (**A**) number of eggs collected, (**B**) number of dead adult females collected weekly from cages, and (**C**) proportion of cage progeny that were transgenic re-entering the cages (fluorescence proportions), over the experimental period. With female moths present in the cages being only wild-type, transgenic progeny (including those re-introduced) were heterozygotes (restrictive conditions). Fluorescence proportions thus equate to twice the MS transgene allele frequency in the cage population at that time point. Solid lines represent OX4319L-treated populations (Cages 1 and 2, circular and square data-points, respectively). Dashed lines represent untreated control populations (Cages 3 and 4, circular and square data-points, respectively). In week 9, return of pupae into treatment and control cages was made proportional and release of OX4319L males into treatment cages began (marked with asterisk)

Insecticide resistance management

This experiment investigated the predicted benefits to both resistance management and population control of combining *Bt* transgenic crops and MS males with *Bt*-susceptible genetic backgrounds. To achieve this, treatments were designed such that each pest control method on its own (*Bt* only and OX4319L release at low levels) would likely be insufficient for population control. The effect of the combination of treatments could then be compared with that of each one applied singly. In the *Bt* only treatment, the presence of resistance alleles in the founder population, and strong selection for these alleles in subsequent generations, were expected to result in a lack of effective control. In the low release rate OX4319L-only treatment, this was achieved by selecting a release rate predicted to be unable to prevent population growth under these conditions. Under this design, all treatment populations would persist for the duration of the experimental period, allowing comparison of their population densities and resistance allele frequencies after multiple generations of treatment effects. In addition, a treatment where OX4319L males were released at a high rate (in the absence of *Bt*) was also conducted to act as an investigation of this strain's suppression potential under more challenging conditions where the target pest population may be expanding rapidly.

In cages with *Bt* broccoli and no OX4319L releases, *P. xylostella* populations were well-controlled in Generation 1 (Figure 2a), presumably due to the initially high frequency of genotypes susceptible to the *Bt* toxin. The subsequent populations in these cages, now highly *Bt*-resistant due to the effects of strong selection in the initial generation, then increased rapidly until Generation 4. Similarly, as expected, low-level releases of OX4319L males into *P. xylostella* populations reared on non-*Bt* broccoli were not effective at preventing population growth, and this treatment was terminated at Generation 3. When *Bt* broccoli and low OX4319L releases were combined, however, the caged populations were well-controlled throughout, only increasing slowly at each generation. In Generation 3, the mean peak population counts recorded for the *Bt* broccoli-only and the low-rate OX4319L-only treatments did not differ significantly (Contrast 1). Concurrently, the mean peak population counts in the combined OX4319L + *Bt* broccoli treatment were significantly lower than those of the low-rate OX4319L-only treatment (Contrast 2), but not significantly different to that of the *Bt* broccoli-only treatment (Contrast 3) (Contrast 1: diff = 337, P = 0.133; Contrast 2: diff = −545, P = 0.0272; Contrast 3: diff = −207, P = 0.315). By Generation 4, however, populations in the *Bt* broccoli-only treatment were significantly larger than those where *Bt* broccoli was combined with low-rate OX4319L releases (t = −4.84, P = 0.0084). The high rate of OX4319L releases performed similarly to the combined OX4319L + *Bt*-broccoli treatment, but by Generation 4 had started to reduce the population in the last remaining cage. Due to the lack of replication in this treatment in Generation 4, it was excluded from statistical analysis.

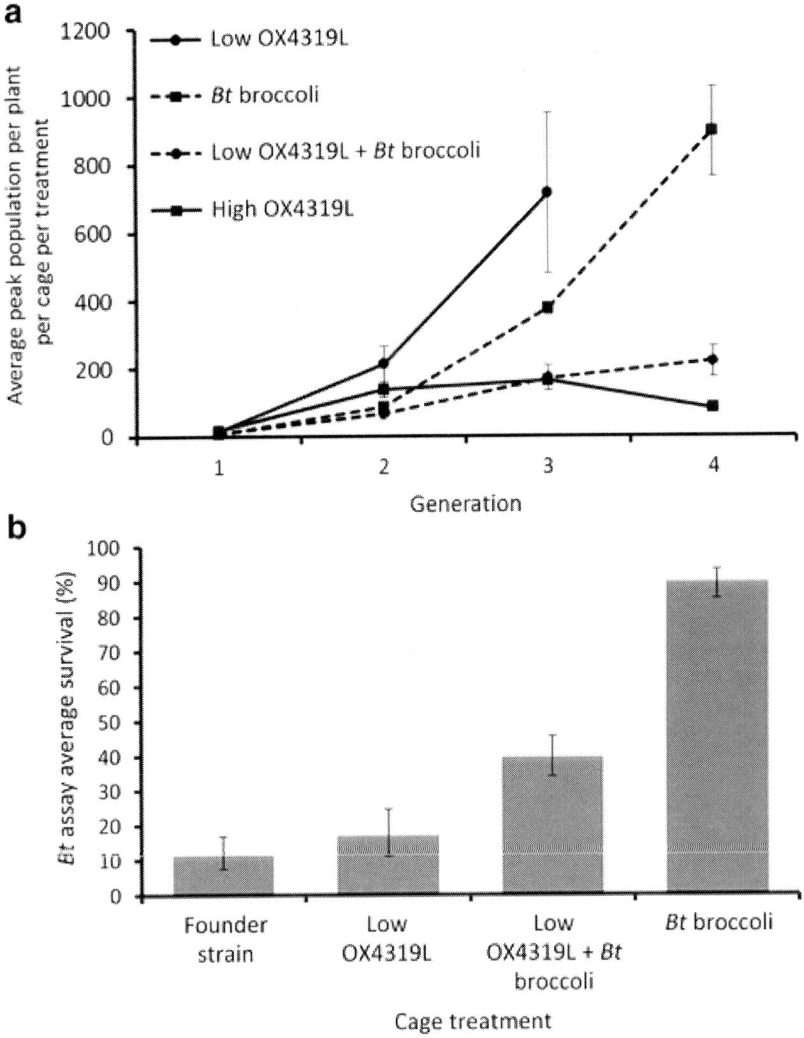

Figure 2. Effects of *Bt* broccoli and OX4319L releases on caged *Plutella xylostella* populations over multiple generations. Caged populations were established with hybrid *Bt*-resistant/wild-type insects – the founder strain – with a low estimated frequency of homozygous-resistant individuals. **A** Graph shows the mean peak population size per plant, per generation, in four experimental treatments over the experimental period: Treatment 1, *Bt* broccoli, no OX4319L releases; Treatment 2, *Bt* broccoli, low-rate weekly OX4319L releases (release rate of 3:1 in Generation 1,

increased to 5:1 in subsequent generations); Treatment 3, non-*Bt* broccoli, low-rate weekly OX4319L releases (identical release rates to Treatment 2); and Treatment 4, non-*Bt* broccoli, high-rate weekly OX4319L releases (release rate of 20:1 in Generation 1, increased to 40:1 in subsequent generations). Means were calculated from three experimental cage replicates, with the exception of Treatments 2 and 4, which were reduced to two and one cage replicates in Generation 3, respectively. Treatment 3 cages were terminated in Generation 3 as the insect populations had reached maximum capacity. Error bars represent standard error of the mean. **B** *Bt* survival assays. Mean survival of third-instar larvae from three experimental cage treatments and the founder strain used to begin these experimental treatments when exposed to a discriminating dose of *Bt* in artificial diet assays. *Bt* dose in this assay is high enough to ensure that only homozygous *Bt*-resistant individuals will survive (as in the high-dose/refuge strategy). This assay therefore indicates the proportion of each population remaining *Bt*-resistant (homozygous) and *Bt*-susceptible (heterozygous or homozygous-susceptible). For each cage, two *Bt* assays and one no-*Bt* control assay were performed. *Bt* assays in each cage were summed and means represent averages of each set of treatment cages corrected for control mortality. The assays took place using individuals from the final generation in which each treatment was run or, in the case of the founder strain, in the generation prior to the start of the experiment. Survival was corrected for control mortality prior to analysis and error bars represent Pearson's exact confidence intervals. Survival of insects from the low OX4319L release cages (Treatment 2) was not significantly different from the founder strain; other pairwise comparisons are significantly different (Table 1)

Larval *Bt* survival assays conducted in the last generation of each treatment, and on the founder strain in Generation 0, showed a significantly reduced proportion of resistant individuals in treatments that combined *Bt* broccoli and low-rate OX4319L releases compared to those where *Bt* broccoli was used alone (Figure 2b). Survival rates on *Bt* were low and not significantly different between the founder strain (11.5 %; CI, 7.4–16.8 %) and the low-rate OX4319L-only caged populations (17 %; CI, 11–24.5 %; Table 1). Both treatments in which larvae were exposed to *Bt* selection (Treatments 1 and 2) displayed significantly higher levels of survival on *Bt* compared to those where non-*Bt* broccoli plants were used, indicating high levels of selection against susceptible genotypes in these cages. However, significant differences between treatments with *Bt* plants were apparent. In the *Bt* broccoli-only treatment, *Bt* resistance rapidly increased in frequency in the population, as indicated by the high percentage survival rate (89.7 %; CI, 85–93.3 %). In the combined OX4319L and *Bt* broccoli treatment, the survival rate was significantly lower (39.5 %; CI, 33.9–45.5 %), implying a significantly higher frequency of *Bt*-susceptible alleles (and thus individuals subject to *Bt* selection pressure) in the populations where low-level releases of OX4319L males had taken place.

Table 1. Pairwise comparisons of survival assay data (from experiments described in Figure 2b)

Comparison		*Z* value	*P* value
Founder strain	Low OX4319L	1.212	0.605
Founder strain	Low OX4319L + *Bt* broccoli	5.016	<0.001
Founder strain	*Bt* broccoli	8.557	<0.001
Low OX4319L	Low OX4319L + *Bt* broccoli	3.323	0.0048
Low OX4319L	*Bt* broccoli	−6.682	<0.001
Low OX4319L + *Bt* broccoli	*Bt* broccoli	−5.594	<0.001

Output was generated using an omnibus logit model for categorical data analysis, followed by post-hoc analysis using Generalized Linear Hypothesis Testing

Output was generated using an omnibus logit model for categorical data analysis, followed by post-hoc analysis using Generalized Linear Hypothesis Testing

Allele frequencies derived from these mean survival assay results indicated that the *Bt* resistance allele had increased in frequency over the experimental period in populations under *Bt* selection, from 0.36 in Generation 0 (founder strain) to an estimated 1.0 in the *Bt* broccoli treatment and 0.71 in the combined OX4319L and *Bt* broccoli treatment (both Generation 4). Fluorescence proportions determined for treatments in which OX4319L males were introduced estimated that 56.0 % (CI, 50.4–61.5 %) of the combined OX4319L and *Bt* broccoli treatment populations (Generation 4); 55.6 % (CI, 47.2–63.7 %) of the low OX4319L treatment populations (Generation 3); and 87.5 % of the high release rate OX4319L treatment population (Generation 3 – single replicate) carried the MS transgene. As in the population suppression experiment, these fluorescence proportions are equal to twice the MS transgene allele frequency in these populations [48].

DISCUSSION

We found that sustained releases of male *P. xylostella* moths carrying a transgene which allows only males to survive to adulthood can have a direct suppressing effect on a target pest population, as previously shown for dipteran pests [18, 20, 27]. We also demonstrate that releases of such insects can significantly delay the spread of resistance to *Bt* in target pest populations by introgression of susceptibility alleles through male progeny, even when initial resistance levels are high. With the increasing number of cases of field resistance to *Bt* plants, transgenic systems such as the one examined here may play an important role in IRM programs aimed at delaying or overcoming the evolution of resistance to this valuable technology.

In stable target populations reared on artificial diet and an initial OX4319L over-flooding ratio of approximately 10:1, reduced numbers of females and suppression of the populations' reproductive capacity became evident approximately two generations after OX4319L male releases were initiated and these populations fell rapidly to extinction thereafter (after approximately three generations). This timeframe until extinction compares well with previous cage experiments exploring the pest suppression potential of female-lethal transgenes in *A. aegypti* (4.66 ± 0.88 mean generations until extinction; 8.5–10:1 over-flooding rate) [27], *B. oleae* (3 ± 0.0; 10:1) [18], and *C. capitata* (3 ± 0.0; 10:1) [20]. With such MS strains applied against isolated populations, as reported here, local eradication of the target pest is achievable as the over-flooding ratio increases as the target population falls. This effect would be lessened when applied in the field against non-isolated pest populations or when pests migrate between areas. However, even though *P. xylostella* is known to travel large distances in some circumstances [49, 50, 51, 52], it does not typically travel far when host plants are available [53, 54]. Although complete isolation of *P. xylostella* and other pest species would be unlikely in most scenarios, efficient control of target populations is likely achievable with some degree of population isolation and/or with sufficient scale of application to reduce edge effects. Moreover, the SIT has previously demonstrated efficient control of other lepidopteran pests – codling moth (*Cydia pomonella*) and pink bollworm (*Pectinophora gossypiella*) – that are able to travel relatively large distances [55, 56].

In our experiments on broccoli, target populations were free to expand in protected environments devoid of limiting factors such as predators and adverse climatic effects. Here, OX4319L male release, with a high over-flooding ratio of up to 40:1, maintained a high level of control. As expected, low rates of OX4319L male releases alone – over-flooding ratio of up to 5:1 – did not prevent rapid population growth under these conditions. Whether a management strategy is able to prevent the expansion of a pest population is a function of that strategy's ability to counteract the reproductive rate (R_0) of the target species [57]. In the case of the lower rate OX4319L treatment, the proportion of matings won by these released insects was predicted to be low enough that the population's reproductive capacity would remain well above 1. Pests with a high R_0, such as Lepidoptera, can represent a challenge to mating-based pest management approaches, and in some cases higher release rates may be required to counter this. However, in a controlled greenhouse setting, R_0 is artificially elevated, while in the field, mortality rates due to parasitism, predation, pathogen infection and environmental factors, especially in pre-adult stages, are likely to be high [40]; it is possible that these factors would reduce the release rate of OX4319L males required to control a target population, relative to release rates used in these experiments.

Populations maintained on *Bt* broccoli alone expanded rapidly after strong selection for *Bt* resistance. However, low rates of OX4319L male release in combination with *Bt* broccoli significantly reduced population growth, controlling the target populations well when each treatment on its own failed to do so. This apparent synergistic effect supports model predictions that introgression of susceptibility alleles into a target population by released transgenic males could prevent or reverse the spread of resistance, thereby preserving susceptibility to *Bt* toxins [29, 30].

Consistent with the hypothesis underlying these models, results from our larval *Bt* survival assays showed significantly higher survival (frequencies of resistant homozygotes and, consequently, resistance alleles) in the *Bt* broccoli-only treatment compared with cages where low levels of OX4319L males were introduced in addition to *Bt* plants. Mean fluorescent protein marker frequencies in these combined OX4319L + *Bt* broccoli cages (56.0 %) displayed a high level of agreement with the proportions of these populations which were *Bt*-susceptible, once survival assay results were calibrated for homozygous-resistant mortality (100–43.7 = 56.3 %). These results are what would be expected if mating by released MS males were the source of *Bt*-susceptibility alleles in these populations, diluting the frequency of resistant (homozygous) genotypes and counteracting *Bt* resistance at a rate predicted by their mating success. Under this hypothesis, no such agreement would be expected if resistance alleles were not undergoing positive selection in the target population, as pre-existing susceptibility alleles would persist independent of those introgressed by MS males. Supporting this argument, populations treated with low-dose OX4319L releases, which received identical release rates of transgenic males to the low OX4319L + *Bt* broccoli treatment (resulting in a very similar population-level fluorescence ratio, 55.6 %) but were not exposed to *Bt* selection, showed a level of *Bt* susceptibility independent of their fluorescence proportions (mean proportion of individuals not surviving *Bt* assay = 100–18.4 = 81.6 %). Taken together, these results provide an empirical demonstration of the mechanism by which MS transgenic insects carrying insecticide-susceptible genetic backgrounds can maintain insecticide efficacy, even in the face of continuous, strong selection for resistance.

These resistance management effects were explored using a well-established model system utilized for investigating IRM strategies for *Bt* crops and risks to non-target organisms [58, 59, 60]. Thus, although *Bt* crucifers are not yet commercially cultivated, the mechanism we demonstrate is broadly relevant to *Bt* crops. In addition, depending on the nature and inheritance of resistance, similar effects would be expected for *Bt* sprays or indeed any other insecticidal intervention against *P. xylostella* and other pests. The efficiency of a transgenic strain in providing these effects will likely depend on performance of its males in the field. Previous studies on OX4319L indicate that this transgene insertion is associated with a small fitness cost in the laboratory [48], but its field performance has not yet been tested.

Theoretical population modeling studies have investigated the release of fertile non-transgenic insecticide-susceptible insects into *Bt* crop areas [29]. At present, the only feasible option for achieving sex separation in lepidopterans, at the scales required for release, is the MS system described herein, so in practice non-transgenic releases would comprise both males and females. In these models, introgression of susceptibility alleles from such releases also provided a degree of resistance management. However, released fertile females would lead to higher pest populations than MS releases, at least initially, and may necessitate increased insecticide usage in neighboring conventionally sprayed areas. For these reasons, an IRM program involving releases of fertile pest females would likely be unacceptable to growers and regulators.

The release of radiation-sterilized *P. gossypiella* has also been described as preventing development of resistance to *Bt*-expressing cotton [61]. However, these two systems (SIT and MS) provide fundamentally different ways of managing resistance. Whereas SIT may prevent increasing resistance by killing resistant insects before they have a chance to reproduce (any alternative population-suppression method would have the same effect), the MS phenotype additionally manages resistance by actively introgressing susceptibility alleles into the target population through the survival of male progeny. In areas of 100 % *Bt* crop cultivation these effects on resistance management may be similar, albeit with these benefits likely to be provided at lower release rates using MS males [22, 23]. However, if combined into existing *Bt*-IRM programs, MS releases may also be advantageous to the maintenance of susceptible populations within refuges as male MS heterozygotes within these areas will survive and continue to pass on susceptibility alleles in subsequent generations. Although some suppression of these refuge areas is expected under these conditions, our results demonstrate that, in the absence of other control measures such as *Bt*, low-level releases of MS males will have a minimal effect on population growth. Moreover, this mechanism may be of benefit to growers utilizing sprayed insecticides as introgressed susceptibility alleles will be spread through the natural mating behavior of the pest population between insecticide sprays. Future modeling studies may help to elucidate the relative benefits of these two potential alternatives to refuges in combating resistance to *Bt*.

CONCLUSIONS

The results described herein demonstrate that releases of MS transgene-carrying, insecticide-susceptible insects can provide two simultaneous effects on pest populations – direct population suppression through death of female progeny and IRM by introgression of susceptibility alleles into target populations through surviving male progeny. Further laboratory experiments are required in order to disentangle the relative contributions of these two effects in suppressing pest populations on *Bt* plants. Nonetheless, our findings strengthen the argument that MS systems, such as OX4319L, could function as effective stand-alone tools against given pest species or as a highly compatible component of an integrated pest management program. If used as an alternative to broad-spectrum insecticides within such an IPM framework, the species-specific action of MS transgenes would reduce negative effects on non-target organisms including natural enemies which can further delay the evolution of resistance to *Bt* crops [60, 62, 63]. For successful IPM programs against *P. xylostella*, conservation of these natural enemies is important [40] and can provide more effective control than the use of broad-spectrum insecticidal sprays [64]. These conservation benefits are also likely to apply to other groups such as pollinators and birds which are currently threatened by the use of some agricultural insecticides [65].

Existing recommendations on minimum refuge size in some *Bt* crops, and compliance with these recommendations, may be inadequate for robust resistance management [32, 35]. MS insects carrying *Bt*-susceptible genetic backgrounds could help to reduce refuge requirements, or potentially replace or supplement refugia as a resistance management tool in *Bt* crops [29, 30]. Area-wide releases of transgenic MS insects in place of refugia could offer area-wide protection to growers and seed producers, for whom such compliance can be expensive and impractical. In addition, our results demonstrate that use of MS systems in combination with other pest management methods would reduce the required number of insects to be produced and released to achieve a given level of control.

The utility of the MS system has been demonstrated here in a lepidopteran pest and elsewhere in dipteran pests [15, 16, 18, 20, 66], so further development in pests in these and other taxonomic groups appears feasible and should be evaluated. This technology represents a promising and sustainable pest management tool offering significant benefits to future IPM strategies.

METHODS

A summary of the two experimental designs used is provided in Figure 3. More detailed explanations of these designs and their aims are provided below.

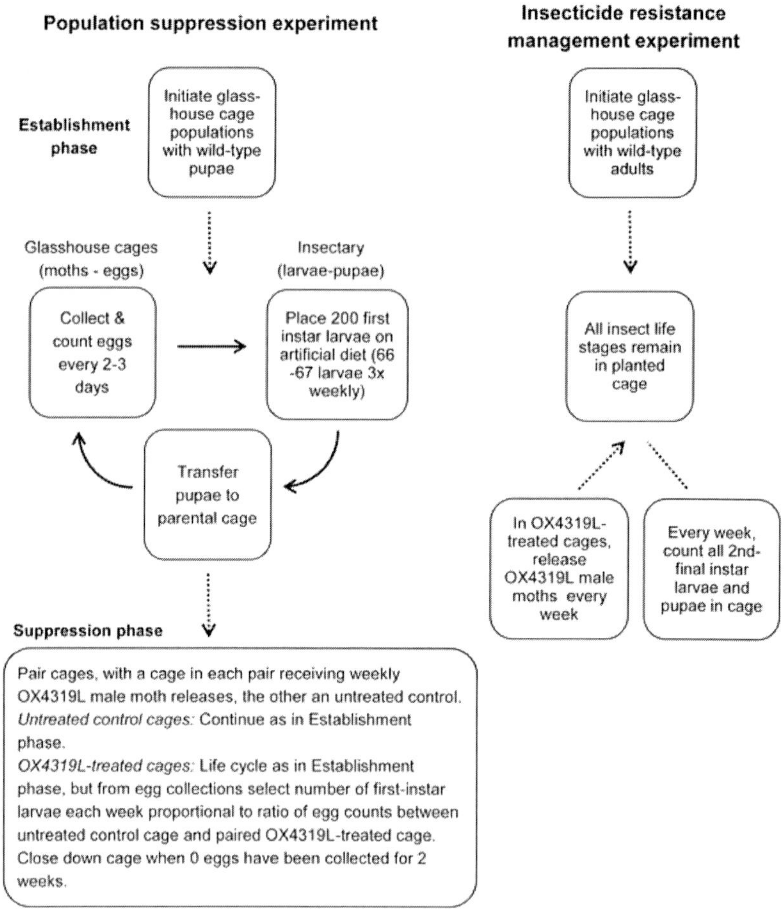

Figure 3. Schematic showing design of the population suppression and insecticide resistance management experiments

Summary of experimental designs

The glasshouse cage population suppression experiment was designed to simulate a field scenario in which biotic and abiotic mortality factors maintain a stable pest population, with adult moths mating and laying eggs in glasshouse cages, and larval rearing being conducted in a separate temperature-controlled room [18, 20, 27]. When weekly OX4319L male moth releases started, OX4319L-treated cages were paired with an untreated counterpart (control cage) and the number of pupae subsequently reintroduced into each treatment cage was made proportional to the ratio of eggs counted between the treatment cage and its paired control cage: this allowed the expected reduced number of females in OX4319L-treated cages to be reflected in reduced numbers of pupae entering the population in the next generation.

In contrast, the subsequent glasshouse cage IRM experiment was conducted on broccoli plants – some on *Bt* broccoli – as they offered a realistic model for many agricultural systems with transgenic crops. Artificially maintaining stable populations on plants, by accurately manipulating the number of progeny surviving, was deemed impractical, and the populations were therefore allowed to expand freely in the near-absence of biotic and abiotic mortality factors, with all life stages of each experimental population residing in their respective cage. This second experiment provided a more rigorous test of the pest management potential of the OX4319L strain as the intrinsic growth rate of target populations was not artificially controlled and competitors from these target populations had the advantage of being reared on host plant material, as opposed to an artificial diet medium.

Experimental population structures

Generations in the population suppression experiment were continuous while those in the resistance management experiment were discrete, at least within the experimental period. In the population suppression experiment a continuous generational structure was achieved by introducing pupae into cages three times per week. As the lifecycle of *P. xylostella* in this experiment was 3 weeks, this created a stable, mixed-age adult population with eggs being collected every 2–3 days over the experimental period. In the resistance management experiment conducted on broccoli, populations were founded by a single introduction of adult moths. Adults in this founding generation, and those in subsequent generations, were allowed to mate freely in their respective cages with females from the same generation ovipositing over a number of days. As such, within the experimental period generations remained discrete and were discernable by peaks in weekly population counts which, as expected, reached a maximum level once per generation (approximately every 3 weeks; see below). Generations did, however, begin to overlap over time.

The generation time of *P. xylostella* in the insecticide resistance experiment was approximately 20–23 days (incrementally: egg, 3–4 days; larvae, 13–14 days; pupae, 3–4 days; adult time to sexual maturity, 1 day). As experiments were conducted in a glasshouse, these time periods varied depending on outdoor weather conditions.

Population suppression experiment

Experiments were conducted in quarantine facilities at Rothamsted Research, Hertfordshire, UK in accordance with legislation concerning the contained use of GM organisms in the UK. Larval rearing took place in a temperature-controlled room and experimental cages were housed within a temperature-controlled glasshouse (both 25 °C and 16:8 light:dark photoperiod). As an efficient, easily replicable proxy for host crop infestation, larval rearing was conducted on artificial diet medium and followed the methods of Martins et al. [67]. All experimental populations were reared on non-tetracycline diet. Transgene-homozygous OX4319L males were produced on diet with chlortetracycline hydrochloride (100 μg/mL) and sexed manually prior to introduction. Experimental cages measured 120 × 100 × 120 cm (width × depth × height), with a zipped entrance at the front. Insects were added to the cages as pupae.

The experiment comprised two phases: establishment and suppression. During establishment, stable mixed-age populations of wild-type *P. xylostella* (wild-type background strain to OX4319L isolated in Vero Beach, Florida, USA) were established in each of the four cages. During suppression, weekly introductions of transgene-homozygous OX4319L males were made into two of the cages to investigate whether engineered female-specific lethality resulted in suppression of these populations. Throughout the experiment, cabbage-extract-baited Parafilm (Bemis Company Inc., Oshkosh WI, USA) pieces were hung from the roof of the cage to act as an artificial leaf oviposition substrate. These were replaced three times per week (Monday, Wednesday, and Friday) and eggs collected on each sheet were counted. During each egg collection, dead adult moths and uneclosed pupae were also collected from the cages, sexed, and counted. Eclosed adults were provided with sugar water-saturated cotton wool reservoirs, changed every 2–3 days.

Establishment phase

Cage populations were initiated by placing 200 unsexed wild-type pupae into each cage. Stable non-expanding populations were maintained in each cage to mimic the stabilizing effects of predation and other limiting factors in the wild. This was achieved by introducing a constant number of pupae back into the cages each week. A total of 200 first-instar larvae were selected each week to carry on the population, taken from the three weekly collections (67 larvae chosen from Monday collection, 67 from Wednesday, and 66 from Friday). These larvae were reared in plastic beakers on non-tetracycline diet, and after pupation were sexed and transferred back into the cage from which they had been collected as eggs, 2 weeks earlier. Tri-weekly egg collections (and subsequent tri-weekly pupal reintroductions) maintained stable, mixed-age populations as might be expected in the field, where adult moths would be continuously entering the population. The first two introductions of wild-type moths into each cage (weeks 1 and 2) originated from an independent laboratory colony. From this point onwards each cage population was self-sustaining.

Suppression phase

Once egg counts had stabilized (indicating stable populations) in week 9, two of the four cages were chosen as OX4319L treatment cages (Cages 2 and 4) and two as control cages (Cages 1 and 3). Cages were designated in a blocked design to minimize bias caused by uncontrolled abiotic factors. In non-treatment control cages, the protocol from the establishment phase was continued. Each OX4319L treatment cage was randomly paired with a control cage for the remainder of the experiment (Cages 1 and 2; Cages 3 and 4). The reintroduction of pupae into each OX4319L treatment cage followed the same protocol as for the establishment phase; however, the total number of early-instar larvae selected for rearing that week was made proportional to the ratio between the number of eggs collected for that treatment cage and its paired control cage for the week when these larvae were collected as eggs. This method ensured that the population suppression effect of female death of transgenic larvae, reflected later by reduced number of eggs collected, resulted in reduced numbers of pupae re-entering the OX4319L-treated cages relative to an untreated population.

After cages had been paired, weekly introductions of homozygous OX4319L males into the treatment cages began. The target over-flooding ratio (OX4319L to wild-type males entering the population) was set at 10:1. This release rate is in line with previous studies investigating the effect of female-lethal transgenes on caged populations [18, 20, 27] and much lower than the sterile:wild ratios that successful SIT programs have aimed to achieve against other Lepidoptera: *P. gossypiella* (60:1) [68]; *C. pomonella* (40:1) [69]; and painted apple moth, *Orgyia anartoides* (100:1) [70, 71]. This over-flooding ratio was calculated as 10× the mean male recruitment rate for the 3 weeks preceding OX4319L male introduction, and the numbers released were held constant for the remainder of the experiment (Cage 2: 990 OX4319L males; Cage 4: 980 OX4319L males). Males were released as pupae into the cages once per week (Wednesday).

Pupae reintroduced into treatment cages were screened for the DsRed2 fluorescent protein transformation marker and fluorescence proportions (proportion of the population which carried the transgene) recorded. Under the restrictive conditions of this experiment and the highly penetrant female-lethal phenotype of OX4319L [15], population-level transgene allele frequencies were equal to half the fluorescence proportion [48]. Each treatment cage's end-point – extinction of each population – was pre-defined as two consecutive weeks of zero eggs collected.

Insecticide resistance management (IRM) experiment

Experiments were conducted at Cornell University, New York State Agricultural Experiment Station, Geneva, NY, in accordance with legislation concerning the contained use of GM organisms in the USA. All *P. xylostella* experimental populations in this experiment were reared on broccoli plants and allowed to expand freely. Eleven cages, each 1.83 m × 0.91 m × 1.83 m (length × width × height), were placed in a temperature-controlled glasshouse with supplementary lights (23–27 °C, 16:8h light:dark photoperiod, and uncontrolled relative humidity). The 11 cages were assigned to four treatments: Treatment 1, *Bt* broccoli plants, no OX4319L release; Treatment 2, *Bt* broccoli plants, low-rate weekly OX4319L releases; Treatment 3, wild-type plants, low-rate weekly OX4319L releases; and Treatment 4, wild-type plants, high-rate weekly OX4319L releases. Treatments 1–3 were assigned three cages (replicates) while Treatment 4 was assigned two cages.

Experimental *Plutella xylostella* strains

Two strains were used: a hybrid *Bt*-resistant/wild-type strain used to generate starting populations in the cages (founder strain) and OX4319L. To generate the founder strain, 25 males from a homozygous Cry1Ac *Bt* toxin-resistant stock colony (showing recessive, monogenic resistance to Cry1Ac toxins) [58] were crossed to 25 females from the homozygous-susceptible 'Geneva 88' colony; 59 of the F1 male progeny from this cross were then mated to 100 females from the Geneva 88 strain. A total of 250 males and 250 females from this cross were then mated to produce the founder strain. These crosses provide an expected resistance allele frequency of 0.25 and an expected homozygous-resistant frequency of 0.0625. At all stages of founder strain production and maintenance, progeny were reared on wild-type broccoli plants in large numbers (>500 pupae collected per generation) to minimize the effects of inbreeding and genetic drift. However, *Bt* survival assay results (Figure 2b) suggest that this frequency had increased to 0.36 ($r = \sqrt{0.1265}$) by the start of the experiment.

Experimental broccoli cultivars

Two strains of broccoli (*Brassica oleracea* L.) plants were used: a wild-type cultivar (Packman) and a transgenic strain engineered to express high levels of the *Bt* toxin Cry1-Ac [41]. Together, this transgenic plant cultivar and the Cry1-Ac-resistant *P. xylostella* strain comprise a well-established model system used to study the dynamics of transgenic crops and their resistance management [58, 59]. Cry1-Ac toxin production was verified by screening the plants (4–5 weeks old) with susceptible Geneva 88 strain neonates. Leaf assays of transgenic plants also killed 100 % of OX4319L/Cry1-Ac hybrids, indicating high levels of *Bt* toxin expression.

Restocking of plant material

All cages started with 20 broccoli plants of their respective cultivar. Plants were replaced after 4 weeks, or when defoliated due to larval feeding, by cutting them at their base and placing them on the replacement plants to allow larvae to transfer. If moth populations grew beyond the capacity of the maximum food supply (estimated by exceeding their cage's plant material within one generation), the cage was terminated.

Establishment of *P. xylostella* populations in cages

In all treatments, replicates were initiated by the release of *P. xylostella* adults from the founder strain into the cages. In treatments involving *Bt* plants, 200 randomly selected adults were released. In treatments involving non-*Bt* plants, seven males and females (total 14 adults) were released into the cages. Due to *Bt* selection, this gave approximately equal starting population densities in Generation 1 in all cages.

Male-selecting transgenic insect treatments

OX4319L-homozygous males for release into the cages were produced by rearing egg collections from a stock colony in the absence of tetracycline in larval feed. Releases in this experiment were proportional: daily estimates of adult male recruitment for each cage were used to calculate a daily male release number for that cage, dependent on the release rate pre-determined for that treatment. A proportional release rate, rather than a constant number, was applied to reduce the likelihood of population extinction and thereby allow exploration of the effect of MS transgene releases on population dynamics and resistance allele frequency.

Release rates were selected in advance based on the outcomes of predictive deterministic models, investigating the effects of each experimental treatment on population size.

In Treatments 2 and 3 (low-OX4319L release treatments), the intended over-flooding ratio was 5:1 (transgenic: wild-type males). However, due to insect rearing limitations this was limited to 3:1 in Generation 1, and increased to 5:1 thereafter. These low release rates, in combination with *Bt* plants, were predicted to maintain a relatively constant population size, but insufficient to cause population suppression when used alone. Similarly, the over-flooding rates in cages with high-OX4319L releases, initially 20:1 and increased to 40:1 in Generation 2 as production capacity increased, were predicted to be sufficient for population suppression when used alone.

Where small populations are expanding in the absence of limiting biotic or abiotic factors, stochastic effects make it difficult to accurately predict the rate of population increase. In response to higher-than-predicted population growth and limited plants available, the cages with *Bt* broccoli plants and low-rate weekly OX4319L releases were reduced to two replicates in Generation 3 (terminated cage selected at random). In Generation 4, the number of plants in each treatment with *Bt* broccoli only (no OX4319L releases) cage was reduced to five (while maintaining per-plant population density levels) in response to limited plant availability. This was achieved by randomly harvesting 25 % of the leaves on each plant within the cage, removing all insect and plant material from the cage, and then restocking the cage with five new plants and returning harvested leaves. As generations were still discrete, this manipulation was timed during a period when the vast majority of insects in the population were present as larvae or pupae on plants, allowing accurate culling of the population. In Generation 3, requirements for OX4319L male moths exceeded production capacity, so the treatment with high-rate weekly OX4319L releases was reduced to one replicate only.

Data collection

Population size estimation

The numbers of second to fourth instar larvae and pupae on each plant in each cage were counted once per week.

Bt resistance assays

With the exception of cages with high-rate OX4319L releases, *Bt* resistance data were collected from all cages in their final generation. As each cage reached maximum egg-laying potential (judged by female recruitment data in each population, collected from eclosion cages), eggs were collected from 8–10 leaves selected at random from each cage. These eggs were transferred to filter paper by paintbrush and resulting larvae were reared on chlortetracycline hydrochloride-augmented artificial diet (100 μg/mL). For each cage, two replicated *Bt* survival assays and one control assay were performed. For *Bt* assays, chlortetracycline hydrochloride-augmented diet was poured into 30 mL plastic pots and 500 μL of 10 ppm *Bt* (Dipel®, Valent BioSciences Corp., Libertyville, IL) – shown previously to discriminate between homozygous-resistant and other genotypes [46] – pipetted onto the dried surface. Non-*Bt* controls were prepared in the same way, with no added *Bt* solution. At third instar, larvae were transferred onto the air-dried diet surface. Exact numbers of larvae per replicate differed between cages according to availability of eggs. A minimum of 33 larvae per pot were used for the first *Bt* replicate; 11 per pot for the second *Bt* replicate; and all control replicates contained 20 larvae. Mortality was assessed 72 h later with surviving larvae taken to be homozygous for the resistance allele. For comparison, the founder strain was subjected to the same assay, in the generation prior to experimental initiation.

Bt resistance allele frequency estimation

Bt assay survival data from the founder strain, *Bt* broccoli alone, and low OX4319L + *Bt* broccoli treatments were used to calculate frequency of the *Bt* resistance allele, as conducted in previous studies utilizing this model system [46, 47, 58]. Under strong *Bt* selection (preliminary assays showed 100 % *Bt* susceptibility in homozygous susceptible and OX4319L/resistant heterozygotes) in the absence of resistance management efforts (as in the *Bt* broccoli-only treatment) it was expected that the *Bt* resistance allele would reach fixation from Generation 1 onwards (all individuals homozygous resistant). Deviation from 100 % *Bt* assay survival in this treatment thus represents mortality to a small proportion of homozygous-resistant individuals under these conditions. This reduction in survival was used to calibrate the results of the assay for other treatments prior to calculation of allele frequencies. Once calibrated, the square of the proportion surviving gave an estimate of the *Bt* allele frequency in the founder strain (under Hardy-Weinberg equilibrium). In the low OX4319L + *Bt* broccoli treatment, the calibrated survival rate represents the proportion of homozygous resistant individuals in the population, with the remainder of (non-surviving) individuals assumed to be the (heterozygous) offspring of resistant females and OX4319L males (as the only source of susceptibility alleles in these populations). The validity of this assumption was explored by comparing the percentage of non-surviving individuals with DsRed2 fluorescence proportions. If the mechanism of resistance management proposed in this experiment were functioning as predicted (released MS males providing susceptibility alleles through mating, maintaining *Bt* efficacy), we would expect to see high levels of agreement between population-level *Bt* susceptibility and fluorescence proportions in the low OX4319L + *Bt* broccoli treatment (where OX4319L males were hypothesized to be the sole providers of susceptibility alleles) but no such agreement in the low OX4319L only treatment populations, which received identical doses of MS transgenic males (and thus would be expected to show similar fluorescence proportions to the combined OX4319L + *Bt* treatment) but where *Bt* resistance alleles were not under selection. The *Bt* resistance allele frequency in the low OX4319L + *Bt* broccoli treatment was calculated as [X + 0.5(1–X)], where X is the calibrated survival rate.

Fluorescence proportion assays

The population-level fluorescence proportion (proportion of the population carrying the transgene) was estimated for all cages into which OX4319L males were released, in the final generations they were conducted. The exception to this was the high OX4319L release treatment where this assay was conducted in the penultimate (third) generation due to the small population size of this treatment cage in Generation 4. Collection of eggs followed procedures described for *Bt* resistance assays with larvae being reared to pupation on chlortetracycline hydrochloride-augmented artificial diet (100 μg/mL) in 500 mL Styrofoam pots. At pupation, individuals were screened for presence/absence of the DsRed2 fluorescent protein marker. Each cage acted as a replicate with a minimum of 75 larvae per assay pot.

Male recruitment rate estimation

For those cages requiring OX4319L male introductions, daily male recruitment rate was assessed by the removal of four randomly selected plants/cage/week. These were placed into smaller cages in an adjacent glasshouse maintained in the same environmental conditions. The number and sex ratio of the adults eclosing from these plants were recorded daily, and eclosed adults were then returned to their respective experimental cages. From this data an estimate of the eclosion rates in the main cage was made, from which the number of OX4319L males required to achieve the respective over-flooding ratio for each cage was calculated.

Statistical analyses

For comparisons between treatment peak population densities in Generation 3, ANOVA was used, followed by Tukey's HSD for pair-wise comparisons. For Generation 4, a *t*-test was used as only two treatments were compared. For *Bt* assay results, results from the two *Bt* assays within each cage were summed to avoid pseudo-replication and then corrected for control mortality using a Henderson-Tilton correction. For comparisons between treatments, a Logit model for Categorical Data Analysis was used followed by pair-wise comparisons using Generalized Linear Hypothesis Testing with each cage acting as a replicate. Error estimates for mean proportions were calculated using Pearson's exact confidence intervals. Data were analyzed using R (Version 2.15.0).

DECLARATIONS

Acknowledgments

The authors thank Debs Granville and Nerys Humphreys-Jones for their invaluable assistance with insect production for the UK experiments and Hongfei Gong for his advice on experimental population modeling. Rothamsted Research receives grant-aided support from the Biotechnology and Biological Sciences Research Council (BBSRC), UK. The work at Rothamsted was carried out under Plant Health License 112593/209885 issued by DEFRA, the UK Plant Health Service. The work at Cornell University was conducted under the requirements of the Code of Federal Regulations (7 CFR 340) and insects were imported from the UK under USDA APHIS BRS permit 12-227-102m. *Bt* broccoli plants used in experiments were developed in ED Earle's program at Cornell University. NA was supported by the BBSRC (grant number BB/H01814X/1). TH-S has been supported by a BBSRC Industrial CASE PhD studentship and thanks his co-supervisor, Dr Sebastian Shimeld, for assistance and advice.

Competing interests

TH-S, NM, AW, TM, KG, SW, and LA are, or were, employees or sponsored students of Oxitec Ltd., which therefore provided salary or stipend and other support for the research program. Also, all these employees have shares or share options in Oxitec Ltd. NA has an indirect interest in shares in the company Oxitec Ltd. Both Oxitec Ltd and the University of Oxford have one or more patents or patent applications related to the subject of this manuscript.

AUTHORS' CONTRIBUTIONS

All authors were involved in experimental design. TH-S, NM AW, TM, JY, and HC performed experiments. TH-S analyzed results. TH-S, NM, NA, SW, TS, and LA wrote the paper. All authors read and approved the final manuscript.

REFERENCES

1. Lounibos LP. Invasions by insect vectors of human disease. Annu Rev Entomol. 2002;47:233–66.
2. Kenis M, Auger-Rozenberg M-A, Roques A, Timms L, Péré C, Cock M, et al. Ecological effects of invasive alien insects. J Biol Inv. 2009;11:21–45.
3. Oerke E-C. Crop losses to pests. J Agric Sci. 2006;144:31–43.
4. Knipling E. Possibilities of insect control or eradication through the use of sexually sterile males. J Econ Entomol. 1955;48:459–69.
5. Dyck VA, Hendrichs J, Robinson AS, editors. Sterile insect technique: principles and practice in area-wide integrated pest management. The Netherlands: Springer; 2005.
6. Bakri A, Mehta K, Lance DR. Sterilizing insects with ionizing radiation. In: Dyck VA, Hendrichs J, Robinson AS, editors. Sterile insect technique principles and practice in area-wide integrated pest management. The Netherlands: Springer; 2005. p. 233–68.
7. Rull J, Brunel O, Mendez M. Mass rearing history negatively affects mating success of male Anastrepha ludens (Diptera: Tephritidae) reared for sterile insect technique programs. J Econ Entomol. 2005;98:1510–6.
8. Davich T, Lindquist D. Exploratory studies on gamma radiation for the sterilization of the boll weevil. J Econ Entomol. 1962;55:164–7.
9. Sharma V, Razdan R, Ansari M. Anopheles stephensi: effect of gamma-radiation and chemosterilants on the fertility and fitness of males for sterile male releases. J Econ Entomol. 1978;71:449–52.
10. Helinski M, Knols B. Mating competitiveness of male Anopheles arabiensis mosquitoes irradiated with a semi- or fully-sterilizing dose in small and large laboratory cages. J Med Entomol. 2008;45:698–705.

11. Hagler J, Jackson C. Methods for marking insects: current techniques and future prospects. Annu Rev Entomol. 2001;46:511–43.
12. Hagler JR, Miller E. An alternative to conventional insect marking techniques: detection of a protein mark of pink bollworm by ELISA. Entomol Exp et Appl. 2002;103:1–9.
13. Robinson AS, Hendrichs J. Prospects for the future development and application of the sterile insect technique. In: Dyck VA, Hendrichs J, Robinson AS, editors. Sterile insect technique principles and practice in area-wide integrated pest management. The Netherlands: Springer; 2005. p. 727–60.
14. Thomas DD, Donnelly CA, Wood RJ, Alphey LS. Insect population control using a dominant, repressible, lethal genetic system. Science. 2000;287:2474–6.
15. Jin L, Walker AS, Fu G, Harvey-Samuel T, Dafa'alla T, Miles A, et al. Engineered female-specific lethality for control of pest Lepidoptera. ACS Syn Biol. 2013;2:160–6.
16. Fu G, Condon KC, Epton MJ, Gong P, Jin L, Condon GC, et al. Female-specific insect lethality engineered using alternative splicing. Nat Biotechnol. 2007;25:353–7.
17. Fu G, Lees RS, Nimmo D, Aw D, Jin L, Gray P, et al. Female-specific flightless phenotype for mosquito control. Proc Natl Acad Sci U S A. 2010;107:4550–4.
18. Ant T, Koukidou M, Rempoulakis P, Gong HF, Economopoulos A, Vontas J, et al. Control of the olive fruit fly using genetics-enhanced sterile insect technique. BMC Biol. 2012;10:51.
19. Schliekelman P, Gould F. Pest control by the release of insects carrying a female-killing allele on multiple loci. J Econ Entomol. 2000;93:1566–79.
20. Leftwich PT, Koukidou M, Rempoulakis P, Gong HF, Zacharopoulou A, Fu G, et al. Genetic elimination of field-cage populations of Mediterranean fruit flies. Proc Biol Sci. 2014;281(1792).
21. Walters M, Morrison NI, Claus J, Tang G, Phillips CE, Young R, et al. Field longevity of a fluorescent protein marker in an engineered strain of the pink bollworm, Pectinophora gossypiella (Saunders). PLoS One. 2012;7, e38547.
22. Rendón P, McInnis D, Lance D, Stewart J. Comparison of medfly male-only and bisexual releases in large scale field trials. In: Tan K, editor. Area-wide control of fruit flies and other insect pests. Penang: Penerbit Universit Sains Malaysia; 2000. p. 517–25.
23. Rendón P, McInnis D, Lance D, Stewart J. Medfly (Diptera:Tephritidae) genetic sexing: large-scale field comparison of males-only and bisexual sterile fly releases in Guatemala. J Econ Entomol. 2004;97:1547–53.
24. Franz G. Genetic sexing strains in Mediterranean fruit fly, an example for other species amenable to large-scale rearing for the sterile insect technique. In: Dyck VA, Hendrichs J, Robinson AS, editors. Sterile insect technique principles and practice in area-wide integrated pest management. The Netherlands: Springer; 2005. p. 427–51.
25. Cáceres C. Mass rearing of temperature sensitive genetic sexing strains in the Mediterranean fruit fly (Ceratitis capitata). Genetica. 2002;116:107–16.
26. Robinson A, Franz G, Fisher K. Genetic sexing strains in the medfly, Ceratitis capitata: development, mass rearing and field application. Trends Entomol. 1999;2:81–104.
27. de Valdez MRW, Nimmo D, Betz J, Gong HF, James AA, Alphey L, et al. Genetic elimination of dengue vector mosquitoes. Proc Natl Acad Sci U S A. 2011;108:4772–5.
28. Bloem KA, Bloem S, Carpenter JE. Impact of moth suppression/eradication programmes using the sterile insect technique or inherited sterility. In: Dyck VA, Hendrichs J, Robinson AS, editors. Sterile insect technique principles and practice in area-wide integrated pest management. The Netherlands: Springer; 2005. p. 677.
29. Alphey N, Bonsall M, Alphey L. Combining pest control and resistance management: synergy of engineered insects with Bt crops. J Econ Entomol. 2009;102:717–32.
30. Alphey N, Coleman PG, Donnelly CA, Alphey L. Managing insecticide resistance by mass release of engineered insects. J Econ Entomol. 2007;100:1642–9.

31. James C. Global status of commercialized biotech/GM crops. In: ISAAA Briefs, vol. 46. Ithaca: International Service for the Acquisition of Agri-biotech Application; 2013.
32. Tabashnik BE, Brevault T, Carriere Y. Insect resistance to Bt crops: lessons from the first billion acres. Nat Biotechnol. 2013;31:510–21.
33. James C. Global status of commercialized transgenic crops. In: ISAAA Briefs, vol. 49. Ithaca: International Service for the Acquisition of Agri-biotech Applications; 2014.
34. Bates SL, Zhao JZ, Roush RT, Shelton AM. Insect resistance management in GM crops: past, present and future. Nat Biotechnol. 2005;23:57–62.
35. Tabashnik BE, Gould F. Delaying corn rootworm resistance to Bt corn. J Econ Entomol. 2012;105:767–76.
36. Zalucki MP, Shabbir A, Silva R, Adamson D, Shu-Sheng L, Furlong MJ. Estimating the economic cost of one of the world's major insect pests, Plutella xylostella (Lepidoptera: Plutellidae): just how long is a piece of string? J Econ Entomol. 2012;105:1115–29.
37. Tabashnik B, Cushing N, Finson N, Johnson M. Field development of resistance to Bacillus thuringiensis in diamondback moth (Lepidoptera: Plutellidae). J Econ Entomol. 1990;83:1671–6.
38. Shelton A, Robertson J, Tang J, Perez C, Eigenbrode S, Preisler H, et al. Resistance of diamondback moth (Lepidoptera: Plutellidae) to Bacillus thuringiensis subspecies in the field. J Econ Entomol. 1993;86:697–705.
39. Whalon M, Mota-Sanchez D, Hollingworth R. Global pesticide resistance in arthropods. Wallingford: CABI; 2008.
40. Furlong M, Wright D, Dosdall L. Diamondback moth ecology and management: problems, progress, and prospects. Annu Rev Entomol. 2013;58:517–41.
41. Metz TD, Roush RT, Tang JD, Shelton AM, Earle ED. Transgenic broccoli expressing a Bacillus thuringiensis insecticidal crystal protein: Implications for pest resistance management strategies. Mol Breed. 1995;1:309–17.
42. Cao J, Shelton AM, Earle ED. Sequential transformation to pyramid two Bt genes in vegetable Indian mustard (Brassica juncea L.) and its potential for control of diamondback moth larvae. Plant Cell Report. 2008;27:479–87.
43. Cao J, Shelton A, Earle ED. Development of transgenic collards expressing a cry1Ac or cry1C Bt gene for control of the diamondback moth. Crop Protect. 2005;24(9):804–13.
44. Cao J, Zhao J, Tang JD, Shelton A, Earle ED. Broccoli plants with pyramided cry1Ac and cry1C Bt genes control diamondback moths resistant to Cry1A and Cry1C proteins. Theor Appl Genet. 2002;105:258–64.
45. Cao J, Tang JD, Strizhov N, Shelton A, Earle ED. Transgenic broccoli with high levels of Bacillus thuringiensis Cry1C protein control diamondback moth larvae resistant to Cry1A or Cry1C. Mol Breed. 1999;5:131–41.
46. Shelton AM, Tang JD, Roush RT, Metz TD, Earle ED. Field tests on managing resistance to Bt-engineered plants. Nat Biotechnol. 2000;18:339–42.
47. Tang JD, Collins HL, Metz TD, Earle ED, Zhao J, Roush RT, et al. Greenhouse tests on resistance management of Bt transgenic plants using refuge strategies. J Econ Entomol. 2001;94:240–7.
48. Harvey-Samuel T, Ant T, Gong HF, Morrison NI, Alphey L. Population-level effects of fitness costs associated with repressible female-lethal transgene insertions in two pest insects. Evol Appl. 2014;7:597–606.
49. Harcourt D. Population dynamics of the diamondback moth in southern Ontario. In: Talekar N, editor. Diamondback moth and other crucifer pests, First International Workshop: 11–15 March 1985. Tainan: Asian Vegetable Research and Development Centre; 1986. p. 3–16.
50. Smith D, Sears M. Evidence for dispersal of diamondback moth, Plutella xylostella (Lepidoptera: Plutellidae), into southern Ontario. Proc Entomol Soc Ont. 1982;113:21–8.
51. Honda K. Hibernation and migration of the diamondback moth in north Japan. In: Talekar NS, editor. Management of Diamondback Moth and Other Crucifer Pests:

Proceedings of the Second International Workshop. Shanhua: Asian Vegetable Research and Development Center; 1992. p. 43–50.

52. Chapman J, Reynolds DS, Riley J, Pedgley D, Woiwod I. High-altitude migration of the diamondback moth, Plutella xylostella, to the UK: a study using radar, aerial netting and ground trapping. Ecol Entomol. 2002;27:641–50.
53. Mo J, Baker G, Keller M, Roush R. Local dispersal of the diamondback moth (Plutella xylostella (L.)) (Lepidoptera: Plutellidae). Environ Entomol. 2003;32:71–9.
54. Cameron P, Walker G, Penny G, Wigley P. Movement of potato tuberworm (Lepidoptera: Gelechiidae) within and between crops, and some comparisons with diamondback moth (Lepidoptera: Plutellidae). Popul Ecol. 2002;31:65–75.
55. Bariola L, Keller J, Turley D, Farris J. Migration and population studies of the pink bollworm in the arid West. Environ Entomol. 1973;2:205–8.
56. Mani E, Wildbolz T. The dispersal of male codling moths (Laspeyresia pomonella L.) in the Upper Rhine Valley. Zeitschrift für Angewandte Entomologie. 1977;83:161–8.
57. Knipling E. Agriculture Handbook no. 512: The basic principles of insect population suppression and management. Washington, DC: USDA; 1979.
58. Zhao JZ, Cao J, Li Y, Collins HL, Roush RT, Earle ED, et al. Transgenic plants expressing two Bacillus thuringiensis toxins delay insect resistance evolution. Nat Biotechnol. 2003;21:1493–7.
59. Zhao JZ, Cao J, Collins HL, Bates SL, Roush RT, Earle ED, et al. Concurrent use of transgenic plants expressing a single and two Bacillus thuringiensis genes speeds insect adaptation to pyramided plants. Proc Natl Acad Sci U S A. 2005;102:8426–30.
60. Liu X, Chen M, Collins HL, Onstad DW, Roush RT, Zhang Q, et al. Natural enemies delay insect resistance to Bt crops. PLoS One. 2014;9, e90366.
61. Tabashnik BE, Sisterson MS, Ellsworth PC, Dennehy TJ, Antilla L, Liesner L, et al. Suppressing resistance to Bt cotton with sterile insect releases. Nat Biotechnol. 2010;28:1304–7.
62. Johnson M, Gould F, Kennedy G. Effect of an entomopathogen on adaptation of Heliothis virescens populations to transgenic host plants. Entomol Exp Appl. 1997;83:121–35.
63. Onstad DW, Liu X, Chen M, Roush R, Shelton A. Modeling the integration of parasitoid, insecticide and transgenic insecticidal crops for the long-term control of an insect pest. J Econ Entomol. 2013;106:1103–11.
64. Bommarco R, Miranda F, Bylund H, Björkman C. Insecticides suppress natural enemies and increase pest damage in cabbage. J Econ Entomol. 2011;104:782–91.
65. Hallmann CA, Foppen RP, van Turnhout CA, de Kroon H, Jongejans E. Declines in insectivorous birds are associated with high neonicotinoid concentrations. Nature. 2014;511:341–3.
66. Tan A, Fu G, Jin L, Guo Q, Li Z, Niu B, et al. Transgene-based, female-specific lethality system for genetic sexing of the silkworm, Bombyx mori. Proc Natl Acad Sci U S A. 2013;110:6766–70.
67. Martins S, Naish N, Walker AS, Morrison NI, Scaife S, Fu G, et al. Germline transformation of the diamondback moth, Plutella xylostella L., using the piggyBac transposable element. Insect Mol Biol. 2012;21(4):414–21.
68. Walters M, Staten R, Roberson R. Pink bollworm integrated management using sterile insects under field trial conditions, Imperial Valley, California. In: Tan K, editor. International Conference on Area-Wide Control of Insect Pests, and the 5th International Symposium on Fruit Flies of Economic Importance: 28 May - 5 June 1998. Penang: Penerbit Universiti Sains Malaysia, Pulau Pinang, Malaysia; 2000. p. 201–6.
69. Proverbs M, Newton J, Campbell C. Codling moth: a pilot program of control by sterile insect release in British Colombia. Can Entomol. 1982;114:363–76.
70. Suckling D, Hackett J, Barrington A, Daly J. Sterilisation of painted apple moth Teia anartoides (Lepidoptera: Lymantriidae) by irradiation. N Z Plant Protect. 2002;55:7–11.

71. Wee S, Suckling D, Burnip G, Hackett J, Barrington A, Pedley R. Effects of substerilizing doses of gamma radiation on adult longevity and level of inherited sterility in Teia anartoides (Lepidoptera: Lymantriidae). J Econ Entomol. 2005;98:732–8.

CHAPTER 17

Efficient Management of Fruit Pests by Pheromone Nanogels

Deepa Bhagat[1] *{ , Suman K. Samanta[1] * & Santanu Bhattacharya[1,2,3]

[1]Department of Organic Chemistry, Indian Institute of Science, Bangalore 560 012, India,

[2] J. C. Bose Fellow, Department of Science and Technology, New Delhi,

[3] Also at JNCASR, Bangalore 560 012, India.

ABSTRACT

Environment-friendly management of fruit flies involving pheromones is useful in reducing the undesirable pest populations responsible for decreasing the yield and the crop quality. A nanogel has been prepared from a pheromone, methyl eugenol (ME) using a low-molecular mass gelator. This was very stable at open ambient conditions and slowed down the evaporation of pheromone significantly. This enabled its easy handling and transportation without refrigeration, and reduction in the frequency of pheromone recharging in the orchard. Notably the involvement of the nano-gelled pheromone brought about an effective management of *Bactrocera dorsalis*, a prevalent harmful pest for a number of fruits including guava. Thus a simple, practical and low cost green chemical approach is developed that has a significant potential for crop protection, long lasting residual activity, excellent efficacy and favorable safety profiles. This makes the present invention well-suited for pest management in a variety of crops.

INTRODUCTION

The pest management practices throughout the world are increasingly leaning toward more environment friendly agriculture without disturbing the balance of the eco-system1. Thus, the evolution of modern organic agriculture depends on biological control agents to disrupt and control pest attacks which impart serious adverse effects on the agricultural economy2.

Guavas (*Psidium guajava* Linn.) comprise plants that contain nearly hundred species of tropical shrubs and trees. Guavas are rich in dietary fibers, minerals, vitamins A and C and folic acid3. The fruit is native to central and northern South America, and cultivated throughout in South-east Asia, Florida, Hawaii, the Caribbean and Africa3. Fruitfly (*Bactrocera dorsalis* Hendel, Figure 1a) is considered to be a harmful key pest of guava. It causes significant yield losses and quality degradations of the fruits particularly during the rainy season4. This results in a poor commercialization of the crops in domestic markets and imposes quarantine restrictions from the international markets5. *Bactrocera dorsalis* (*B. dorsalis*) are polyphagous6 pests and it breeds on a number of other fruits and vegetables as well. For example, fruits such as, apple (*Annona squamosa*), banana (*Musa paradisiaca*), mango (*Mangifera indica*), orange (*Citrus sinensis*), peach (*Prunus persica*), plum (*Prunus domestica*) *etc.* are also affected by this fruit fly7,8. The fruit flies are attracted toward the concerned fruits and infect them (Figure 1b). Attacked fruits show signs of ovipositional damage in the form of minute depressions which can be even seen from outside (Figure 1c, d). The affected fruits soften at the infested positions, then disintegrate and eventually drop down prematurely (Figure 1e). Eggs of *B. dorsalis* are laid below the skin of the host fruit and the maggots feed on the pulp of the fruits (Figure 1f). These hatch within ~1–3 days and the larvae feed for another 9–35 days. The pupariation process occurs in the soil near the host plant and the adult flies emerge after ~1–2 weeks and persist throughout the year9.

Figure 1.(a) Photograph of an adult *B. dorsalis* fruit fly; (b) infection of the fruit guava by *B. dorsalis*; (c, d) eggs of *B. dorsalis* laid below the skin of the host fruit and the attacked fruit shows the signs of the ovipositional damage in the form of minute depressions; (e) the affected fruits soften at the site of infestation which then rot and drop down prematurely and (f) the maggots feed on the pulp of the fruits.

For the pest management of *B. dorsalis,* several methods have been developed in the last few decades. These pest management practices include destruction of the fallen and infested fruits[10], cold treatment[11], heat teatment[12,13], bait sprays method containing protein hydrolysate[14], biological control method[15], sterile insect techniques[16,17] *etc.* However, most of these treatments fail due to one or another type of drawbacks. Use of insecticides provides another alternative. But it affects both the agricultural fields, habitation and even the locality resulting in affecting non-target species such as humans, mammals, arthropods and other natural enemies[18]. In addition, the pest populations have increasingly become resistant to the chemical pesticides compelling the farmers to resort to an use of even larger amounts of pesticides[19,20,21]. These pose significant risk on the human health and the environment in a bid to achieve appropriate control of the pest.

To alleviate such problems highly viscous hydrogels have been used to deliver pheromones[22]. However, hydrogels often swell and shrink with the changes in humidity and temperature. Therefore, such hydrogels cannot be used in all seasons and in particular during the rainy season. Male annihilation technique has been used in the pest management of *B. dorsalis* due to the attraction of males to methyl eugenol (ME) at very low concentrations[23,24]. However, there is a significant limitation in that these solutions contain pesticides and have a limited shelf-life.

To overcome these problems a new yet simple approach is essential which should provide viable solutions in the pest management arena. Pheromones are naturally occurring volatile semiochemicals and are considered as eco-friendly biological control agents[25]. They induce impaired sexual communication and mating disruptions among pests[26]. High specificity of pheromones towards a given species further enables pest control without weakening the beneficial organisms[27]. However, due to the unstable nature of pheromones, such as photo-oxidation, auto-oxidation, isomerization and volatility *etc.*[28], the formulation development for the field trials involving pheromones needs a major refinement specifically in terms of the slow-release and protection from decomposition under ambient conditions[29]. ME in this context may be useful since it belongs to the class of a para-pheromone which mimics the role of a male attractant[30,31]. Using supramolecular self-assembly principles and nanotechnology we have been able to achieve an increased shelf-life of the pheromone (ME) when immobilized in a nanogel. We demonstrate here that such pheromone nanogels exhibit high residual activity, excellent efficacy in open orchard even during adverse seasons and are also environment friendly.

RESULTS

Nanotechnology plays a pivotal role in the increase of the effectiveness of pheromones in terms of their increased shelf-life through their entrapment by suitable host matrix[32]. Low-molecular mass gelators (LMMGs) have attracted much attention because of their interesting properties and potential uses in various fields of research and industry[33,34]. LMMGs such as all-*trans* tri(*p*-phenylenevinylene) bis-aldoxime (**1**)[35], self-assemble through weak intermolecular non-covalent interactions such as hydrogen bonding, π-π stacking and van der Waals leading to the gelation of specific solvents (Figure 2)[36,37,38,39]. In the present work, we demonstrate the nanogel formation upon entrapment and immobilization of ME into the entangled three-dimensional nano-sized supramolecular networks of **1**. Such nanogels provide high pheromone retention capacity, enhanced shelf-life and protection of ME from environmental decompositions (from exposure to ambient air, water and sunshine). ME alone however, is quite volatile on its own and evaporates under ambient conditions significantly rapidly during the summer (see below).

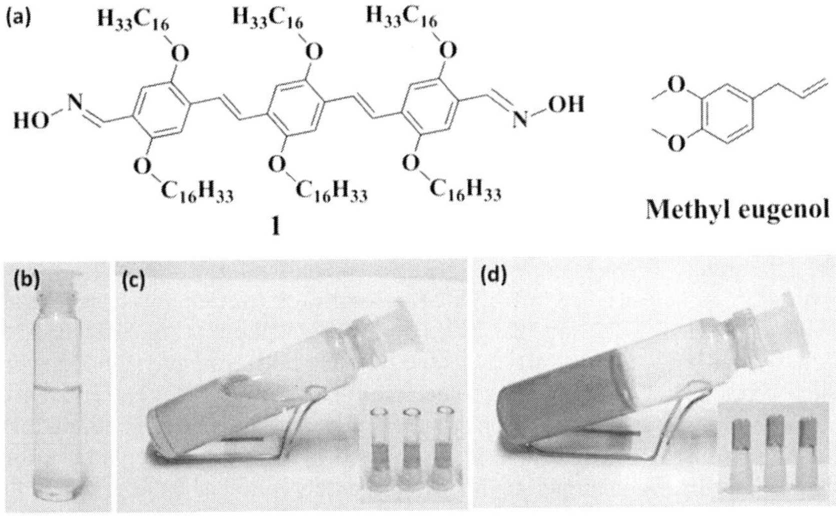

Figure 2.(a) Molecular structures of the gelator, **1** and methyl eugenol (ME).Photographs showing (b) colorless ME, (c) a 'sol' after warming 12 mg/mL of **1** in ME at 70°C; inset shows glass vials holding 0.2 mL of the 'sol' and (d) a 'gel' at room temperature (25°C) after cooling the previous solution without any external perturbation; inset shows the inverted vials holding the gel suitable for the field study.

IMMOBILIZATION OF METHYL EUGENOL IN NANOGELS

The present invention describes a simple and practical method by which pheromones could be immobilized in the nanogels. ME alone is a colorless semi-viscous liquid (Figure 2b) which turns yellow (Figure 2c) on solubilization of the gelator **1** (2.5 mg/mL) upon brief warming of the mixture to ~65°C. In this condition, the pheromone did not undergo any decomposition. When cooled to room temperature without any external perturbation for a period of 10–15 min, the yellow solution turned into a thick, viscous orange colored mass, which did not flow under the influence of gravity when the vial was inverted upside down demonstrating the onset of gelation (Figure 2d). The gelation was retained unperturbed even after several consecutive heating and cooling cycles (from room temperature to 70°C). The nanogel was thermoreversible in nature, a property which was found to be useful for its long term use in the field. Thus the preparation of the nanogels (**1** + ME) is indeed convenient without requiring addition of any organic solvents or chemical cross-linking agents etc. Also, after several heat-cool cycles, or upon exposure to open orchard for a number of weeks, neither a thermal or photo-chemical decomposition of ME occurred as evidenced from GC-MS analysis (see, Supplementary Information, [SI] Figure S1 and S2). This ensured good chemical stability of the resulting nanogels.

The minimum amount of the gelator (**1**) required for the formation of the gel is known as the minimum gelator concentration (MGC). This was estimated to be ~2.5 mg/mL indicating high ME content in the nanogel. Thus as little as ~2.5 mg of **1** could immobilize about 1 mL of ME. The freeze-dried gel was further characterized using both transmission electron microscopy (TEM) and scanning electron microscopy (SEM). The presence of three-dimensional nano-sized fibrillar networks having fiber diameters ranging from 100–200 nm was evidenced. These fibers collate upon themselves in the nanogel to form aggregated bundles having the fiber diameters ranging from 500 nm to 1 μm (Figure 3).

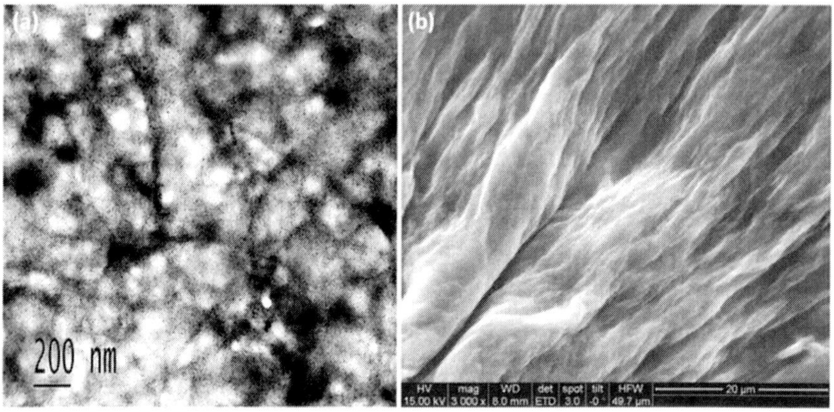

Figure 3.(a) TEM and (b) SEM images of the nanogel showing the existence of nano-fibrillar networks.

THERMAL STABILITY OF NANOGELS

The melting temperature of the nanogels, *i.e.* gel-to-sol transition (T_{gel}) increased progressively with increasing concentration of **1**, a phenomenon that is general for many LMMGs (Figure 4a)[34,35]. This indicates that, with increasing concentration of the gelator, the density of the gel assembly increases ensuring participation of a larger number of gelator molecules per unit volume in the thermally induced transitions. At the concentration of 12 mg/mL of **1** in ME the T_{gel}reached ~63°C which was well above the ambient temperature even during peak summer in India. Thus the thermal stability of the sample should be adequate for the agricultural field trials even in hot climates around different geographical regions of the world.

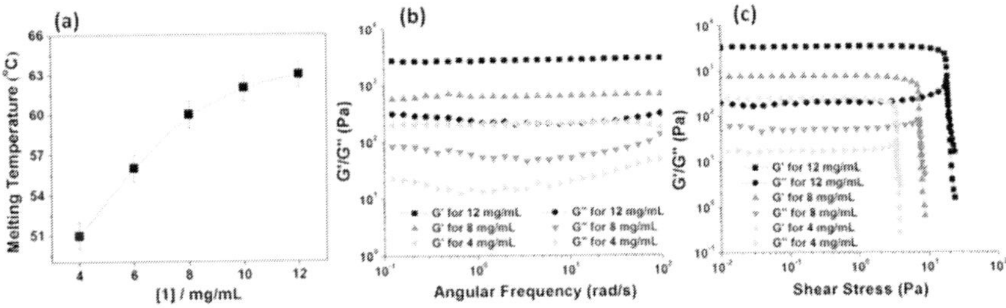

Figure 4.(a) Plot of gel-to-sol melting temperatures as a function of the concentration of **1** indicating an improved thermal stability of the nanogels with increasing amount of **1**.(b) Rheological studies of the nanogels show oscillatory frequency sweep and (c) stress amplitude sweep dependencies. With increasing amount of **1**, the viscoelasticity of the gel increased providing greater mechanical stability.

Rheological properties and mechanical stability of gels

The oscillatory frequency response of the nanogel revealed that the elastic modulus G′ is independent of frequency and much higher than the viscous modulus G″ over the entire angular frequency range (0.1–100 rad/s) at 0.01% strain amplitude (Figure 4b). A ~13-fold higher G′ over G″ suggested a substantial elastic response of the gels that did not relax over the long timescale. This is a typical characteristic of the nanogels34,35. The viscoelastic solid-like nature of the gel increases with increasing concentration of **1** and it becomes sufficiently strong (G′ ~2800 Pa) at 12 mg/mL. Under oscillatory stress, the nanogels show G′ > G″ by an order of magnitude indicating the dominant elastic behavior (stiffness) of the system and show the resistance of the mass to flow under applied stress (Figure 4c). An applied stress above which the gel starts to flow is called the yield stress (σ_y) and its value (~ 20 Pa at 12 mg/mL) is sufficiently high to provide adequate mechanical stability to prohibit the samples being relaxed (dropped) during its transportation or shipment.

SLOW RELEASE OF ME FROM THE NANOGELS

The release pattern of the volatile pheromone in terms of its relative rates of evaporation at a particular temperature was determined both for the liquid ME alone and ME immobilized in nanogel (12 mg/mL) under identical conditions. The extent of evaporation was investigated by monitoring the weight loss from both the samples kept at four different temperatures (10, 20, 30 and 50°C) in a temperature gradient chamber. Plots of %-weight losses *vs.* the number of weeks showed that the rates of evaporation were always greater from the ME alone compared to the ME immobilized in nanogel (Figure 5). The calculated slopes from the linear fit of the data at various temperatures showed that the evaporation rate of ME increased significantly with increasing temperatures (Figure 5e). ME in nanogel afforded only ~2–3% weight loss after 10 weeks at 30°C (Figure 5) although it showed ~50% weight loss at 50°C indicating that the weight loss occurred ~13 times slower at 30°C compared to that at 50°C. On the other hand, at ~50°C (a temperature at the peak summer in Indian subcontinents), ME alone evaporated entirely within 3 weeks and in contrast it took ~30 weeks for the near complete evaporation of ME from the nanogel. The extrapolated graphs for the nanogel in their weight loss profiles further indicate that, it would take ~7.8 years at 30°C and 12.4 years at 20°C for the complete evaporation of ME from the nanogel (SI, Figure S3). Thus, the sustainable nature of the pheromone in the nanogel for a significant period of time makes the nanogel particularly useful for the agricultural pest management applications.

Thus, the nanogel was found to be chemically, thermally and mechanically stable. The self-assembly of individual gelators is stabilized by hydrogen bonding donor (–OH) and acceptor (= N–) sites, van der Waals interactions through the long aliphatic hydrocarbon chains and π-π stacking through the extended aromatic π-conjugated backbones of the gelator **1**35. The pheromone molecules immobilized in the nanofiber networks were found to be particularly resistant to evaporation at room temperature which confirms its increased shelf-life under these conditions.

Evidence of enhanced shelf-life of ME in the nanogel

As the weight loss experiments indicated a significantly reduced evaporation rate of ME from the nanogels compared to the ME alone, a thin film of the prepared ME nanogel was cast (plate 'A', Figure 6) on glass plates to study the extent of shelf-life of the pheromone. Control experiments were performed with the cast film of **1** solubilized in toluene (plate 'B') and ME alone (plate 'C'). All the three plates were exposed to a guava orchard for a number of hours. The results showed that the fruit flies were attracted

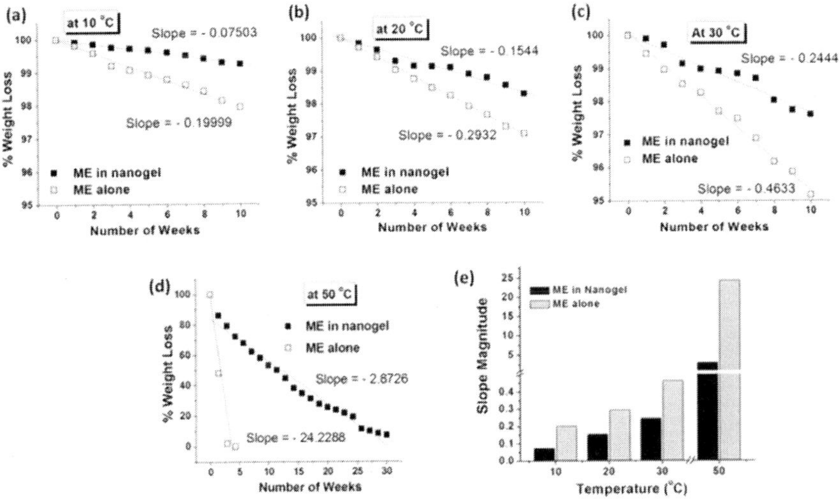

Figure 5. Plots of percent weight losses as a function of time confirm sustained 'release' pattern of the nano-gelled pheromone (ME) studied at (a) 10°C, (b) 20°C, (c) 30°C and (d) 50°C. (e) Histograms show a comparison of the slopes of the release patterns between ME in nanogel and ME alone.The individual weight loss at any week is an average of three replications performed in identical conditions.

specifically to the plate A and to the plate C throughout the period and not to the plate B which held the gelator **1** in toluene (a non-pheromone solvent, see Movie S1). This indicates that the pheromone is biologically active in the nanogel and is responsible for attracting the pest although the compound **1** itself is inactive toward the fruit flies. The accumulation of the pests increased with time around both the plates A and C. The same plates were preserved at room temperature (~30°C) and were exposed again in the same guava orchard after 21 days. Interestingly this time the fruit flies were attracted to the plate A only which contained the nanogel and not to the plate B or C for the entire time period of observation. This indicates that the nanogel laced with ME still retained the pheromone for sustained release and has also retained significant pest attractant property due to the higher shelf-life of ME in nanogel than the pure ME alone which evaporates away lot faster. This experiment also revealed that the fruit flies upon being attracted by ME released from the nanogels, gathered in the vicinity or on top of the plate. However, they did not die during this experiment (see Movie S1). This indicates that the nanogel (**1** + ME) is non-toxic in nature and thus acts only as an attractant source for the fruit flies alone.

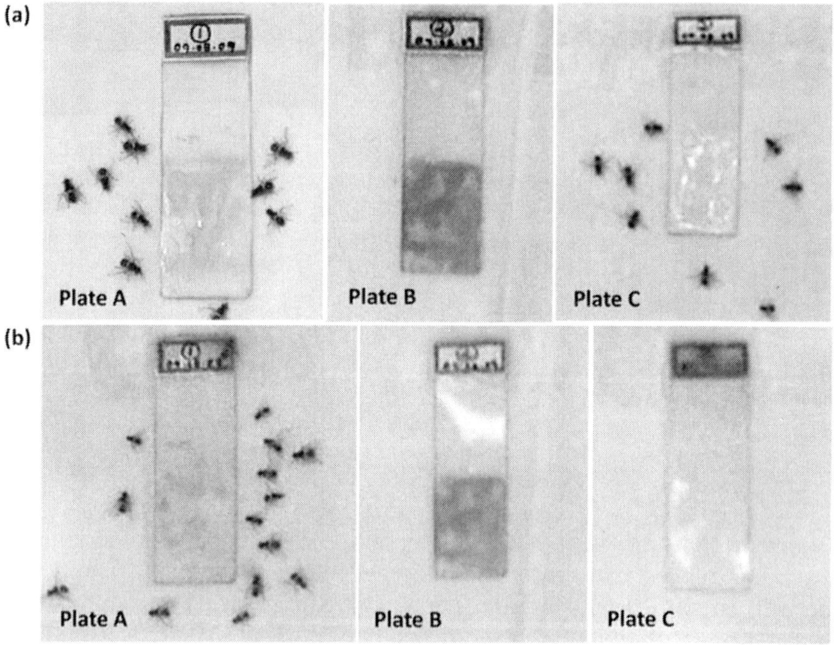

Figure 6. (a) Thin films prepared on 09th Aug'2009; plate A: **1** + ME gel, plate B: gels of **1** in toluene, plate C: ME alone and the plates were exposed to a guava orchard.(b) The same plates were exposed to the guava orchard after three weeks on 31st Aug'09.

Field trials with the nanogels

The nanogel containing ME was then utilized for the efficient pest management of *B. dorsalis* as a bait trap along with a technique to kill the pest. The total area where the experiments conducted was 67274 sq. ft. (0.625 hectare) with the distance of 15 ft. in between two adjacent trees. The samples containing ME alone and ME in nanogel were applied in the guava orchard each with six replications in such a way that the horizontal and vertical distance between the two traps was 60 ft. and 75 ft. respectively as depicted in Figure S4 (SI). The experiment was set up during the ripening stage of guava (9th August, 2009) when the environmental temperature varied within ~35–40°C and this continued till the crops were harvested (7th September, 2009). The nanogels (12 mg/mL) were taken in glass vials (5 mm inner-diameter, inset of Figure 2d) and placed inside plastic bottles (Figure 7a, b). Two holes were made in the bottles and were filled with water upto the white arrow level as shown in Figure 7b (also *c.f.* Methods for the detailed Sampling Process). Control experiments were also performed with ME by directly adding it into water in the plastic bottles as convenient to the previously existing method40. Due to the presence of pheromone, the pests were attracted towards the hanging vials containing the nanogel only. During this event a significant number of fruit flies fell into water, which followed by their drowning and eventual death. Dead pests were finally collected by filtration and the total number of killed flies were counted (Figure 7c, d) and summarized against events recorded each day in Table S1 (see SI). The number of trap catches for the control containing only ME was significantly lower in comparison to the nanogel throughout the experiment.

Figure 7. (a) A simple and low cost sampling technique in which nanogels are placed just above the water level contained inside plastic bottles.This arrangement was distributed throughout the guava orchard. (b) A view of the sampling process where in the plastic bottle, two holes were made oppositely up (rear-side shown by yellow rear-arrow) and down (front-side yellow arrow) to each other and filled with water near the level of the down hole (shown by white arrow). The nanogel was placed inside the bottles just above the water level. Control experiments were performed with ME added directly into the water ~2 cm below the level of the down hole. (c) Owing to the presence of pheromone, the fruit flies were attracted towards the bottles and fell down into the water and drowned to eventual death. (d) Dead flies were filtered and their numbers were counted. (e) Plot represents averages (±SE, n = 6) of the number of dead flies for the nanogel and the control (ME alone).

The number of catches for the control bottle containing ME alone also attracted pests initially. However, this showed a progressive decrease in comparison to the bottle containing nanogel and from the 8th day onward no fly was attracted to this bottle. This indicates that the effectiveness of ME alone towards the *B. dorsalis* without the gelator **1** is at best limited to one week in the real field environment (Figure 7e). On the contrary significant counts of *B. dorsalis* catches were continually recorded almost throughout the month with peak trap catches occurring during the first three weeks as compared to the control. After harvesting all the guavas, the infestation of *B. dorsalis* was found to be remarkably lower as shown in the results after three weeks of experiment since the number of catches got reduced. However, the nanogels were still active in releasing pheromone as revealed from the weight loss experiments which were conducted upto ten weeks. The procedure is therefore reproducible and particularly attractive due to the low cost, simplicity, selectivity and efficiency of the sampling technique. This also produced high yield and damage free crops.

Statistical Analysis of Variance (ANOVA) were performed on the collected data containing the number of trap catches in each day throughout the experiment (30 days) for all the replications (6 for both ME alone and ME in nanogel) of the pheromone containing samples (Table S1, SI). The two-factor ANOVA calculations considering the factor 1 as treatment (number of trap catches) and factor 2 as the period of observations (days), were carried out on the transformed data $[(x + 0.5)^{1/2}$, x = number of trap catches in each day] to homogenize the variances (Table S2, SI)41. The interaction between the treatments (t) and the period of observations (d) has been calculated from the data presented in Table 1 to obtain the extent of significance of the experiments.

Table 1. Mean values of the trap catches from the six replications per treatment (ME alone and ME in nanogel)[a]

Days	ME alone (t1)	ME in nanogel (t2)	Mean	Days	ME alone (t1)	ME in nanogel (t2)	Mean
1	105 (10.18)	120.5 (10.95)	112.75 (10.57)	16	0 (0.71)	64.33 (8.01)	32.17 (4.36)
2	101.66 (10.06)	104.33 (10.21)	103 (10.14)	17	0 (0.71)	91.33 (9.47)	45.67 (5.09)
3	105 (10.18)	109.83 (10.46)	107.42 (10.32)	18	0 (0.71)	88.83 (9.4)	44.42 (5.05)
4	67.33 (8.19)	115 (10.69)	91.17 (9.44)	19	0 (0.71)	105.5 (10.19)	52.75 (5.45)
5	47.83 (6.87)	97.66 (9.87)	72.75 (8.37)	20	0 (0.71)	74.33 (8.62)	37.17 (4.66)
6	23.66 (4.81)	100 (9.99)	61.83 (7.4)	21	0 (0.71)	32 (5.42)	16 (3.06)
7	10.33 (2.96)	89.5 (9.45)	49.92 (6.20)	22	0 (0.71)	8.17 (2.69)	4.08 (1.7)
8	0.33 (0.85)	99.16 (9.94)	49.75 (5.39)	23	0 (0.71)	7.17 (2.43)	3.58 (1.57)
9	0 (0.71)	98.83 (9.89)	49.42 (5.3)	24	0 (0.71)	5.83 (2.16)	2.92 (1.43)
10	0 (0.71)	103.33 (10.15)	51.67 (5.43)	25	0 (0.71)	7.66 (2.34)	3.83 (1.52)
11	0 (0.71)	105.83 (10.24)	52.92 (5.47)	26	0 (0.71)	15.16 (3.69)	7.58 (2.20)
12	0 (0.71)	82.16 (9.07)	41.08 (4.89)	27	0 (0.71)	5.83 (2.21)	2.92 (1.46)
13	0 (0.71)	95.5 (9.75)	47.75 (5.23)	28	0 (0.71)	2.5 (1.54)	1.25 (1.12)
14	0 (0.71)	98.33 (9.91)	49.17 (5.31)	29	0 (0.71)	1 (1.09)	0.5 (0.9)
15	0 (0.71)	87.5 (9.32)	43.75 (5.02)	30	0 (0.71)	1 (1.13)	0.5 (0.92)
Grand Mean					15.37 (2.32)	67.27 (7.34)	41.32 (4.83)
Critical Difference (CD) between each treatments (t)					0.193*	0.255**	
Critical Difference between periods of observation (d)					0.749*	0.987**	
Critical Difference between the interactions ($t \times d$)					1.059*	1.396**	

[a]This was analyzed from the data shown in Table S1 and Table S2 in the supplementary information. Figures in the parentheses are the mean values for the transformed data [$^*p < 0.05$, $^{**}p < 0.01$].

The two-factor ANOVA calculations showed that the computed F value was larger than the tabular F value both at probability (p) < 0.05 and $p < 0.01$ level of significance for the treatments (t), period of observations (d) as well as the interaction between the two (t × d) (Table 2). This suggests that the difference between the treatments, the period of observations pertaining to fruit flies' visitation and the interaction between the treatments (t1 and t2) with the period of observations is highly significant. The coefficient of variation (CV) is 19.39% (< 20%) (Table 2) indicating that the degree of precision with which the treatments were compared, showed a good index of reliability of the experiments42.

Table 2. Two-factor ANOVA calculations of the transformed data of guava field trials

Source of Variation	Degree of Freedom	Sum of Squares	Mean Squares	Computed F value	Probability	Tabular F value	
						$p < 0.05$	$p < 0.01$
Treatment (t)	1	2268.84	2268.84	2610	0.000**	3.87	6.72
Days (d)	29	2937.61	101.29	116.53	0.000**	1.51	1.78
Interactions (t × d)	29	1171.92	40.41	46.49	0.000**	1.51	1.78
Replications	5	1.45	0.291	0.33		2.24	3.08
Experimental error	295	256.44	0.87	1			
Total	359	6636.26	18.48	21.26			

**Significant at $p < 0.05$ and $p < 0.01$; Coefficient of variation (CV) = 19.39% [(CV) = {(Error Mean square)$^{1/2}$/Grand Mean} × 100]

The least significant difference (LSD) calculated (with AgRes software) for the treatments (ME alone and ME in nanogel), the period of observations (days) and the interaction between the two (t × d) considering the mean values are all presented in the Table 1. LSD results clearly indicate that the treatment 2 (t2, ME in nanogel) is more efficient than the treatment 1 (t1, ME alone) in trapping fruit flies (see AgRes results in Table S5, Supplementary Information) as the critical difference (CD) between the treatments is 0.193 ($p < 0.05$). This further indicates that there is significant difference between the mean for the treatment 1 (2.32) and treatment 2 (7.34). The differences between the mean values of the periods of observations (days) were estimated using LSD considering the CD = 0.749 ($p < 0.05$) which showed a descending order of distribution among days (Table S3, SI). The maximum number of fruit flies catches was recorded on the first three days. As the days progressed and guava fruits were harvested fruit flies infestation in field became gradually less significant and thus the lowest number of trap catches took place from the day 23 to day 30. LSD calculations were also performed from the mean values for the comparison of the interactions between the treatments and the period of observations (t × d) considering CD = 1.059 ($p < 0.05$) and the relevant data are recorded in Table S4 (SI). Treatment 1 (ME alone) and treatment 2 (ME in nanogel) were at par on the first three days and from day 8, the treatment 2 showed more effectiveness than the treatment 1 (Table S4, Group o). Therefore, ME immobilized in nanogel acted as an effective formulation compared to ME alone for the pest management of *B. dorsalis* in the field trial studies.

DISCUSSION

The present invention provides a simple and effective route to a slow delivery of pheromones from a nanogel. This does not require any use of environmentally harmful and toxic chemicals such as, cross-linkers, antioxidants, antimicrobials, pesticides, volatile organics *etc.* While other existing methods pose risk of human contacts where ME is added to water in the bait, the present method avoids any direct contact with the fruit crop and the agricultural workers with pheromone. This keeps the crop absolutely clean from the chemical contaminations unlike the prevalent practice of spraying toxic pesticides on the orchard. Moreover, the aggressive techniques like destruction of fallen and infested fruits, cold or vapor-heat treatments, hot-water treatment, forced hot-air treatment *etc.* are avoided in the present approach which makes this economically viable as well.

The nanogel of pheromone ME described herein is insoluble in water, which makes it superior to hydrogels and microcapsules. The nanogel does not significantly swell and shrink in presence of water, confirming that the humidity plays no effect whatsoever on this material. This pheromone nanogel is particularly useful for the pest management even in rainy season as *B. dorsalis* shows maximum activity during the rainy season resulting in the most significant yield loss of crops. It may be further noted that, use of microcapsules encounters problems associated with capsule bursts or potential leaks. These critical drawbacks have been avoided in the present method.

The flexibility in using the nanogel in any season at any temperature (<60°C) are feasible due to the oxidative, photochemical and thermal stability of the nanogel. Also the transportation of the nanogel based samples is trouble free due to their significant mechanical strength. This is a very general method and the present finding could be potentially useful in preparing new nanogel type formulations with other pests such as *Helicoverpa armigera* (Hubner) (Lepidoptera, Noctuidae), *Scirphophaga incertulas* (Walker) (Lepidoptera, Pyralidae), *Leucinodes orbonalis* (Guenee) (Lepidoptera: Pyralidae) *etc.* with suitable LMMGs and appropriate pheromones that attract these species. This strategy may also be applicable even for kairomones for the attraction of predators and parasitoids (natural enemies). Such an approach could therefore be efficiently carried forward from the laboratory to the agricultural field.

In summary, this is a simple and straightforward method where a nanogel is prepared as a controlled release system to deliver fruit fly pheromone that remains active over a sustained period in an environmentally safe manner. In view of the demonstrated efficiency of this pheromone nanogel in a fruit orchard, in rainy season and also in humid, sunny peak summer months, this approach should be useful in pest control in most parts of the world.

METHODS

Materials

Commercial ME was obtained from Sisco Research Laboratories Pvt. Ltd., India and was used without further purification. Gelator **1** has been synthesized as reported earlier35.

Scanning electron microscopy (SEM)

The nanogels were carefully scooped onto the brass stubs and were freeze-dried. The samples were then coated with gold vapour and analyzed on a Quanta 200 SEM operated at 15 kV.

Transmission electron microscopy (TEM)

A solution of **1** + ME (1×10^{-4} M) was loaded onto a carbon coated copper grid (200 mesh size) and the TEM images were recorded on a TECNAI F30 instrument.

GC-MS studies

Samples were analyzed on a gas-chromatograph (GC 17A, Shimadzu) coupled with a mass-spectrometer (GC-MS QP 5050 Shimadzu) with an electron impact ionization at 70 eV. The GC was equipped with a WCOT fused silica stationary phase CP-Sil 5CB open tubular capillary column DB-5 ms (30 m × 0.25 mm, I.D. 0.25 μM) using helium as carrier gas (flow rate 1.3 ml/min). The temperature was programmed from 60°C (hold 5 min) to

250°C at 10°C/min. The samples were introduced at the injector temperature of 250°C. Tentative identification of each compound was achieved based on the comparison of matching of the respective mass spectrum as recorded in the Wiley 139 computer databases. Identification of ME was ensured from a comparison with the mass spectra and the GC retention data with those of the known standards.

Determination of gel melting temperatures (T_{gel})

The melting temperature of the nanogels from gel-to-sol was determined by an 'inverted vial' method. A sealed glass vial containing 0.5 mL of nanogel was gradually heated (2°C/min), and the temperature at which the gel started to flow under gravity was recorded as the gel melting temperature. Each experiment was performed in duplicate to determine the reproducibility.

Weight loss studies

Mini glass containers of fixed (5 mm) diameter containing 0.2 mL of nanogel samples (in three replications for each sample) were placed inside a temperature gradient chamber (NK system, model: TG-180 CCFL-5LE) and were allowed to evaporate slowly. The weights were recorded on a weekly basis.

Rheological studies

An Anton Paar 100 rheometer using a cone and plate geometry (CP 25-2) with a gap distance fixed at 0.05 mm was utilized. The rheometer has a built-in computer which converts the torque measurements into either G′ (the storage modulus) and G″ (the loss modulus) in oscillatory shear experiments. The nanogel was first scooped on the plate of the rheometer and then a stress amplitude sweep experiment was performed at a constant oscillation frequency of 1 Hz for the strain range of 0.01 to 500% at 20°C. Oscillatory frequency sweep experiments were performed in the linear viscoelastic region (0.01% stain amplitude) by varying the angular frequency range (0.1–100 rad/s) to ensure that the calculated parameters correspond to an intact network structure.

Sampling process

We have taken a ~25 cm long plastic bottle of ~7 cm diameter with a closed screw-cap. This is kept hanging from the branch of a tree with the help of a hook in an upside down orientation. Two circular holes (diam. ~4.5 cm) have been made on the bottle in such a way that they face up and down opposite to each other. Two holes are useful for the facile passage of fruit flies that are attracted to pheromone. Water is introduced into the bottle through the lower hole and maintained at nearly the same level. The pheromone nanogel sample is kept in a hanging vial and its opening faces downwards just a few cm above the water level. With this arrangement, even when it rains, the excess of water gets automatically drained away from the lower hole allowing it to maintain the water level. As the opening of the nanogel vial is oriented downwards, rain water cannot enter into it. Also at the end of each day, the contents from the plastic bottle may be released by opening the screw-cap. This allows collection of dead flies after the day-long exposure in the orchard. The same bottle may be reused following the above procedure throughout the season.

Field studies

The experiment was conducted at Amrud Odyan Ropan under Rastriya Audhanik Mission at Gosaipur Mohab, Varanasi, Uttar Pradesh, India. Total area of the orchard was 0.625 hectare (67274 sq. ft) with the distance of 15 feet in between two adjacent trees. The experiment was conducted during rainy season for a period of 30 days from 9th August to 7th September, 2009. We took 6 replications for each sample containing ME in nanogel and ME alone. Dead pests were counted everyday for each bottle and were averaged from six bottles under each category. Thus the data containing ME samples represents a randomized block design (RBD model)42.

STATISTICAL ANALYSIS

The data were analyzed with the AgRes software (version 7.01) ANOVA (Analysis of Variance) Package for Researchers 1994, Pascal Intl software solutions. The data entry was executed using AGDATA AgRes (version 3.01) in the data entry module for the AgRes Statistical Software 1994 (Pascal Intl software solutions). Significance is assessed at 95% ($p < 0.05$) and 99% ($p < 0.01$) level of confidence. F values were calculated using the method as described in http://www.biokin.com/tools/fcrit.html.

ACKNOWLEDGEMENTS

DB thanks the present and past Directors of NBAII, Bangalore for their encouragements. We thank DST, Government of India (J C Bose Fellowship to SB), INI, IISc for the SEM and TEM studies, DBT, Government of India for funding this work. We thank Dr. K. J. David for the identification of *Bactrocera dorsalis* (Hendel).

REFERENCES

1. Tilman, D., Cassman, K. G., Matson, P. A., Naylor, R. & Polasky, S. Agricultural sustainability and intensive production practices. Nature 418, 671–677 (2002).
2. Copping, L. G. & Menn, J. J. Biopesticides: a review of their action, applications and efficacy. Pest Manage. Sci. 56, 651–676 (2000).
3. Nutrition data, know what you eat; guavas, common, raw. http://nutritiondata.self.com/facts/fruits-and-fruit-juices/1927/2 (retrieved on 11th 2012 April).
4. Allwood, A. J. et al. Host plants for fruit flies (Diptera; Tephritidae) in South East Asia. Raff. Bull. Zoology, Supplement 1–92 (1999).
5. Vayssieres, J.-F., Wharton, R., Adandonon, A. & Sinzogan, A. Preliminary inventory of parasitoids associated with fruit flies in mangoes, guavas, cashew pepper and wild fruit crops in Benin. BioControl 56, 35–43 (2011).
6. Shelly, T. et al. Capture probability of released males of two Bactrocera species (Diptera: Tephritidae) in detection traps in California. J. Econ. Entomol. 103, 2042–2051 (2010).
7. Koyama, J. Pest status; south-east Asia and Japan. In: World Crop Pests 3(A). Fruit flies; their biology, natural enemies and control, 63–66 (1989).
8. Iwaizumi, R. Species and host record of the Bactrocera dorsalis complex (Diptera: Tephritidae) detected by the plant quarantine of Japan. Appl. Entomol. Zool. 39, 327–333 (2004).
9. Christenson, L. D. & Foote, R. H. Biology of fruit flies. Ann. Rev. Entomol. 5, 171–192 (1960).
10. Liquido, N. J. Reduction of oriental fruit fly populations in papaya orchards by field sanitation. J. Agr. Entomol. 10, 163–170 (1993).
11. Burikam, I., Sarnthoy, P., Charensom, K., Kanno, T. & Homma, H. Cold temperature treatment for mangosteens infested with the oriental fruit fly. J. Econ. Entomol. 85, 2298–2301 (1992).
12. USDA Treatment manual. USDA/APHIS, Frederick, USA (1994).

13. Armstrong, J. W., Hu, B. K. S. & Brown, S. A. Single-temperature forced hot-air quarantine treatment to control fruit flies (Diptera: Tephritidae) in papaya. J. Econ. Entomol. 88, 678–682 (1995).
14. Roessler, Y. Insecticidal bait and cover sprays. In: World Crop Pests 3A&B. Fruit flies: their biology, natural enemies and control. 337–345 (1989).
15. Wharton, R. H. Control; classical biological control of fruit-infesting Tephritidae. In: World Crop Pests 3B. Fruit flies; their biology, natural enemies and control. 303–313 (1989).
16. Dyck, V. A. et al. Sterile insect technique: principles and practice in area-wide integrated pest management. Springer, Dordrecht, 525–545 (2005).
17. Shelly, T. E., Edu, J. & McInnis, D. Pre-Release consumption of methyl eugenol increases the mating competitiveness of sterile males of the oriental fruit fly, Bactrocera dorsalis, in large field enclosures. J. Insect. Sci. 10, 1–16 (2010).
18. PubMedArticle
19. Leroy, P. D. et al. Microorganisms from aphid honeydew attract and enhance the efficacy of natural enemies. Nat. Commum. 2, 348–354 (2011).
20. Witzgall, P., Kirsch, P. & Cork, A. Sex pheromones and their impact on pest management. J. Chem. Ecol. 36, 80–100 (2010).
21. Hsu, J. C., Haymer, D. S., Wu, W. J. & Feng, H. T. Mutations in the acetylcholinesterase gene of Bactrocera dorsalis associated with resistance to organophosphorus insecticides. Insect Biochem. Mol. Biol. 36, 396–402 (2006).
22. Daane, K. M. & Johnson, M. W. Olive fruit fly: Managing an ancient pest in modern times. Annu. Rev Entomol 55, 151–169 (2010).
23. Vargas, R. I. et al. Attraction and mortality of oriental fruit flies to SPLAT-MAT-methyl eugenol with spinosad. Entomol. Exp. Appli. 131, 286–293 (2009).
24. Vargas, R. I. et al. Evaluation of SPLAT with spinosad and methyl eugenol or cue-lure for "attract-and-kill" of oriental and melon fruit flies (Diptera: Tephritidae) in Hawaii. J. Econ. Entomol. 101, 759–768 (2008).
25. Leblanc, L., Vargas, R. I., MacKey, B., Putoa, R. & Pinero, J. C. Evaluation of cue-lure and methyl eugenol solid lure and insecticide dispensers for fruit fly (Diptera: Tephritidae) monitoring and control in Tahiti. Florida Entomol. 94, 510–516 (2011).
26. Thomson, D. R., Gut, L. J. & Jenkins, J. W. Biopesticides: use and delivery, Ed. Hall, F. R. & Menn, J. J. Humana Press Inc., New Jersey., 385–412 (1999).

27. Campion, D. G., Critchley, B. R. & McVeigh, L. J. Mating disruption. In: Insect Pheromones in Plant Protection; Ed. Jutsum, A. R. & Gordon, R. F. S. Wiley: New York, 89–119 (1985).
28. Evans, T. A., Dawes, T. Z., Ward, P. R. & Lo, N. Ants and termites increase crop yield in a dry climate. Nat. Commun. 2, 262–268 (2011).
29. Cork, A. Pheromone manual. Chatham Maritime, UK: Natural Resources Institute, Chapter 4, 13–17 (2004).
30. Heuskin, S., Verheggen, F. J., Haubruge, E., Wathelet, J.-P. & Lognay, G. The use of semiochemical slow-release devices in integrated pest management strategies. Biotechnol. Agron. Soc. Environ. 15, 459–470 (2011).
31. White, I. M. & Elson-Harris, M. M. Fruit flies of economic significance: their identification and bionomics. CAB International, Wallingford, UK, 601 (1992).
32. Shelly, T. E. et al. Capture of Bactrocera fruit flies (Diptera: Tephritidae) in traps baited with liquid versus solid formulations of male lures. J. Asia-Pacific Entomol. 14, 463–467 (2011).
33. Jadhav, S. R. et al. Molecular gels-based controlled release devices for pheromones. Soft Matter 7, 864–867 (2011).
34. Bhattacharya, S. & Samanta, S. K. Soft functional materials induced by fibrillar networks of small molecular photochromic gelators. Langmuir 25, 8378–8381 (2009).
35. Terech, P. & Weiss, R. G. Low molecular mass gelators of organic liquids and the properties of their gels. Chem. Rev. 97, 3133–3159 (1997).
36. Samanta, S. K., Pal, A. & Bhattacharya, S. Choice of the end functional groups in tri(p-phenylenevinylene) derivatives controls its physical gelation abilities. Langmuir 25, 8567–8578 (2009).
37. Samanta, S. K. & Bhattacharya, S. Wide range light harvesting donor-acceptor assemblies through inter-gelator interactions via self-assembly. Chem. Eur. J. 18, 15875–15885 (2012).
38. Samanta, S. K., Subrahmanyam, K. S., Bhattacharya, S. & Rao, C. N. R. Composites of Graphene and other Nanocarbons with Gelators Assembled through Supramolecular Interactions. Chem. Eur. J. 18, 2890–2901 (2012).
39. Bhattacharya, S. & Samanta, S. K. Unusual salt induced color modulation through aggregation-induced emission switching of bis-cationic phenylenedivinylene based π-hydrogelator. Chem. Eur. J. 18, 16632–16641 (2012).
40. Samanta, S. K. & Bhattacharya, S. Aggregation induced emission switching and electrical properties of chain length dependent π-gels derived from phenylenedivinylene bis-pyridinium salts in alcohol-water mixtures. J. Mater. Chem. 22, 25277–25287 (2012).

41. Ibrahim, A. G., Singh, G. & King, H. S. Trapping of the Fruit-flies, Dacus spp. (Diptera: Tephritidae) with methyl eugenol in Orchards. Pertanika 2, 58–61 (1979).
42. CAS Yamamura, K. Transformation using (x + 0.5) to stabilize the variance of populations. Res. Popul. Ecol. 41, 229–234 (1999).
43. Gomez, K. A. & Gomez, A. A. Statistical procedures for Agricultural research. Second edition, pp. 17. (1984).

CHAPTER 18

Developing the Arsenal Against Pest and Vector Dipterans: Inputs of Transgenic and Paratransgenic Biotechnologies

Christian E. Ogaugwu[1*] and Ravi V. Durvasula[2]

[1] *Department of Animal and Environmental Biology, Federal University Oye-Ekiti, Ekiti State, Nigeria*

[2] *Department of Internal Medicine, University of New Mexico, Albuquerque, USA*

ABSTRACT

Insects are the most numerous of all animals and are found in almost every inhabitable place on earth. The order Diptera (true flies) contains many members that are notorious agricultural pests, nuisance or vectors of diseases. The list is long: mosquitoes, tsetse flies, screw worms, fruit flies, sand flies, blow flies, house flies, gall and biting midges, black flies, leaf miners, horse flies, and so on. Efforts to combat some of these pests and vectors have resulted in control measures such as the chemical, physical, and cultural control methods. These methods, though largely beneficial, have disadvantages and limitations, which sometimes seem to outweigh the problems initially sought to be controlled. The chemical method, for example, is not environment-friendly since it negatively affects many nontarget organisms and disrupts ecosystem balance. Development of insecticide resistance by pests/vectors is another concern. Molecular biotechnology has introduced vast arrays of novel ways to tackle pests and disease vectors, as well as improve the potency of existing control methods. This chapter looks at transgenic and paratransgenic biotechnologies and how they have been applied so far to develop and expand the arsenal against dipteran pests and disease vectors. Further, we discuss the advantages, disadvantages, and limitations of these technologies.

Keywords: insects, dipterans, crop pests, disease vectors, transgenesis, paratransgenesis

1. INTRODUCTION

Insects are highly abundant and are the most numerous classes of all described living animals. They account for about half or more of all living animals and are found in almost every inhabitable place on earth [1, 2]. Their success and abundance could be attributed to their ability to adapt and colonize diverse habitats. Among insects, the order Diptera is one of the largest orders with an estimated 120 families and 250,000 described species [1]. They are generally regarded as the two-winged insects or true flies [2, 3]. The main characteristics for members of this order include larvae that lack legs (apodous maggots), pupae enclosed in a thick larval cuticle (puparium), and adults that possess a pair of membranous forewings, vestigial hindwing modified into halters, as well as a tubular sucking or sponging mouthparts [2]. Dipterans are longtime foes and arguably considered the insect archenemy of man. This stems from the fact that many members of this order constitute pests of cultivated crops, are major causes of annoyance or are highly notorious as vectors of human diseases either in their larval or adult stages. Examples of pest, annoyance causing and vector dipterans are given in **Table 1** [4–8].

Table 1. Dipteran crop pests, nuisance or vectors of diseases [4–8].

Family	Genus/species involved	Problematic or damage-causing stage; problem caused
Agromyzidae (leaf miners)	*Phytomyza angelicastri; Melani agromyza*	Larva; damage to leaves of crops
Anthomyiidae	*Antherigona* spp; *Delia radicum*	Larva; damage to stems of crops like cauliflower and sorghum causing disease like "dead heart"
Calliphoridae (blow flies)	*Callitroga* spp.; *Cordylobia anthroporhaga; Lucilia* spp; *Chrysomya bezziana*	Larva; myiasis or flesh infesting damage to man and livestock
Cecidomyiidae (gall midges)	*Contarinia sorghicola*	Larva; damage to leaves of crops such as rice, pear, sorghum, etc.
Ceratopogonidae (biting midges)		Adult; blood sucker from man and livestock
Chloropidae (chloropid flies)		Larva; damage to leaves of crops like rice and cereals
Culicidae (mosquitoes)	*Anopheles* spp.; *Aedes* spp.; *Culex* spp.; *Mansonia* spp.; *Psorophora* spp.; *Stegomyia* spp	Adult; blood sucker from man and livestock transmitting parasites that cause various diseases like malaria, dengue fever, West Nile fever, yellow fever, encephalitis, O'nyong nyong fever, Bancroftian filariasis, chikungunya, Igbo-Ora, Zika, etc.; nuisance, major cause of disturbance and annoyance to man at night
Drosophilidae	*Drosophila suzukii*	Larva; damage to fruits
Glossinidae (tsetse flies)	*Glossina* spp.	Adult; blood sucker from man and livestock transmitting the causative agent of Trypanosomiasis (sleeping sickness)
Muscidae (house flies)	*Musca domestica*	Adult; transmits microorganisms that cause cholera and amoebic dysentery; nuisance, major cause of annoyance to man during the day
Oestridae (warble or bot flies)	*Oestrus ovis; Gasterophilus* spp.; *Hypoderma bovis; Dermatobia hominis*	Larva; myiasis or flesh infesting damage to man and livestock
Psychodidae (sand flies and moth flies)	*Phlebotomus* spp.	Adult; blood sucker from man transmitting the parasite causing disease leishmaniasis
Sarcophagidae (flesh flies or screw worms)	*Cochliomyia hominivorax; Sarcophaga* spp; *Wohlfahrtia* spp.	Larva; myiasis or flesh infesting damage to man and livestock
Simuliidae (black flies)	*Simulium* spp.	Adult; blood sucker from man transmitting the causative agent of the disease onchocerciasis (river blindness)
Tabanidae (horse flies)	*Tabanus* spp.; *Haematopota* spp.; *Chrysops* spp.	Adult; blood sucker from man and livestock transmitting the causative agent of diseases like trypanosomiasis (sleeping sickness) and loasis
Tephritidae (fruit flies)	*Anastrepha* spp.; *Bactrocera* spp.; *Ceratitis* spp.; *Dacus* spp.; *Rhagoletis* spp.; *Tephritis* spp.	Larva; serious damage to fruits and vegetables

Efforts by man in the fight against dipteran pest and vector insects have resulted in the generation of an arsenal with several weapons. These range from chemical method which involves the use of insecticides to cultural methods such as sanitation, physical interference or destruction of breeding sites, and cropping methods. However, many of these methods have major disadvantages and (or) limitations that sometime seem to outweigh their benefits. For example, the chemical method is very widely used, but has the disadvantages of environmental consequences such as pollution, health challenges on man and livestock, killing of nontarget insect species, as well as the challenge of the targeted insect species developing resistance to the insecticides applied [9, 10]. Most cultural methods applied against dipteran pest or vector control are labor-intensive and can only be most suitably applied on a small scale.

Biologically-based approaches are generally friendlier to the environment, more sustainable and cost-effective than many other methods used for dipteran control. Here, control methods such as the use of natural enemies like predators and parasitoids are environment-friendly with varying levels of success, but the major limitation is the fact that it is unpredictable as chances are usually low on finding a suitable parasitoid or predator that can survive the weather and conditions wherever the pest or vector dipteran is and continue to effectively eat or parasitize the host [10]. The time it takes to find a good parasitoid may be so long that farmers or entomologists concerned may opt for other control methods, in addition to the fact that the process of actual control by a parasitoid or predator itself is slow. The biological method of using pathogens (microbial or biopesticides) has been quite promising, but recently there has been concerns of insect resistance as is the case with *Bacillus thuringiensis*, and also the disadvantage that the applied pathogen may infect other nontarget insects, livestock, or man himself. Major limitations of biopesticides are usually that one may need to find an efficient way to get the pathogens to their host and that the pathogens may be negatively affected by environmental conditions such as weather.

Another biologically-oriented and environment-friendly method for controlling dipterans is the use of pheromones or suitable attractants. However, the scale of its application and area that it covers is also limited, while the potency of the attractants does reduce gradually with time or could easily be influenced by environmental factors such as rainfall or masked by other chemicals within the vicinity.

Genetic methods such as the sterile insect technique (SIT) majorly use radiation to sterilize male insects and thus reduce the fertility of females that mate with them [11, 12]. This method works well with sexually reproducing insects and has so far has great success among many dipterans [13]. Its major problem is the fact that the gamma radiation used for sterilization also reduces the fitness of the males and makes them less competitive than the wild males.

Obviously, no present method of pest control is devoid of disadvantages or limitations. As such, an integration of different control methods that are compatible is the recent paradigm. Integrated pest management (IPM) has offered a way to augment control methods to achieve a more efficient and sustainable management of pests and vectors.

The new millennium has witnessed advanced progress in genetic biotechnology which in turn has greatly influenced insect control. Biotechnology approaches have been used and are continually been pursued as a means to develop novel ways or improve some of the methods used to fight pest and vector dipterans. For example, new strains of reproductively sterile insects or strains exhibiting other desired traits could be engineered to control a population or designed to fit into control methods like SIT, entomopathogens or biopesticides that are adversely affected by weather conditions where a pest or vector is located or have environmental concerns regarding nontarget insects could be encapsulated in materials that will release the pathogens only in a desired condition, while nonharmful microorganisms could be engineered to deliver therapeutic or antiparasitic molecules to pathogens in their environment. Many of these new biotechnology approaches could also be used as a part of IPM programs which is suitable for other methods. In this chapter, we focus on how transgenic and paratransgenic biotechnologies have been applied to expand the array of weapons in man's arsenal against dipteran crop pests and vectors of diseases.

2. TRANSGENIC BIOTECHNOLOGY

Transgenesis aims at the transformation of an organism by altering its genetic composition and the final outcome is the generation of a transgenic or a genetically modified organism (GMO). Basically, desired genes or genes-of-interest from a different organism(s) are inserted into the genome of a wild type organism majorly with the aid of "jumping genes" called transposable elements or transposons and the transgenic organism generated carry these desired genes (transgenes), while exhibiting characters or traits encoded by the transgenes as well (**Figure 1**). For insects, germline transformation is sought and microinjections are performed to achieve it, allowing the genome modification to be passed on from generation to generation [14]. To enable detection of successfully transformed organisms, fluorescent proteins such as the green fluorescent protein (GFP), the red fluorescent protein (RFP), and fluorescent proteins of other colors are used as markers [15–18]. Consequently, GM dipterans harboring a transgene that incorporates a fluorescent protein gene cassette as marker would express the fluorescent protein and can be visualized under a fluorescent microscope (**Figure 2**).

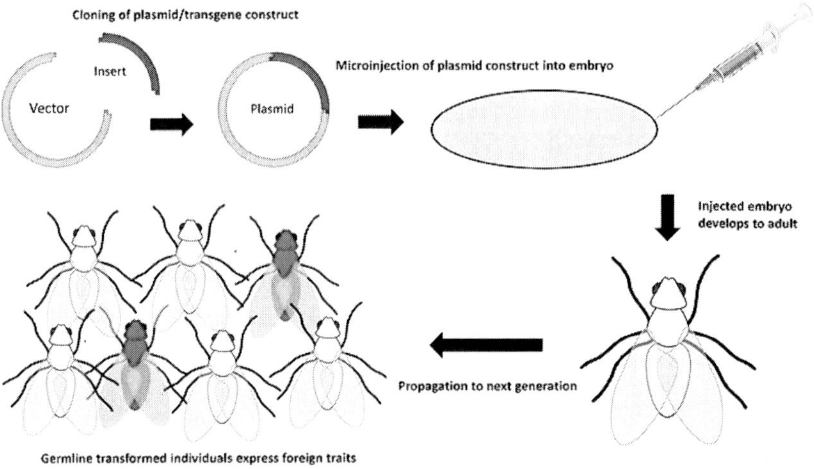

Figure 1. Schematic representation of transgenesis: desired gene-of-interest is cloned into a vector to generate a plasmid or transgene construct which is then microinjected into embryos. Adults developing from the injected embryos are outcrossed to non-injected ones and their progeny are screened. Progeny that are stably germline transformed express traits encoded by the genes in the plasmid construct injected, for example a red fluorescent protein, and as such are distinguished from untransformed ones.

2.1. Transgenic Strategies Against Agricultural Crop Pest Dipterans

2.1.1. Drosophilids

The family Drosophilidae consists of many members including the well-known model fly *Drosophila melanogaster*, but only the spotted wing drosophila, *Drosophila suzukii*, is considered a major pest of cultivated crops [19]. However, *D. melanogaster* has been immensely beneficial in genetic studies and many proofs-of-principle of transgenic strategies against dipteran population control, or even for other insect orders, have been developed in this model insect.

A proof-of-principle transgene-based, embryo-specific lethality system for insect control was developed by Horn and Wimmer [20]. The system used embryo stage-specific promoters such as serendipity alpha (sryα) to regulate the expression of a hidAla5 lethal effector placed under the control of a tetracycline-response element [20]. Such a strain would effectively achieve reproductive sterility in insect populations because the offspring die during the embryo stage and could replace radiation sterilization of insects as is the case for conventional SIT.

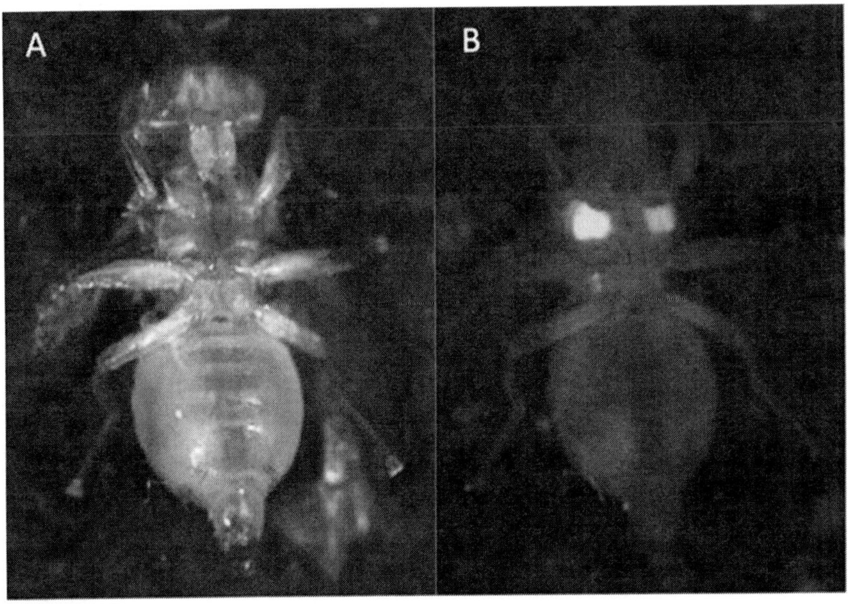

Figure 2. A transgenic strain of the Mediterranean fruit fly Ceratitis capitata: (A) visualized under cold light, (B) visualized under fluorescent light, the same transgenic fly shows a pattern of expression of green fluorescent protein GFP in its thorax and legs.

Two proofs-of-principle for transgenic sex-specific lethality systems for insect population control were developed: (i) using female-specific enhancers of yolk protein 1 (yp1) gene to drive expression of a hid effector under control of tetracycline-responsive element [21]. (ii) using a female-specific yolk protein and fat-body enhancer Yp3 to regulate the expression of a Ras64BV12 effector under control of tetracycline-responsive element, as well as using a Hsp26 promoter to regulate expression of a dose-compensation gene, mutant male-specific lethal 2 (msl-2NOPU), under the control of a tetracycline-response element [22]. These kinds of systems limit lethality or death of offspring to only female individuals and could be used for efficient sex separation of dipterans prior to field release during area-wide dipteran pest control programs such as SIT.

Besides these afore-mentioned transgenic lethality systems which were all based on the tetracycline-repressible binary expression system (**Figure 3**) [23], a gene-driven system capable of driving population replacement was also developed in Drosophila [24]. Basically, a gene-driven system such as a maternal-effect dominant embryonic arrest (Medea) system use a combination of two genes that encode for a toxin and its antidote, respectively, to create a condition whereby a heterozygote female would express only the maternal toxin in half of her oocytes without the antidote resulting in death of those offspring. The Medea strain which was developed by Chen et al. in Drosophila used microRNA-mediated silencing of a maternally expressed embryonic development gene, my88, as its toxin and early zygotic expression of a rescuing transgene as the antidote. A more complex Medea system employing additional mechanisms such as targeting signaling pathways like the Notch pathway has since been also demonstrated in *D. melanogaster* [25].

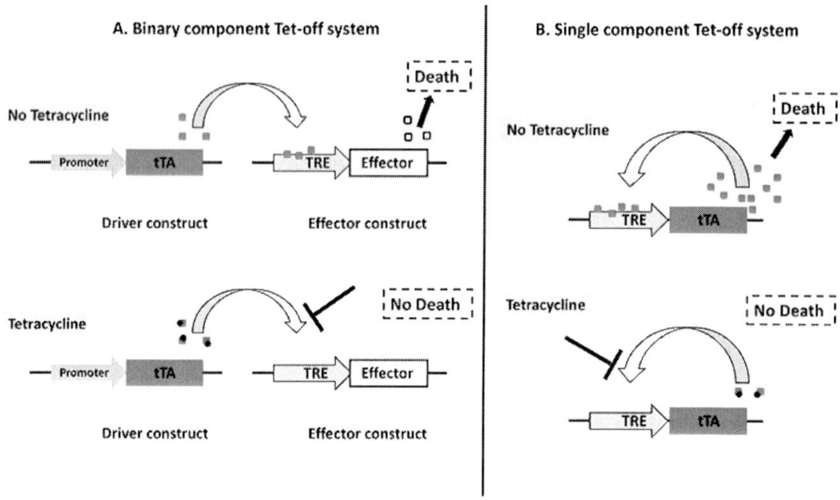

Figure 3. Diagrams showing different versions of tetracycline-repressible expression system: (A) binary component tet-off system, (B) single component tet-off system [**23, 29**]. In the absence of tetracycline, tetracycline-represssible transactivator (tTA) is produced and goes on to bind to the tetracycline-response element (TRE) to activate expression of a downstream gene. Both systems are turned off in the presence of tetracycline which bind to the tTA and stops expression of a downstream gene. The binary system needs an effector for lethality, while the single component uses tTA which is toxic at high concentration.

For the crop pest Drosophilid, the spotted wing drosophila *D. suzukii*, germline transformation has recently been performed and transgenic strains for control of this strain may soon be generated [26].

2.1.2. Tephritid Fruit Flies

Tephritids are very important pests of fruits and vegetables and majority of transgenic strategies for crop pests have been developed against members of this group. Lethality systems that their proofs-of-principle have earlier been developed in Drosophila have also been successfully transferred to many Tephritids. Among these are the conditional embryonic lethality strains transferred from *D. melanogaster* to both the Mediterranean fruit fly, *Ceratitis capitata* and the Caribbean fruit fly, *Anastrepha suspensa*, and using the tetracycline-regulated binary expression system, embryonic promoters/enhancers and proapoptotic hid effector [27, 28]. In addition, the lethality strains not previously established in Drosophila was developed for *C. capitata* using a simplification of the tetracycline-regulated binary expression system to a single expression component that relies on auto feedback-driven overexpression of a version of the tetracycline-repressible transactivator (tTA) for its lethality (**Figure 3**) [29].

For sex separation of Tephritids, transgenic sexing strains were developed for different fruit fly genera: (i) an RNA interference (RNAi) system developed for *C. capitata* based on knockdown of transcripts of the sex determination gene transformer (tra) [30], (ii) lethality systems relying on a simplified single component tetracycline expression system and developed for *C. capitata* and the olive fruit fly, *Bactrocera oleae* [31, 32], (iii) lethality systems relying on a tetracycline-regulated binary expression system [23], embryonic promoters/enhancers and proapoptotic hid effector, and developed for *C. capitata, A. suspensa*, and the Mexican fruit fly, *Anastrepha ludens* [33–35]. Unlike the two proof-of-principle transgenic sexing systems based on lethality earlier developed in Drosophila, all the afore-mentioned transgenic sexing systems based on lethality in Tephritids employed the sex-specifically spliced intron of the gene transformer (tra) to confer lethality only to the female individuals. However, only those systems employing the tetracycline-regulated binary expression system and embryonic promoters or enhancers achieved female-specific lethality in the embryo stage [33–35]. Another type of transgenic sex-specific lethality system has recently been developed for the Oriental fruit fly, *Bactrocera dorsalis* [36]. This system combined the mechanism of alternative splicing of the double sex (dsx) gene and the toxicity of expressed ricin to ensure female-specific lethality and kill off the female progeny in *B. dorsalis* [36].

Since area-wide dipteran pest control strategies like SIT involve release of sterile males, a way to monitor the released males is also as important as the sterilization and sex separation of the males. Scolari et al. developed a transgenic strain that would facilitate such monitoring in C. capitata by using the promoter of a sperm-specifically expressed gene β2-tubulin (β2t) to regulate the expression of RFP and GFP to only male testis. Males of this sperm-marked fly strain were shown to still have brightly glowing fluorescent testis for several months after they had died [37]. As such, the released males could easily be monitored if caught in traps or found dead in the field in the case they were used in any SIT control program.

2.2. Transgenic Strategies Against Dipterans of Medical and Veterinary Importance

2.2.1. Mosquitoes

The battle against any other dipteran insects has perhaps never been as intense as it is for mosquitoes due to the wide range of diseases they vector and transmit. Almost every kind of approach that is imaginable is under development or has been developed in the effort to win the battle against mosquitoes. Ever since the first germline transformation of an Anopheles mosquito [38], several transgenic strategies have been constructed including gene drive systems, lethality, flightless, sperm-monitoring, as well as spermless systems, and mosquito strains that have been impaired in their ability to transmit a parasite.

Among the gene-driven systems include a maternally-regulated transposition system in the yellow fever mosquito, *Aedes aegypti*, which utilized regulatory elements of a maternal gene Nanos to control events in mosquito embryos [39]. A synthetic gene drive system developed in the human malaria mosquito, *Anopheles gambiae*, exploited I-SceI which is a selfish genetic element known as a homing endonuclease gene (HEG) to drive rapid invasion of mosquito population genomes by the engineered gene of interest [40]. The recently developed clustered, regularly interspaced, short palindromic repeats (CRISPR) and CRISPR-associated protein (Cas), better known as the CRISPR-Cas9 system that enables flexible genome editing in both prokaryotic and eukaryotic cells [41–43] using a guide-RNA to direct the nuclease to its target has also been exploited to develop a gene drive system with high transmission rate to progeny of up to 99.6% in *An. gambiae* [44].

Strains exhibiting dominant lethality or major incapacitation in the form of a flightless phenotype have also been generated. A transgenic strain based on the expression of dominant lethality in *Ae. aegypti* was constructed using similar components (a single expression component that relies on auto feedback-driven overexpression of a version of tTA) as was used for the Mediterranean fruit fly *C. capitata* [29, 45]. Fu et al. also generated a flightless strain of *Ae. aegypti* for dengue fever control by using its Actin4

(Act4) gene promoter in the single component tet-off expression system (**Figure 3**) to regulate the expression of tTA [46]. Act4 is female-specifically expressed in the indirect flight muscles [47], and as such, the tTA-mediated lethality regulated by its promoter is obtained predominantly in the female indirect flight muscles rendering them flightless and providing a way to genetically separate the sexes or enable possible male-only mosquito release in an SIT program [46].

For transgenic sexing and sperm monitoring, Catteruccia et al. established a strain that exhibited fluorescent sperms in the Asian malaria mosquito *Anopheles stephensi*, employing the promoter/enhancer elements of the β2 tubulin gene to control and ensure expression of enhanced green fluorescent protein (EGFP) in male testis [48]. This has been followed by further sperm manipulation whereby to study possible roles of sperms in regulation of postmating female responses in the major malaria mosquito *An. gambiae*, a spermless strain was engineered by RNAi-mediated silencing of a developmental gene required for early germ cell differentiation, zero population growth (zpg) [49]. Mosquito control programs may benefit from the interesting report that female mosquitoes mated to the spermless males become refractory to further mating [49]. Moreover, such a spermless mosquito strain also possesses reproductive sterility and could also find application in SIT programs for mosquito population control. A very recent sex distorting system developed in An. gambiae employs the CRISPR/Cas9 endonuclease to shred the X chromosome and lead to male bias in progeny without significantly reducing the adult's fertility [50].

Another strategy that has also been pursued is to transgenically impair the ability of mosquitoes to transmit malaria Plasmodium parasites. To this end, transgenic Anophelines were developed that were unable to vector Plasmodium parasites as they express an antiparasitic peptide, the salivary gland and midgut peptide 1 (SM1) in their midgut epithelia under regulation by a carboxypeptidase promoter [51]. In the wake of insecticide and drug resistance by both vector and parasite, respectively, this approach offered an avenue to curtail transmission while not removing the vector and could easily be spread to wild mosquito population using some of the gene drive systems developed. Several other researchers have followed this strategy and developed transgenic mosquitoes that cannot transmit their parasites. Bee venom phospholipases, synthetic antimalaria proteins like vida3, single chain antibodies (scFv) targeting malaria parasites, as well as an antimicrobial peptide cecropin A have been used as effectors and mosquitoes engineered to express them lack the ability to effectively transmit parasites [52–56]. RNAi-based resistance to dengue virus has also been engineered in *Ae. aegypti* mosquitoes by using inverted-repeat RNA (IR-RNA) from the premembrane protein coding region of the DENV-2 RNA genome whose expression was regulated by a carboxypeptidase promoter to suppressed viral replication in the midgut [57].

2.2.2. Blow Flies

Veterinary pests such as Blow flies that inflict enormous damage to sheep and other livestock have also received attention lately. Transgenic sexing strains that allow male-only production for control of the Australian sheep blow fly Lucilia cuprina were developed using both the single and binary component tetracycline-repressible expression system. An initial single component female-specific lethality system showing lethality in pupa used a heat shock promoter Hsp70 and the transformer (tra) intron from *Cochliomyia hominivorax* to limit lethality to only females [58, 59]. A later strain which used the binary component of the tetracycline-repressible expression system and showed lethality in embryos utilized promoters of cellularization genes to drive expression of tTA and the transformer (tra) intron from *C. hominivorax* placed inside a hidAla2 effector gene to confer the lethality to only female individuals [60].

2.2.3. Screwworms

Though the very successful strategy of SIT had originally been developed against the New World Screwworm *C. hominivorax* [11], this has not translated into success in development of transgenic strains of this insect. Transformation of *C. hominivorax* is much more challenging than other dipterans and efforts have resulted in few transgenic strains that allow genetic marking with fluorescent proteins for management and control of screwworms [61, 62]. It is expected that similar transgenic sterile male strains, sexing strains, sperm marked, and organismal lethal strains will be developed for screwworms in the near future as had already been done for other dipterans [62].

3. PARATRANSGENIC BIOTECHNOLOGY

Similar to transgenesis, paratransgenesis also involves the genetic transformation of organisms. However, paratransgenesis targets to achieve the genetic transformation or transgenesis of the symbionts that live inside an insect instead of the insect itself and cause the symbionts to express or secrete substances that act against parasites and pathogens that are transmitted by the insect (**Figure 4**). Consequently, paratransgenesis is suitably applied against disease vectors. Originally developed against the triatomine bug vector of Chagas disease, Rhodinus prolixus using its symbiont Rhodococcus rhodnii and the antimicrobial peptide cecropin A as an effector [63], this strategy has been adopted for many dipterans that vector diseases of humans and livestock.

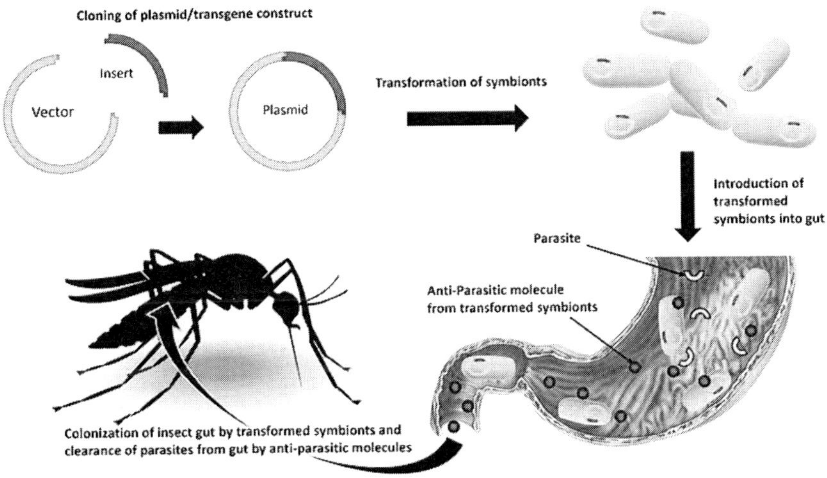

Figure 4. Schematic representation of paratransgenesis: transgenes encoding anti-parasitic molecules are cloned into a plasmid which is used to transform suitable symbionts of a vector insect. When the transformed symbiont is introduced into the gut of the vector, they colonize the gut whereas the anti-parasitic molecules they produce act against the parasites and clear them off.

3.1. Paratransgenic Strategies Against Medical and Veterinary Important Dipterans

3.1.1. Mosquitoes

Since mosquitoes transmit several disease-causing pathogens, many paratransgenic studies have been conducted on it. Inhibition of vectorial competence in mosquitoes via bacterial symbiont paratransgenesis were demonstrated in the malaria mosquito *An. stephensi* using genetically modified strains of the popular gram negative bacteria Escherichia coli to express scFvs that block development of the parasite *Plasmodium berghei* [64] or an anti-Plasmodium molecule such as SM1 [65]. Other bacteria such as Asaia spp and Pantoea agglomerans (formerly Enterobacter agglomerans) have also been used. Favia et al. showed that the Asaia associates stably with *An. stephensi* and that transgenic strain of this bacteria expressing GFP are able to colonize the gut and salivary gland of females of this mosquito [66]. In another study using Asaia spp, paternal transmission of recombinant strains expressing the green fluorescent protein GFP or the red fluorescent protein DsRed to progeny through mating of paratransgenic males with wild females was obtained in *An. stephensi* showing that it is possible to utilize nonbiting male mosquitoes in malaria transmission [67]. Working with *Pantoea agglomerans*, Wang et al. were able to express several anti-Plasmodium molecules such as SM1 peptide, scFv, mutated phospholipase (mPLA2), Plasmodium enolase-plasminogen interaction peptide (EPIP), synthetic antiparasitic lytic peptide Shiva1, etc., in both *An. gambiae* and *An. stephensi* and successfully suppressed transmission of *Plasmodium falciparum* and *P. berghei*, respectively [68].

Besides bacteria, fungi and viruses have also been utilized in mosquito paratransgenesis. The entomopathogenic fungi Metarhizium anisopliae was engineered by Fang et al. to express the anti-Plasmodium molecules SM1, a sporozoite-agglutinating scFv, as well as an antimicrobial toxin scorpine in An. gambiae [69]. Using the densonucleosis virus (DNV) in a proof-of-concept viral paratransgenesis work in *An. gambiae*, the potential of virions in paratransgenesis was demonstrated by ability of transgenic DNV expressing GFP to infect larvae, persist to the adult stage and disperse to vital tissues such as fat bodies, ovaries and midgut, and be transmitted subsequently to other generations [70]. This shows that viruses such as DNV could be used to express antiparasitic molecules not only in one but several mosquito developmental stages and subsequent generations and could effectively mitigate and eliminate malaria transmission. Another virus, the Sindbis virus, has also been exploited in paratransgenesis and used to express scFv that acts against *Plasmodium gallinaceum* sporozoites in *Ae. aegypti* [71]. As it appears, the Sindbis virus has great potentials for control of various viruses transmitted by Aedes mosquitoes [72–74].

3.1.2. Sandflies

Efforts on control of sandfly vectors that transmit Leishmania parasites which cause the disease Leishmaniasis has been done mainly with chemical insecticides. To develop a more environment-friendly strategy, Hurwitz et al. recently demonstrated the feasibility of paratransgenesis for sandflies in a proof-of-principle work in *Phlebotomus argentipes* in which they used recombinant *Bacillus subtilis* fed to larvae to express GFP in the gut lumen of emerging adults (**Figure 5**) [75]. This proof-of-principle study has paved the way for future development of strains that will express anti-Leishmania molecules and block transmission of the parasites by the sandfly vector.

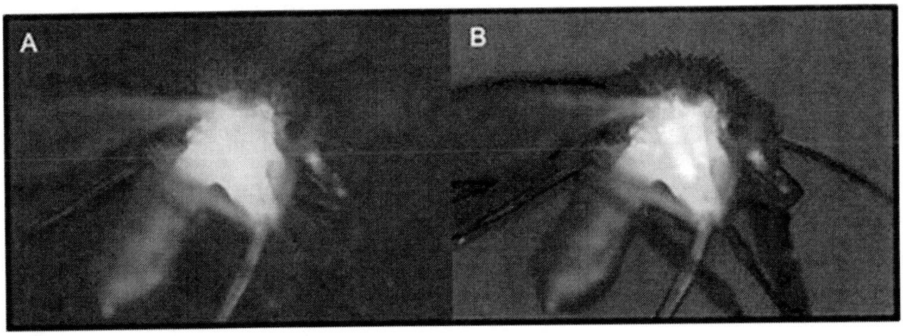

Figure 5. Paratransgenic sandfly Phlebotomus argentipes: (A) auto fluorescence of the outer carapace of the sand fly is seen amidst the presence of GFP expressed by the symbiont, (B) subtilis in the sand fly's midgut B. visualization of GFP specifically localized in the sand fly's midgut chamber upon uncoupling of the GFP signal from the background [**75**].

3.1.3. Tsetse Flies

Although several control strategies including SIT have been applied against tsetse flies, continual effort is made to develop other methods that would not have limitations of the existing methods, be more sustainable, more cost-effective or suitable for use in IPM control. To investigate the possibility of paratransgenesis in tsetse flies, transgenic *Sodalis glossinidius* were introduced into adult females where they were able to express GFP and interestingly passed on subsequently to the progeny of those females [76, 77]. Actual utilization of antitrypanosomal molecules to block parasite development and transmission by tsetse flies could be achieved in the near future.

4. ADVANTAGES, DISADVANTAGES AND LIMITATIONS OF TRANSGENIC AND PARATRANSGENIC TECHNOLOGIES

4.1. TRANSGENIC TECHNOLOGY

4.1.1. Advantages of transgenic technology

The main advantage of transgenic biotechnology is its ability to generate strains that possess traits that are unique and special, and accurately designed or tailored to be specific as desired. Also, the flexibility of transgenic technology allows generation of such desired strains in many species which would have been very difficult or impossible to achieve by other means. Transgenic strains are usually generated after one generation (**Figure 1**), and hence take less time to generate compared to other methods like classical genetics. Moreover, generation of strains possessing desired traits in one species can almost always be reproduced and transferred to related species with relative ease [78]. Quite unlike earlier genetic methods such as SIT where the use of radiation generates unknown and uncharacterized genetic mutations, transgenic technologies generate known and characterizable genetic modifications. Transgenes could easily be thoroughly characterized, and same goes for genomic positions in the dipteran insect where desired transgenes had got inserted. Also, most transgenic strategies are environment-friendly, sustainable and target-specific. For example, while chemicals developed against fruit flies may kill pollinator insects, transgenic strains developed for control of fruit flies are not likely to have any negative effect on pollinators that their wild counterparts do not already exhibit. Also, the development of resistance against control agents is less likely to occur when using transgenics.

In terms of costs, transgenic technologies as well as many other control strategies are not so cheap to develop. But it is difficult to say with all certainty whether transgenic (and paratransgenic technology) is cheaper than many other methods as there have not been any such economic studies to the best of our knowledge. Nevertheless, transgenic (and paratransgenic) approaches are considered less expensive with regards to the farmer or public beneficiaries as they are usually area-wide-oriented and implemented by big organizations at overall little or no cost to the individual farmers or the public.

4.1.2. Disadvantages of Transgenic Technology

Probably due to the fact that transgenic technology is just beginning to move from laboratory to the field [79, 80], there are yet no scientifically proven disadvantages. Despite this, many public negative concerns already exist on the use of transgenics [78], mostly environmental and social, as well as safety and ethical issues. These are mainly due to speculations and the uncertainty as to what might happen in nature following field use of transgenics, and whether unintentional and unforeseen mutations could lead to harmful consequences (though these can potentially occur also in nontransgenics). There are also thoughts on how field use of transgenics could interfere with diversity and evolution due to possible loss of genetic material of original insects and the associated future downstream events. Potential horizontal transfer of transgenes could also be a potential disadvantage that could be associated with the use of transgenic technology. However, a "self-limiting" transgenic approach such as use of transgenic lethality dipteran strains should not present some of the aforementioned environmental problems since they are most likely to remove themselves from the environment with time unlike "self-sustaining" strains [81]. Though not always the case, there may be fitness costs that might arise in dipterans and other insects due to the various loads of transgenes they carry [82, 83]. When well assessed, the fitness costs could be determined and measures taken to eliminate them if necessary or avoid using strains that suffer such lack of fitness. New transgenic strains that may not have the observed fitness cost could also be developed and utilized instead. Proper assessment should be done to determine the associated risks and benefits before any GMO can be utilized [84].

4.1.3. Limitations of Transgenic Technology

Transposable elements or "jumping genes" have been the main tool relied on to achieve germline transformation and generate transgenic dipterans. However, most of the transposons used in dipterans (and other insects too) are insect-derived [85] and a major concern is that a transposon could potentially be remobilized from its integrated genomic position in the insect if transposases required for its activity is encountered in the field. The consequence of such transgene-transposase exposure could be the remobilization of a transgene to another genomic location or total loss of a transgene from an insect's genome. Measures to avoid potential transgene remobilization in engineered dipterans such as postintegrational transgene modification to alter the transposon and achieve nonmobilization or stability has been demonstrated in *D. melanogaster* and *C. capitata* [86–88]. Other strategies that offer transgene stability are becoming available. The recently developed genome editing tool, CRISPR/Cas9, which allows RNA-guided modification of target DNA locations [41] has been utilized to achieve stable germline transformation in *D. melanogaster* [43]. Unlike transposon-mediated germline transformation, CRISR/Cas9-mediated germline transformation is seamless and should not be prone to subsequent remobilization.

Transgenesis is not yet possible in all dipterans as not all members are amenable to it. Since the development of a transgenic insect strain involves germline transformation (**Figure 1**), it is therefore important that the biology of a target insect must be in such a way that allows the necessary manipulations to achieve genetic transformation. Tsetse flies are yet to be genetically transformed due to their viviparity which makes it difficult to obtain embryos needed for microinjections and subsequent germline transformation [77].

4.2. Paratransgenic Technology

4.2.1. Advantages of paratransgenic technology

While similar to transgenic approach in terms of its ability to generate within a short time strains that possess unique and special traits designed specifically as desired, paratransgenesis also has an additional advantage of leaving the insect itself genetically unmodified and rather targets the parasites transmitted. This gives paratransgenic approaches a major plus in the sense that it has a more positive public perception than transgenic approaches as many of the disadvantages with use of transgenics would not be present [78, 89]. In addition, this technology has a high potential to be transferred between different species [78]. Moreover, paratransgenic biotechnology mostly employs microorganisms that live within the target dipterans (symbionts) and as a result also has a high likelihood of field success. Another advantage of paratransgenics is the absence of fitness cost of genetic manipulation compared to transgenics or other control strategies [90].

4.2.2. Disadvantages of paratransgenic technology

Field application of paratransgenic strategies is yet to be actualized and any potential disadvantage of this technology is still to be proven scientifically. Nevertheless, safety concerns and risk assessments have become necessary requirements that need to be addressed to ensure that the benefits outweigh the risks of utilized genetically modified organisms [84]. One concern for paratransgenics is the potential exposure of engineered symbionts to the environment and likely consequences such as horizontal gene transfer. Measures such as symbiont encapsulation to ensure regulated release are being taken to address some of these regulatory concerns [91].

4.2.3. Limitations of paratransgenic technology

Despite the known advantages of the paratransgenic approach, a major limitation is that it is not suitable for most dipteran crop pests and has been developed mostly for those dipterans (and other insects) that transmit disease pathogens. Symbiont choice and utilization in a paratransgenic expression approach depend not only on availability of symbionts that can be isolated, cultured, reintroduced, and survive well in the targeted host, but also on the ability of the symbiont to be genetically transformed and to possibly express antiparasitic molecules [77]. The lack of some of these requirements would render several good symbionts unusable for paratransgenic control. The bacteria symbiont Wolbachia is one such microorganism that is promising for paratransgenic application, but the lack of success in genetically transforming it has hindered its further utilization for expression of antiparasitic molecules.

5. FUTURE OF TRANSGENIC AND PARATRANSGENIC TECHNOLOGIES

In the near future, transgenic and paratransgenic pest/vector control strategies may become common place and more widely applied than it is now. Some of the novel approaches of these technologies are promising and offer great hopes for control of several human diseases and could be implemented in the near future if regulatory and ethical issues are satisfied [92, 93]. This could usher in a new era where cases major dipteran-vectored diseases of man such as malaria and dengue, as well as agricultural pest like Tephritid fruit flies become much reduced or even eradicated.

The arrays of weapons in man's arsenal against his dipteran enemies are also expected to continue to expand. Continuous improvement will be made to existing control strategies, while new and better strategies are expected to be developed in the future as more advances are made in genetics and molecular biology. The RNA-guided genome editing tool, the CRISPR/Cas9 endonuclease recently developed from bacteria such as *Streptococcus pyogenes*and *Neisseria meningitidis* [41, 94] has equally enabled genome modification and generation of transgenic control strategies in dipterans [43, 44, 50, 95]. More recently, a DNA-guided genome editing which makes use of an argonaute from the bacteria *Natronobacterium gregoryi* [96] has also been developed and it is expected that this new tool, as well as others that may soon be developed, will definitely lead to the generation of new transgenic or paratransgenic approaches to better control pest, nuisance or vector dipterans.

REFERENCES

1. Gullan PJ, Cranston PS. The Insects: An Outline of Entomology. 1st ed. London: Chapman and Hall; 1994. 491 p.
2. Encyclopedia of Life [Internet]. [Updated: 2016]. Available from: http://eol.org/pages/421/details [Accessed: 2016-09-26]
3. Wiegmann BM, Yeates DK. Diptera—True Flies [Internet]. [Updated: 2007]. Available from: http://tolweb.org/Diptera in the Tree of Life Web project [Accessed: 2016-07-03]
4. Igbinosa IB. Parasitic Arthropod. In: Ezigbo JC, editor. Parasitology for Medical Students. Lagos: New Frontiers Publishers Limited; 1990. pp. 41–66.

5. Discover Life. Diptera [Internet]. [Updated: 2016]. Available from: http://www.discoverlife.org/mp/20q [Accessed: 2016-09-26]
6. Wikipedia. Biology of Diptera [Internet]. [Updated: 2015]. Available from: https://en.wikipedia.org/wiki/Biology_of_Diptera [Accessed: 2016-07-03]
7. Pialoux G, Gauzere BA, Jaureguiberry S, Strobel M. Chikungunya, an epidemic arbovirosis. Lancet, Infectious Disease. 2007;7:319–327.
8. Olaleye OD, Omilabu SA, Fagbami AH. Igbo-Ora virus (an alphavirus isolated in Nigeria): a serological survey for haemagglutination inhibiting antibody in humans and domestic animals. Transactions of the Royal Society of Tropical Medicine and Hygiene. 1988;82:905–906.
9. Kumar R. Insect Pest Control. ELBS ed. London: Edward Arnold Publishers Ltd; 1984. 298 p.
10. van Emden HF. Pest Control. 2nd ed. Cambridge: Cambridge University Press; 1989. 117 p.
11. Knipling EF. Possibilities in insect control or eradication through the use of sexually sterile males. Journal Economic Entomology. 1955;48:459–462.
12. Robinson AS. Genetic basis of the Sterile Insect Technique. In: Dyck VA, Hendrichs J, Robinson AS, editors. Sterile Insect Technique: Principles and Practice in Area-Wide Integrated Pest Management. Dordrecht, The Netherlands: Springer; 2005. pp. 95–114.
13. Franz G. Genetic sexing strains in Mediterranean fruit fly, an example for other species amenable to large-scale rearing for the Sterile Insect Technique. In: Dyck VA, Hendrichs J, Robinson AS, editors. Sterile Insect Technique: Principles and Practice in Area-Wide Integrated Pest Management. Dordrecht, The Netherlands: Springer; 2005. pp. 427–451.
14. Handler AM. An introduction to the history and methodology of insect gene transfer. In: Handler AM, James AA, editors. Insect Transgenesis: Methods and Applications. Boca Raton: CRC Press LLC; 2000. pp. 3-26.
15. Chalfie M, Tu Y, Euskirchen G, Ward WW, Prasher DC. Green fluorescent protein as a marker for gene expression. Science. 1994;263(5148):802–805.
16. Matz MV, Fradkov AF, Labas YA, Savitsky AP, Zaraisky AG, Markelov ML, Lukyanov SA. Fluorescent proteins from nonbioluminescent Anthozoa species. Nature Biotechnology. 1999;17:969–973.
17. Campbell RE, Tour O, Palmer AE, Steinbach PA, Baird GS, Zacharias DA, Tsien RY. A monomeric red fluorescent protein. Proceedings of the National Academy of Science USA. 2002;99:7877–7882.
18. Shaner NC, Campbell RE, Steinbach PA, Giepmans BN, Palmer AE, Tsien RY. Improved monomeric red, orange and yellow fluorescent proteins derived from Discosoma sp. red fluorescent protein. Nature Biotechnology. 2004;22:1567–1572.
19. Goodhue RE, Bolda M, Farnsworth D, Williams JC, Zalom FG. Spotted wing Drosophila infestation of California strawberries and raspberries: economic analysis of potential revenue losses and control costs. Pest Management Science. 2011;67:1396–1402.
20. Horn, C, Wimmer EA. A transgene-based, embryo-specific lethality system for insect pest management. Nature Biotechnology. 2003;21:64–70.
21. Heinrich JC, Scott M. A repressible female-specific lethal genetic system for making transgenic insect strains suitable for a sterile-release program. Proceedings of the National Academy of Science USA. 2000;97:8229–8232.
22. Thomas DD, Donnelly CA, Wood RJ, Alphey LS. Insect population control using a dominant, repressible, lethal genetic system. Science. 2000;287:2474–2476.
23. Gossen M, Bujard H. Tight control of gene expression in mammalian cells by tetracycline-responsive promoters. Proceedings of the National Academy of Science USA. 1992;89(12):5547–5551.
24. Chen C-H, Huang H, Ward CM, Su JT, Schaeffer LV, Guo M, Hay BA. A Synthetic Maternal-Effect Selfish Genetic Element Drives Population Replacement in Drosophila. Science. 2007;316(5824):597–600.

25. Akbari OS, Chen CH, Marshall JM, Huang H, Antoshechkin I, Hay BA. Novel synthetic Medea selfish genetic elements drive population replacement in Drosophila; a theoretical exploration of Medea-dependent population suppression. ACS Synthetic Biology. 2014;3(12):915–928.
26. Schetelig MF and Handler AM. Germline transformation of the spotted wing drosophilid, Drosophila suzukii, with a piggyBac transposon vector. Genetica. 2013;141:189–193.
27. Schetelig MF, Caceres C, Zacharopoulou A, Franz G, Wimmer EA. Conditional embryonic lethality to improve the sterile insect technique in Ceratitis capitata (Diptera: Tephritidae). BMC Biology. 2009a;7:4.
28. Schetelig MF, Handler AM. Strategy for enhanced transgenic strain development for embryonic conditional lethality in Anastrepha suspensa. Proceedings of the National Academy of Science USA. 2012a;109(24):9348–9353.
29. Gong P, Epton MJ, Fu G, Scaife S, Hiscox A, Condon KC, Condon GC, Morrison NI, Kelly DW, Dafa'alla T, Coleman PG, Alphey L. A dominant lethal genetic system for autocidal control of the Mediterranean fruit fly. Nature Biotechnology. 2005;23(4):453–456.
30. Saccone G, Pane A, De Simone A, Salvemini M, Milano A, Annunziata L, Mauro U, Polito LC. New sexing strains for Mediterranean fruit fly Ceratitis capitata: transforming females into males. In: Vreysen MJB, Robinson AS, Hendrichs J, editors. Area-wide control of insect pests: from research to field implementation. Dordrecht, The Netherlands: Springer; 2007. pp. 95–102.
31. Fu G, Condon KC, Epton MJ, Gong P, Jin L, Condon GC, Morrison NI, Dafa'alla TH, Alphey L. Female-specific insect lethality engineered using alternative splicing. Nature Biotechnology. 2007;25(3):353–357.
32. Ant T, Koukidou M, Rempoulakis P, Gong H-F, Economopoulos A, Vontas J, Alphey L. Control of the olive fruit fly using genetics-enhanced sterile insect technique. BMC Biology. 2012;10:51.
33. Schetelig MF, Handler AM. A transgenic embryonic sexing system for Anastrepha suspensa (Diptera: Tephritidae). Insect Biochemistry and Molecular Biology. 2012b;42(10):790–795.
34. Ogaugwu CE, Schetelig MF, Wimmer EA. Transgenic sexing system for Ceratitis capitata (Diptera: Tephritidae) based on female-specific embryonic lethality. Insect Biochemistry and Molecular Biology. 2013;43:1–8.
35. Schetelig MF, Targovska A, Meza JS, Bourtzis K, Handler AM. Tetracycline-suppressible female lethality and sterility in the Mexican fruit fly, Anastrepha ludens. Insect Molecular Biology. 2016;25(4):500–508.
36. Chang C, Huang CY, Dai SM, Atlihan R, Chi H. Genetically engineered ricin suppresses Bactrocera dorsalis (Diptera: Tephritidae) based on demographic analysis of group-reared life table. Journal of Economic Entomology. 2016;
37. Scolari F, Schetelig MF, Bertin S, Malacrida AR, Gasperi G, Wimmer EA. Fluorescent sperm marking to improve the fight against the pest insect Ceratitis capitata (Wiedemann; Diptera: Tephritidae). New Biotechnology. 2008;25:76–84.
38. Catteruccia F, Nolan T, Loukeris TG, Blass C, Savakis C, Kafatos FC, Crisanti A. Stable germline transformation of the malaria mosquito Anopheles stephensi. Nature. 2000;405:959–962.
39. Adelman ZN, Jasinskiene N, Onal S, Juhn J, Ashikyan A, Salampessy M, MacCauley T, James AA. Nanos gene control DNA mediates developmentally regulated transposition in the yellow fever mosquito Aedes aegypti. Proceedings of the National Academy of Science USA. 2007;104(24):9970–9975.
40. Windbichler N, Menichelli M, Papathanos PA, Thyme SB, Li H, Ulge UY, Hovde BT, Baker D, Monnat RJ Jr., Burt A, Crisanti A. A synthetic homing endonuclease-based gene drive system in the human malaria mosquito. Nature. 2011;473(7346): 212–215.

41. Jinek M, Chylinski K, Fonfara I, Hauer M, Doudna JA, Charpentier E. A programmable dual-RNA-guided DNA endonuclease in adaptive bacterial immunity. Science. 2012;337:816–821.
42. DiCarlo JE, Norville, JE, Mali P, Rios X, Aach J, Church GM. Genome engineering in Saccharomyces cerevisiae using CRISPR Cas systems. Nucleic Acids Research. 2013;41:4336–4343.
43. Gratz SJ, Cummings AM, Nguyen JN, Hamm DC, Donohue LK, Harrison MM, Wildonger J, O'Connor-Giles KM. Genome Engineering of Drosophila with the CRISPR RNA-Guided Cas9 Nuclease. Genetics. 2013;194:1029–1035.
44. Hammond A, Galizi R, Kyrou K, Simoni A, Siniscalchi C, Katsanos D, Gribble M, Baker D, Marois E, Russell S, Burt A, Windbichler N, Crisanti A and Nolan T. A CRISPR-Cas9 gene drive system targeting female reproduction in the malaria mosquito vector *Anopheles gambiae*. Nature Biotechnology. 2016;34:78–83.
45. Phuc HK, Andreasen MH, Burton RS, Vass C, Epton MJ, Pape G, Fu G, Condon KC, Scaife S, Donnelly CA, Coleman PG, White-Cooper H and Alphey L. Late-acting dominant lethal genetic systems and mosquito control. BMC Biology. 2007;5:11,
46. Fu G, Lees RS, Nimmo D, Aw D, Jin L, Gray P, Berendonk TU, White-Cooper H, Scaife S, Phuc HK, Marinotti O, Jasinskiene N, James AA, Alphey L. Female-specific flightless phenotype for mosquito control. Proceedings of the National Academy of Science USA. 2010;107(10):4550–4554.
47. Muñoz D, Jimenez A, Marinotti O, James A. The AeAct-4 gene is expressed in the developing flight muscles of females *Aedes aegypti*. Insect Molecular Biology. 2004;13(5):563–568.
48. Catteruccia F, Benton JP, Crisanti A. An Anopheles transgenic sexing strain for vector control. Nature Biotechnology. 2005;23:1414–1417.
49. Thailayila J, Magnussona K, Godfray HCJ, Crisanti A, Catteruccia F. Spermless males elicit large-scale female responses to mating in the malaria mosquito Anopheles gambiae. Proceedings of the National Academy of Science USA. 2011;108(33):13677–13681.
50. Galizi R, Hammond A, Kyrou K, Taxiarchi C, Bernadini F, O'Loughlin SM, Papathanous P-A, Nolan T, Windbichler N, Crisanti A. A CRISPR-Cas9 sex-ratio distortion system for genetic control. Scientific Reports. 2016;6:31139.
51. Ito J, Ghosh A, Moreira LA, Wimmer EA, Jacobs-Lorena M. Transgenic anopheline mosquitoes impaired in transmission of a malaria parasite. Nature. 2002;417(6887):452–455.
52. Moreira LA, Ito J, Ghosh A, Devenport M, Zieler H, Abraham EG, Crisanti A, Nolan T, Catteruccia F, Jacobs-Lorena M. Bee venom phospholipase inhibits malaria parasite development in transgenic mosquitoes. Journal of Biological Chemistry. 2002;277:40839–40843.
53. Rodrigues FG, Santos MN, de Carvalho TX, Rocha BC, Riehle MA, Pimenta PF, Abraham EG, Jacobs-Lorena M, Alves de Brito CF, Moreira LA. Expression of a mutated phospholipase A2 in transgenic Aedes fluviatilis mosquitoes impacts Plasmodium gallinaceum development. Insect Molecular Biology. 2008;17(2):175–183.
54. Kim W, Koo H, Richman AM, Seeley D, Vizioli J, Klocko AD, O'Brochta DA. Ectopic expression of a cecropin transgene in the human malaria vector mosquito Anopheles gambiae (Diptera: Culicidae): effects on susceptibility to Plasmodium. Journal of Medical Entomology. 2004;41:447–455.
55. Isaacs AT, Li F, Jasinskiene N, Chen X, Nirmala X, Marinotti O, Vinetz JM, James AA. Engineered resistance to Plasmodium falciparum development in transgenic Anopheles stephensi. PLoS Pathogens. 2011;7:e1002017.
56. Meredith JM, Basu S, Nimmo DD, Larget-Thiery I, Warr EL, Underhill A, McArthur CC, Carter V, Hurd H, Bourgouin C, Eggleston P. Site-specific integration and expression of an anti-malarial gene in transgenic Anopheles gambiae significantly reduces Plasmodium infections. PLoS One. 2011;6:e14587.

57. Franz AW, Sanchez-Vargas I, Adelman ZN, Blair CD, Beaty BJ, James AA, Olson KE. Engineering RNA interference-based resistance to dengue virus type 2 in genetically modified Aedes aegypti. Proceedings of the National Academy of Science USA. 2006;103:4198–4203.
58. Li F, Wantuch HA, Linger RJ, Belikoff EJ, Scott MJ. Transgenic sexing system for genetic control of the Australian sheep blow fly Lucilia cuprina. Insect Biochemistry and Molecular Biology. 2014;51:80–88.
59. Scott MJ. Development and evaluation of male-only strains of the Australian sheep blowfly, Lucilia cuprina. BMC Genetics. 2014;15:S3.
60. Yan Y, Scott MJ. A transgenic embryonic sexing system for the Australian sheep blow fly Lucilia cuprina. Scientific Reports. 2015;5:16090.
61. Allen ML, Handler AM, Berkebile DR, Skoda SR. PiggyBac transformation of the New World screwworm, Cochliomyia hominivorax, produces multiple distinct mutant strains. Medical and Veterinary Entomology. 2004;18:1–9.
62. Handler AM, Allen ML, Skoda SR. Development and utilization of transgenic New World Screwworm, Cochliomyia hominivorax. Medical and Veterinary Entomology. 2009;23:98–105.
63. Durvasula RV, Gumbs A, Panackal A, Kruglov O, Aksoy S, Merrifield RB, Richards FF, Beard CB. Prevention of insect-borne disease: an approach using transgenic symbiotic bacteria. Proceedings of the National Academy of Science USA. 1997;94:3274–3278.
64. Yoshida S, Ioka D, Matsuoka H, Endo H, Ishii A. Bacteria expressing single-chain immunotoxin inhibit malaria parasite development in mosquitoes. Molecular and Biochemical Parasitology. 2001;113:89–96.
65. Riehle MA, Moreira CK, Lampe D, Lauzon C, Jacobs-Lorena M. Using bacteria to express and display anti-Plasmodium molecules in the mosquito midgut. International Journal of Parasitology. 2007;37:595–603.
66. Favia G, Ricci I, Damiani C, Raddadi N, Crotti E, Marzorati M, et al. Bacteria of the genus Asaia stably associate with Anopheles stephensi, an Asian malarial mosquito vector. Proceedings of the National Academy of Science USA. 2007;104:9047–9051.
67. Damiani C, Ricci I, Crotti E, Rossi P, Rizzi A, Scuppa P, Esposito F, Bandi C, Daffochio D, Favia G. Paternal transmission of symbiotic bacteria in malaria vectors. Current Biology. 2008;18(23): R1087-1088.
68. Wang S, Ghosh AK, Bongio N, Stebbings KA, Lampe DJ, Jacobs-Lorena M. Fighting malaria with engineered symbiotic bacteria from vector mosquitoes. Proceedings of the National Academy of Science USA. 2012;109:12734–12739.
69. Fang W, Vega-Rodríguez J, Ghosh AK, Jacobs-Lorena M, Kang A, St Leger RJ. Development of transgenic fungi that kill human malaria parasites in mosquitoes. Science. 2011;331:1074–1077.
70. Ren, X, Hoiczyk, E, Rasgon, JL. Viral paratransgenesis in the malaria vector Anopheles gambiae. PLOS Pathogens. 2008;4(8): e1000135.
71. de Lara Capurro M, et al. Virus-expressed, recombinant single-chain antibody blocks sporozoite infection of salivary glands in Plasmodium gallinaceum-infected Aedes aegypti. American Journal of Tropical Medicine and Hygiene. 2000;62(4):427–433.
72. Higgs S, et al. Engineered resistance in Aedes aegypti to a West African and a South American strain of yellow fever virus. American Journal of Tropical Medicine and Hygiene. 1998;58(5):663–670.
73. Adelman ZN, et al. Sindbis virus-induced silencing of dengue viruses in mosquitoes. Insect Molecular Biology. 2001;10(3):265–273.
74. Coutinho-Abreu IV, Zhu KY, Ramalho-Ortigao M. Transgenesis and paratransgenesis to control insect-borne diseases: Current status and future challenges. Parasitology International. 2010;59(1):1–8.
75. Hurwitz I, Hillesland H, Fieck A, Das P, Durvasula R. The paratransgenic sand fly: A platform for control of Leishmania transmission. Parasites and Vectors. 2011;4:82.

76. Cheng Q, Aksoy S. Tissue tropism, transmission and expression of foreign genes in vivo in midgut symbionts of tsetse flies. Insect Molecular Biology. 1999;8(1):125–32.
77. Aksoy S, Weiss B, Attardo G. Paratransgenesis applied for control of tsetse transmitted sleeping sickness. Advances in Experimental Medecine and Biology. 2008;627:35–48.
78. Bourtzis K, Lees RS, Hendrichs J, Vreysen MJB. More than one rabbit out of the hat: Radiation, transgenic and symbiont-based approaches for sustainable management of mosquito and tsetse fly populations. Acta Tropica. 2016;157:115–130.
79. Harris AF, Nimmo D, McKemey AR, Kelly N, Scaife S, Donnelly CA, Beech C, et al. Field performance of engineered male mosquitoes. Nature Biotechnology. 2011;29:1034–1037.
80. Harris AF, McKemey AR, Nimmo D, Curtis Z, Black I, Morgan SA, Oviedo MN, et al. Successful suppression of a field mosquito population by sustained release of engineered male mosquitoes. Nature Biotechnology. 2012;30: 828–830.
81. WHO-TDR. Report on Planning Meeting 1 Technical Consultations on Current Status and Planning for Future Development of Genetically Modified Mosquitoes for Malaria and Dengue Control. In: Crawford VL, Reza JN, editors. Progress and Prospects for the Use of Genetically Modified Mosquitoes to Inhibit Disease Transmission; 4–6 May 2009; Geneva, Switzerland. World Health Organization; 2010. pp. 1–66.
82. Massonnet-Bruneel B, Corre-Catelin N, Lacroix R, Lees RS, Hoang KP, Nimmo D, Alphey, L, Reiter, P. Fitness of transgenic mosquito Aedes aegypti males carrying a dominant lethal genetic system. PLoS ONE. 2013;8(5): e62711.
83. Harvey-Samuel T, Ant T, Gong H, Morrison NI, Alphey L. Population-level effects of fitness costs associated with repressible female-lethal transgene insertions in two pest insects. Evolutionary Applications. 2014;7(5):597–606.
84. European Food Safety Authority (EFSA). Guidance on the environmental risk assessment of genetically modified animals. EFSA Journal. 2013;11(5):3200.
85. Atkinson PW, O'Brochta DA, Craig NL. The hobo, Hermes and Herves transposable elements of insects. In: Vreysen, MJB, Robinson AS, Hendrichs J, editors. Area-wide control of insect pests: from research to field implementation. Dordrecht, The Netherlands: Springer; 2007. pp. 61–71.
86. Handler AM, Zimowska GJ, Horn C. Post-integration stabilization of a transposon vector by terminal sequence deletion in Drosophila melanogaster. Nature Biotechnology. 2004;22:1150–1154.
87. Dafa'alla TH, Condon GC, Condon KC, Phillips CE, Morrison NE, Jin L, Epton MJ, Fu G, Alphey L. Transposon-free insertions for insect genetic engineering. Nature Biotechnology. 2006;24(7):820–821.
88. Schetelig MF, Scolari F, Handler AM, Kittelmann S, Gasperi G, Wimmer EA. Site-specific recombination for the modification of transgenic strains of the Mediterranean fruit fly Ceratitis capitata. Proceedings of the National Academy of Science USA. 2009b;106(43):18171–18176.
89. Wilke AB, Marrelli MT. Paratransgenesis: a promising new strategy for mosquito vector control. Parasites and Vectors. 2015;8:342.
90. Weiss BL, et al. Interspecific transfer of bacterial endosymbionts between tsetse fly species: infection establishment and effect on host fitness. Applied and Environmental Microbiology. 2006;72(11):7013–7021.
91. Arora A, Forshaw A, Durvasula R. A delivery system for field application of paratransgenic control. BMC Biotechnology. 2015;15:59.
92. New York Times. Panel Endorses 'Gene Drive' Technology That Can Alter Entire Species [Internet]. [Updated: 2016]. Available from: http://www.nytimes.com/2016/06/09/science/national-academies-sciences-gene-drive-technology.html?_r=0 [Accessed: 2016-08-17]

93. Harvard Magazine. Editing an End to Malaria? [Internet]. [Updated:2016]. Available from: http://www.harvardmagazine.com/2016/05/editing-an-end-to-malaria [Accessed: 2016-08-17]

94. Hou Z, Zhang Y, Propson NE, Howden SE, Chu L-F, Sontheimer EJ, Thomson JA. Efficient genome engineering in human pluripotent stem cells using Cas9 from Neisseria meningitides. Proceedings of the National Academy of Science USA. 2013;110(39): 15644–15649.

95. Bassett AR, Tibbit C, Ponting CP, Liu J-L. Highly efficient targeted mutagenesis of drosophila with the CRISPR/Cas9 System. Cell Reports. 2013;4:220–228.

96. Gao F, Shen XZ, Jiang F, Wu Y, Han C. DNA-guided genome editing using the Natronobacterium gregoryi Argonaute. Nature Biotechnology.2016;34:768–773.

Index

A

Amblyseius andersoni, 327, 328, 329, 332, 333, 334, 337, 338
Aspergillus, 262, 264, 281, 282

B

Bacillus thuringiensis, 48, 49, 187, 201, 262, 282, 342, 343, 344, 368, 369, 396
biological control, 2, 7, 8, 11, 279, 283, 319, 325, 337, 339, 343, 372, 373, 374, 389
biopesticide, 169, 170, 180, 261, 279, 280, 281
Botrytis Fruit Rot, 286, 289, 293

C

curative methods, 13
Cylas puncticollis, 190, 203
cytochrome p450, 88

D

detoxifying enzymes, 88, 94, 95, 99, 107, 110, 116, 117, 128
dipterans, 394, 396, 397, 400, 404, 405, 409, 410, 411

E

entomopathogen, 170, 201, 369
entomopathogenic, 46, 169, 261, 262, 264, 277, 280, 281, 282, 283, 287, 406

I

Insecticide resistance, 77, 88, 122, 126, 128, 343, 348, 360
Irrigation, 25, 218, 285, 286, 288, 290, 295, 296

K

kairomones, 130, 131, 132, 139, 144, 149, 384

M

mating disruption, 129, 130, 135, 136, 137, 138, 148, 343
Micro-Sprinkler, 286, 296

O

olfactometer, 130, 134, 140, 141, 142, 143, 144, 145, 146
organic agriculture, 13, 15, 16, 18, 55, 372

P

paratransgenesis, 394, 404, 405, 406, 407, 410, 415
pest control, 2, 3, 4, 5, 8, 10, 280, 282, 298, 301, 302, 312, 374, 385, 397, 400, 401
pheromones, 8, 30, 31, 130, 374, 375, 384, 389, 390, 396
Photorhabdus, 169, 170, 174, 175, 184
Phytoseiidae, 117, 167, 328, 337, 338, 339
Powdery Mildew, 286, 289, 293
preventive, 13, 16, 17, 27, 139

S

Spinosad, 170
Strawberry, 285, 286, 292, 295, 296
Streptomyces, 46, 169, 170, 171, 177, 178, 186
Sweetpotato butterfly, 190

T

Typhlodromus pyri, 302, 327, 328, 329, 331, 332, 337, 338

X

Xenorhabdus, 169, 170, 171, 174, 175, 184